Recent Titles in This Series

118 **Kenkichi Iwasawa,** Algebraic functions, 1993
117 **Boris Zilber,** Uncountably categorical theories, 1993
116 **G. M. Fel′dman,** Arithmetic of probability distributions, and characterization problems on abelian groups, 1993
115 **Nikolai V. Ivanov,** Subgroups of Teichmüller modular groups, 1992
114 **Seizô Itô,** Diffusion equations, 1992
113 **Michail Zhitomirskiĭ,** Typical singularities of differential 1-forms and Pfaffian equations, 1992
112 **S. A. Lomov,** Introduction to the general theory of singular perturbations, 1992
111 **Simon Gindikin,** Tube domains and the Cauchy problem, 1992
110 **B. V. Shabat,** Introduction to complex analysis Part II. Functions of several variables, 1992
109 **Isao Miyadera,** Nonlinear semigroups, 1992
108 **Takeo Yokonuma,** Tensor spaces and exterior algebra, 1992
107 **B. M. Makarov, M. G. Goluzina, A. A. Lodkin, and A. N. Podkorytov,** Selected problems in real analysis, 1992
106 **G.-C. Wen,** Conformal mappings and boundary value problems, 1992
105 **D. R. Yafaev,** Mathematical scattering theory: General theory, 1992
104 **R. L. Dobrushin, R. Kotecký, and S. Shlosman,** Wulff construction: A global shape from local interaction, 1992
103 **A. K. Tsikh,** Multidimensional residues and their applications, 1992
102 **A. M. Il′in,** Matching of asymptotic expansions of solutions of boundary value problems, 1992
101 **Zhang Zhi-fen, Ding Tong-ren, Huang Wen-zao, and Dong Zhen-xi,** Qualitative theory of differential equations, 1992
100 **V. L. Popov,** Groups, generators, syzygies, and orbits in invariant theory, 1992
99 **Norio Shimakura,** Partial differential operators of elliptic type, 1992
98 **V. A. Vassiliev,** Complements of discriminants of smooth maps: Topology and applications, 1992
97 **Itiro Tamura,** Topology of foliations: An introduction, 1992
96 **A. I. Markushevich,** Introduction to the classical theory of Abelian functions, 1992
95 **Guangchang Dong,** Nonlinear partial differential equations of second order, 1991
94 **Yu. S. Il′yashenko,** Finiteness theorems for limit cycles, 1991
93 **A. T. Fomenko and A. A. Tuzhilin,** Elements of the geometry and topology of minimal surfaces in three-dimensional space, 1991
92 **E. M. Nikishin and V. N. Sorokin,** Rational approximations and orthogonality, 1991
91 **Mamoru Mimura and Hirosi Toda,** Topology of Lie groups, I and II, 1991
90 **S. L. Sobolev,** Some applications of functional analysis in mathematical physics, third edition, 1991
89 **Valeriĭ V. Kozlov and Dmitriĭ V. Treshchëv,** Billiards: A genetic introduction to the dynamics of systems with impacts, 1991
88 **A. G. Khovanskiĭ,** Fewnomials, 1991
87 **Aleksandr Robertovich Kemer,** Ideals of identities of associative algebras, 1991
86 **V. M. Kadets and M. I. Kadets,** Rearrangements of series in Banach spaces, 1991
85 **Mikio Ise and Masaru Takeuchi,** Lie groups I, II, 1991
84 **Đào Trọng Thi and A. T. Fomenko,** Minimal surfaces, stratified multivarifolds, and the Plateau problem, 1991

(*Continued in the back of this publication*)

Translations of
MATHEMATICAL
MONOGRAPHS

Volume 118

Algebraic Functions

Kenkichi Iwasawa

Translated by
Goro Kato

American Mathematical Society
Providence, Rhode Island

代数函数論

DAISUU KANSUURON (Algebraic Functions)
by Kenkichi Iwasawa
Originally published in Japanese by Iwanami Shoten, Publishers,
Tokyo in 1952 and 1973

Translated from the Japanese by Goro Kato

1991 *Mathematics Subject Classification*. Primary 14H05, 30F10;
Secondary 33E05.

ABSTRACT. This book gives an introduction to the theory of algebraic functions. The algebraic part of the theory, e.g., the Riemann-Roch theorem for function fields over arbitrary fields of constants, is discussed in the first two chapters, and the analytic part, i.e., the classical theory of analytic functions on compact Riemann surfaces, is described in the latter three chapters.

Library of Congress Cataloging-in-Publication Data

Iwasawa, Kenkichi, 1917–
[Daisū kansūron. English]
Algebraic functions/Kenkichi Iwasawa; translated by Goro Kato.
p. cm.—(Translations of mathematical monographs, ISSN 0065-9282; v. 118)
Translated from the Japanese.
Includes index.
ISBN 0-8218-4595-0 (alk. paper)
1. Functions, Algebraic. I. Title. II. Series.
QA341.I9313 1993 92-39922
512′.74—dc20 CIP

This publication was typeset using $\mathcal{A}_{\mathcal{M}}\mathcal{S}$-TEX,
the American Mathematical Society's TEX macro system

10 9 8 7 6 5 4 3 2 1 98 97 96 95 94 93

Contents

Foreword to the Revised Edition vii

Foreword ix

Preface xi

Chapter 1. Preparation from Valuation Theory 1
- §1. Valuations and prime divisors 1
- §2. The metric induced by a valuation 11
- §3. Extension and projection of prime divisors 19

Chapter 2. Algebraic Theory of Algebraic Function Fields 33
- §1. Algebraic function fields 33
- §2. Prime divisors on algebraic function fields 35
- §3. Divisors on algebraic function fields 42
- §4. Idele and differential 52
- §5. Hasse differential 61
- §6. Applications of the Riemann-Roch theorem 74
- §7. Special function fields 78

Chapter 3. Riemann Surfaces 89
- §1. Riemann surfaces and analytic mappings 89
- §2. Functions on a Riemann surface 94
- §3. Differentials and integrals on Riemann surfaces 97
- §4. Existence of analytic functions 104
- §5. Covering Riemann surfaces 123
- §6. Simply connected Riemann surfaces and canonical forms of Riemann surfaces 130

Chapter 4. Algebraic Function Fields and Closed Riemann Surfaces 147
- §1. The Riemann surface for an algebraic function field 147
- §2. Analytic function fields on closed Riemann surfaces 156
- §3. Topological properties of closed Riemann surfaces 167

Chapter 5. Analytic Theory of Algebraic Function Fields 175
§1. Abelian integrals 175
§2. Additive functions and multiplicative functions 193
§3. Theorems of Abel-Jacobi and Abelian functions 213
§4. Extension fields 231
§5. Elliptic function fields 242

Appendix 265
§1. Finitepotent map and its trace 265
§2. The differential and its residue 269
§3. Differentials and residues in an algebraic function field 276

Subject Index 285

Foreword to the Revised Edition

The theory of algebraic functions had been essentially established by the time this book was originally published. Neither noteworthy developments nor crucial reforms have occurred during the last twenty years. However, there have been important ideas and results in various areas of the theory. Therefore, an appendix has been added as an introduction to such new methods as differentials and their residues. I express my thankfulness to Dr. Genjiro Fujisaki and Mr. Makino of Iwanami Shoten for their cooperation in this publication.

Kenkichi Iwasawa
Spring 1973

Foreword

The classical theory of algebraic functions is explained in this book, and a detailed description of the contents can be found in the preface. To the reader who is learning the theory of algebraic functions for the first time, we recommend reading the preface once again after completing the book.

I sincerely thank Professor Iyanaga for his help in publishing this treatise. I also express my gratitude to Professor Tsuneo Tamagawa for his help in proofreading.

Kenkichi Iwasawa
Fall 1951

Preface

1. As is well known, Descartes is the recognized founder of analytical geometry. He explains in Volume 2 of his *La Géométrie*, Paris, 1637, the interaction and unification of algebra and geometry. Descartes associates a geometric curve with the relation between variables x and y given by a polynomial with real coefficients $f(x, y) = 0$, and conversely, he associates such an equation with a given algebraic curve. Since Descartes studied the dependency between x and y satisfying $f(x, y) = 0$, one may regard the idea of Descartes as the beginning of algebraic function theory.

As calculus was being developed, Newton and Euler studied such a relation $y = \varphi(x)$ between real variables x and y as a plane algebraic curve. However, the essential nature of the algebraic function $y = \varphi(x)$ is revealed only in the framework of complex analysis. That is, the development of algebraic function theory necessitated the founding of complex function theory as was accomplished by Gauss and Cauchy.

Around 1800, Gauss was already studying functions with double periods, i.e., elliptic functions; however, it was Abel who first published a paper on elliptic functions: *Recherches sur les fonctions elliptiques*, Crelle's Jour. **3**, **4** (1827–28). On learning about Cauchy's complex analysis during his visit to Paris, Abel immediately applied Cauchy's theory to his elliptic integral

$$u = \int^{x} \frac{dx}{\sqrt{f_4(x)}},$$

where $f_4(x)$ is a quartic equation in x. Abel successfully obtained an elliptic function $x = x(u)$ as the inverse function of the above integral. He then studied the double periodicity of the elliptic function, complex multiplication, etc. Abel extended the notion of an elliptic integral to an integral of an algebraic function, now called an Abelian integral, in order to obtain the famous theorem of Abel. Abel understood algebraic functions as complex functions and found essential properties of algebraic functions for the first time. Abel should be regarded as the founder of the theory of algebraic functions (Wirtinger).

To generalize the theory, Jacobi began his investigation with an elliptic function, as did Abel. That is, Jacobi studied a hyperelliptic integral generalizing an elliptic integral. As the simplest case, Jacobi obtained the inverse functions $x_1 = x_1(u_1)$ and $x_2 = x_2(u_2)$ of

$$u_1 = \int^{x_1} \frac{dx_1}{\sqrt{f_6(x_1)}} \quad \text{and} \quad u_2 = \int^{x_2} \frac{x_2\, dx_2}{\sqrt{f_6(x_2)}},$$

where $f_6(x)$ is a polynomial in x of degree 6. He found multivalued functions with four periods. However, there were at that time no clear concepts of analytic continuation and multivalued functions, and such functions were discarded as "unvernünftig." Using Abel's theorem, in 1832 Jacobi discovered that, for the sums

$$u_1 = \int^{x_1} \frac{dx_1}{\sqrt{f_6(x_1)}} + \int^{x_2} \frac{dx_2}{\sqrt{f_6(x_2)}},$$
$$u_2 = \int^{x_1} \frac{x_1\, dx_1}{\sqrt{f_6(x_1)}} + \int^{x_2} \frac{x_2\, dx_2}{\sqrt{f_6(x_2)}},$$

the symmetric functions $x_1 + x_2$ and $x_1 x_2$ are single-valued functions of u_1 and u_2 with four periods instead of hyperelliptic integrals. He conjectured that $x_1 + x_2$ and $x_1 x_2$ can be expressed as theta series in u_1 and u_2. This was proved later by Rosenhain (1846) and Göpel (1847).

Inspired by the works of Abel and Jacobi, many mathematicians focused on solving Jacobi's 'Umkehrproblem' as they tried to extend the results to the case of general Abelian integrals, i.e., to construct the general theory of Abelian functions and theta functions. That effort led to the fundamental development of analytic function theory, especially algebraic function theory and multiperiodic function theory in several variables. The complex function theory of Riemann and Weierstrass appeared under these circumstances.

We will now describe Weierstrass' work. As is well known, Weierstrass built his function theory based on power series. As for an algebraic function, Weierstrass considered the following. He named the totality of all the pairs (a, b) of complex values $x = a$ and $y = b$ satisfying an irreducible equation $f(x, y) = 0$ as algebraisches Gebilde. Each (a, b) is called a 'Stelle'. He considered a rational function $R(x, y)$ of x and y as a function on algebraisches Gebilde having the value $R(a, b)$ at the Stelle (a, b). The fundamental method that he chose was the following. For an arbitrary Stelle (a, b), find a pair of power series $(x(t), y(t))$ of the form

$$x(t) = a + a_1 t + a_2 t^2 + \cdots \quad \text{and} \quad y(t) = b + b_1 t + b_2 t^2 + \cdots, \quad |t| < \varepsilon$$

satisfying $f(x(t), y(t)) = 0$. Then his theorem states that the coordinates x and y of Stelle (x, y) sufficiently near (a, b) are given by "Elements" $(x(t), y(t))$, (generally, one Element suffices). Furthermore, Weierstrass developed algebraically the theory of power series in several variables obtaining many results on $R(x, y)$. Among other results, he proved that the sum of residues of $R(x, y)$ is 0. He considered the totality of $R(x, y)$ that have value ∞ at given Stelles to obtain the concept of genus and the Riemann-Roch theorem. Thus, the theory of Weierstrass is algebraic in nature, beginning with $f(x, y) = 0$ and proceeding with power series. When he could, Weierstrass transformed problems even on Abelian integrals to problems of differentials. He seems to have avoided the geometric consideration of the Gebilde as a curve. At that time, the foundation of the theory of Riemann surfaces had not been established. With his critical spirit, he may have intentionally avoided the usage of Riemann surfaces. Nowadays, we have a rigorous foundation for Riemann surfaces, and Weierstrass' approach appears unsuited to us.

It was Weierstrass who first defined an analytic function as a power series; however, the power series expansion of an algebraic function in a neighborhood of a given point was known previously. Namely, Cauchy and Laurent obtained power series expansions at a regular point and at a pole. In 1850, Puiseux showed that an algebraic function $y = \varphi(x)$ has an expansion in fractional powers at a branch point. This Puiseux series essentially coincides with Weierstrass' expansions of $x = x(t)$ and $y = y(t)$. (However, Weierstrass emphasized that the parameter t can be chosen as a rational function of x and y. This is another movement towards his algebraic method.) Puiseux recognized that the multivaluedness of $\varphi(x)$ comes from the analytic continuation of $\varphi(x)$, starting with a power series for $\varphi(x)$. He also recognized the periodicity of the Abelian integral $\int \varphi(x)\,dx$. Even though these findings reflected the essential nature of an algebraic function, it was the genius of Riemann that exposed the very root of these results.

Riemann's contribution to algebraic function theory is decisive. He defined the notion of a Riemann surface as he studied the multivalued nature of analytic functions, especially algebraic functions. Riemann showed that if the complex sphere is replaced by a Riemann surface, then a multivalued function can be treated just as a single-valued function. In particular, he studied in depth the correspondence between an algebraic function determined by $f(x, y) = 0$ and the topological structure of the associated Riemann surface $\mathfrak{R}$. In general, the Riemann surface is a $(2p+1)$-sheeted connected closed surface. That is, when one cuts $\mathfrak{R}$ along $2p$ closed curves, one gets a simply connected surface. In addition to the above, Riemann showed that p, the genus of $\mathfrak{R}$, is given by

$$p = \frac{w}{2} - n + 1,$$

where n is the number of sheets covering the complex x-sphere, i.e., the degree of $f(x, y)$ in y, and w is the number of branch points for the complex x-sphere. Such a topological study enabled one to understand intuitively the reason behind the periodicity of an Abelian integral. In particular, it was made clear that an arbitrary period of an Abelian integral can be written as a sum of $2p$ fundamental periods with rational integral coefficients.

Thus, through the introduction of a geometric concept, a Riemann surface, Riemann provided a clear picture of many difficult topics. His most significant contribution, rather than his construction of a Riemann surface from an algebraic function, was to derive the existence of algebraic functions as he built his theory based on the concept of a Riemann surface. (The construction of a Riemann surface was somewhat achieved by Weierstrass with the notion of an algebraisches Gebilde.) That is, Riemann studied an arbitrary closed Riemann surface $\mathfrak{R}$ which is not a priori associated with an algebraic function, and he proved by the Dirichlet principle the existence of a differential (or a harmonic function) having singularities at given points. Then he proved that the totality K of analytic functions on $\mathfrak{R}$, which he called the "Klasse", is the field of algebraic functions. Namely, he proved the fact that one can choose suitable functions x and y in K so that an arbitrary function in K can be written as a rational expression of x and y with complex coefficients. Moreover, x and y satisfy an irreducible equation $f(x, y) = 0$.

Riemann further proved that the Riemann surface determined by the above algebraic function is nothing but the original Riemann surface. Thus, the concept of an algebraic function went beyond the superficial expression $f(x, y) = 0$ to the notion of analytic function on a closed Riemann surface; thus, revealing the true nature of algebraic functions. In algebraic terms, the fundamental properties of an algebraic function do not depend upon $f(x, y) = 0$ but upon the algebraic function field $K = k(x, y)$, ($k =$ field of complex numbers).

After his careful study of Abelian integrals, Riemann proved the well-known Riemann-Roch theorem, which was later supplemented by Roch. Furthermore, by defining a general theta function, Riemann solved Jacobi's Umkehrproblem (converse problem). Thus, we witness the essential completion by Riemann of classical algebraic function theory. However, various proofs by Riemann are not rigorous enough for current standards. It is natural that he should have lacked rigorousness prior to the establishment of abstract algebra and topology. After Riemann's death (1869), Weierstrass pointed out a theoretical flaw in the proof of the Dirichlet principle used by Riemann. This criticism by Weierstrass caused a stir among mathematicians at that time.

The Dirichlet principle is a method in potential theory for finding the Green's functions used by Gauss, W. Thompson, Dirichlet, etc. The principle is as follows: in order to find a harmonic function $u = u(x, y, z)$ satisfying

$$\Delta u = \frac{\partial^2 u}{\partial x^2} + \frac{\partial^2 u}{\partial u^2} + \frac{\partial^2 u}{\partial z^2} = 0$$

in a 3-dimensional domain G, assuming specified values on the boundary of G, choose a differentiable function, among the functions satisfying the boundary condition, that minimizes the integral

$$\int_G \left(\left(\frac{\partial u}{\partial x}\right)^2 + \left(\frac{\partial u}{\partial y}\right)^2 + \left(\frac{\partial u}{\partial z}\right)^2 \right) dx\,dy\,dz.$$

Riemann ingeniously used the above principle for the two-dimensional case to prove the existence of a differential and a harmonic function having given singularities. He then assumed the existence of a function that minimizes the above integral as an obvious fact. This is the point Weierstrass criticized.

Though Weierstrass shook the foundation of Riemann's theory, Schwarz and C. Neuman proved the existence of such a harmonic function without using the Dirichlet principle, thus saving Riemann's theory. Honoring Riemann's original idea, Klein attempted to prove the Dirichlet principle itself. Hilbert, who was a professor at Göttingen as were Klein and Riemann, finally gave a proof of the Dirichlet principle in 1901. Hilbert's method has been simplified by Courant and Weyl, and in Weyl's book, *Die Idee der Riemannschen Fläche*, Teubner, Leipzig, 1913, in which a generalization of Hilbert's method is beautifully described, combined with an exposition of the topological properties of a Riemann surface. In this book, Riemann's original fundamental thought has been revived in complete form. Hence, Weyl put an end to the classical theory of algebraic functions.

We return to the previous century to describe the further development of our theory after the work of Riemann and Weierstrass.

After the investigation by Descartes of a real coefficient algebraic equation as a plane algebraic curve, and with the birth of projective geometry, the projective geometric properties of an algebraic curve were studied. In addition to Riemann surfaces, Riemann also studied the complex coefficient equation $f(x, y) = 0$ determining an algebraic function as a curve in a complex projective plane. This shed further light on the nature of algebraic functions. Having obtained the algebraic function field $K = k(x, y)$, Riemann paid attention to a birational transformation between $f(x, y) = 0$ and $g(x, y) = 0$. It is natural that he should focus on the invariant properties of algebraic curves under transformations, i.e., the algebraic geometric properties of the curves. Recognizing the effectiveness of the analytic theory of algebraic functions, e.g., Abelian integrals, to solve geometric problems about algebraic curves he applied, as an example, elliptic function theory to an equation of degree four.

Riemann's program was pushed further by Clebsch who produced many results. He not only applied Riemann's analytic theory to algebraic geometry, but with Gordan, from an algebraic geometric view, tried to refine Riemann's theory, especially Abelian function theory, by skillfully using the homogeneous coordinates in a projective space. (See *Theorie der Abelschen Funktionen*, 1866.) It is said that their work popularized Riemann's thought, which was considered esoteric at that time.

Building a theory of algebraic curves independently of the analytic theory, Brill and M. Noether succeeded for the first time in proving, in an algebraic geometric way, the Riemann-Roch theorem and Abel's theorem (1873). Thus, algebraic geometry had become an independent field of mathematics. Its intuitively attractive geometric results provided rich insight into algebraic function theory. In this direction, Halphen, Picard, Poincaré, and in Italy, Cremona, Segre, Enriques, Casternuovo, and Severi extended the research from plane curves to space curves and to more general algebraic varieties. Their methodology, which had become more refined, lead to modern algebraic geometry. It was a quite recent event when the purely algebraic method was established in algebraic geometry without using analytic methods. This development was made by van der Waerden, Zariski, Weil, Chevalley, and others. We will return to this topic later.

In addition to the analytic (classical) and the algebraic geometric methods of algebraic function theory, there is another method, which may be called the algebraic or the arithmetic method.

In addition to Weierstrass' work mentioned earlier, the number theoretic investigations of algebraic number fields by Kummer, Kronecker, and Dedekind contributed the profound understanding of the notion of fields. Their works significantly influenced the development of algebraic function fields. Namely, Dedekind and H. Weber discovered that theorems for "Norm", "Spur", and "Diskriminante" hold parallely between algebraic number fields and algebraic function fields. Always modeling algebraic number fields, they succeeded in erecting the theory of algebraic functions in a purely algebraic (arithmetic) way (see *Theorie der algebraischen Funktionen einer Veränderlichen*, Crelle's Jour. **92** (1882), 181–290). As is well known, Ideal theory was fundamental to their work. Thus, they defined the so-called absolute Riemann surface of an algebraic function field as the set of "Primdivisor's", and defined a "Polygon", i.e., a "Divisor" in current terminology, as a product of points on the absolute Riemann surface. They then obtained the notions of Divisor of a function, "Divisorenklasse", "Differential", "Geschlecht", etc. Finally, they proved the Riemann-Roch theorem in a purely algebraic manner using "Normal basis". One can say that the main part of the algebraic theory of algebraic functions was established by Dedekind and Weber.

The Dedekind-Weber theory was further simplified by Hensel and Landsberg in their voluminous *Theorie der algebraischen Funktionen einer Variabeln und ihre Anwendung auf algebraische Kurven und Abelsche Integrale*, Teubner, Leipzig, 1902. As contrasted with the purely algebraic method of Dedekind-Weber, the method in this book is much closer to Weierstrass' method using the power series expansion of an algebraic function as an analytic function. Henzel improved the power series expansion method even further to obtain his famous theory of $\mathfrak{p}$-adic numbers. With that he gave an algebraic foundation for the power series expansion of an algebraic function independent of analysis. Thus, he confirmed that his method is entirely algebraic.

In the preface of Dedekind-Weber's paper noted above, Dedekind and Weber mention that their theory is also valid for algebraic function fields whose coefficient fields are the field of algebraic numbers. However, Artin seems to be the first to begin the theory of an algebraic function field over a field that differs from the complex field (1924). There, he regards an algebraic function no longer as an analytic function. Artin considered a quadratic extension K over the rational function field $k(x)$ with coefficients in a finite field k. Comparing K with a quadratic field over the field of rational numbers, he proved that both fields have similar number theoretic properties. Furthermore, following Dedekind's ζ-function, Artin defined the ζ-function of K and proved the functional equation for the ζ-function. He also showed through several examples that zeros s of $\zeta(s)$ are located on the line $\mathrm{Re}(s) = 1/2$, ($\mathrm{Re}(s)$ is the real part of s). This is the statement analogous to the famous Riemann hypothesis for the ζ-function of an algebraic number field.

Artin's results were generalized by F. K. Schmidt to general algebraic function fields over a finite field. Namely, Schmidt proved that, just as for algebraic number fields, the main theorems of class field theory (the theory of Abelian extensions) for such algebraic function fields still hold. He also gave a better definition of a ζ-function, and he gave a simple proof of the functional equation for the ζ-function. For his proof, he extended and proved the classical Riemann-Roch theorem to the case of algebraic function fields over a finite field. Schmidt also noticed that his proof is valid not only for the finite field case, but also for the perfect field case. By ingeniously applying valuation theory to his method, he further developed his theory so as to obtain the Riemann-Roch theorem for an algebraic function field over an arbitrary coefficient field (1936). His method, using "Normal basis", fundamentally follows Dedekind-Weber's idea. Meanwhile, Weil obtained the same result by a totally different idea.

Since the time of Artin's work, the ζ-function for an algebraic function field over a finite field, i.e., "Kongruenzzetafunktion", has attracted the attention of many mathematicians. In particular, the proof of the Riemann hypothesis for this ζ-function was the focus of much attention. In particular, Hasse studied this problem by reconstructing the classical elliptic function theory through algebraic (abstract) means. Then he proved the above Riemann hypothesis for elliptic function fields (1936). This work of Hasse was significant because he suggested the importance of the theory of algebraic correspondences of algebraic function fields for the proof of the Riemann hypothesis for the generalization. In fact, through the general theory of his new algebraic geometry, Weil built the theory of algebraic correspondences of algebraic curves in characteristic p. Then he completed the proof of the Riemann hypothesis for the "Kongruenzzetafunktion" by his theory of algebraic correspondences (1940–41).

Weil's paper on the detailed proof of the Riemann hypothesis recently appeared, just as his book on algebraic geometry was published. With Weil's work, the most general and complete foundation of algebraic geometry has been established, and the relationship between the algebraic geometric method and the algebraic method for algebraic function field theory has been clarified. The theory of algebraic functions has been somewhat settled by this approach as well as the analytic approach. This certainly does not mean that nothing needs to be done. For instance, we still have the problem of Hecke on the representation of the Galois group by Abelian integrals, the problem on a complex multiplication, etc. However, we are more equipped with methods and theories to study those problems, and the connection among these methods has been clarified.

Recently, an analytic approach and an algebraic geometric approach to the theory of algebraic functions of several complex variables has been actively studied. The single variable algebraic function theory may be a good model for the study of algebraic functions in several variables.

2. Among the three main approaches to algebraic function fields, we describe the algebraic approach in the first two chapters and the analytic approach in the latter three chapters of this book. We do not mention here the algebraic geometric approach since we expect the volume on algebraic geometry in this series will cover this topic.([1])

Next we will describe each of the chapters, taking into account the above historical comments.

Chapter 1 is a summary of the valuation theory that is needed for Chapter 2. Based on Cantor's theory of real numbers and the $\mathfrak{p}$-adic numbers of Hensel, Kürschák, Ostrowski, Krull, etc., generalized the valuation theory so that it has become indispensable for algebra and number theory. We focus on discrete non-Archimedean valuations for direct use in algebraic function theory. We follow orthodox methods in the proofs of theorems and in the description of the theory, and we emphasize somewhat the topological method by means of the metrics induced by valuations.

In Chapter 2 we explain the algebraic theory of algebraic function fields with the central theme being the Riemann-Roch theorem. Hence, the contents of this chapter coincide formally with the Dedekind-Weber paper. However, since we use valuation theory throughout the chapter, the theoretical structure is clarified and simplified. Since we consider an algebraic function field over an arbitrary field, our treatment is more general than that of Dedekind and Weber. As noted earlier, the general version of the Riemann-

([1]) *Translator's note.* This book is one of the books in the mathematical book series from Iwanami Shoten, Publishers.

Roch theorem was proved by F. K. Schmidt and A. Weil. Our proof of the Riemann-Roch theorem is based on Weil's idea, but we describe Weil's proof with the notion of idéle developed by Chevalley for algebraic number theory. In order to clarify the general Riemann-Roch theorem obtained through Weil's notion of differentials, we explain the more explicit notion of Hasse differentials defined in an algebraic function field over a perfect field. By using Hasse differentials, we aim to connect Chapter 2 with the classical algebraic function theory in the later chapters. Following the Riemann-Roch theorem, we mention a few applications of the Riemann-Roch theorem. Then we discuss simple algebraic results about rational function fields and elliptic function fields as preparation for the classical theory. As we mentioned in the historical note, the algebraic theory of an algebraic function field should include the number theoretic study of an algebraic function field over a finite field and the study of its Kongruenzzetafunktion. We have left these topics to the volumes on number theory and algebraic geometry.

In Chapter 3, as foundation for the analytic theory of algebraic functions, we introduce the notion of a Riemann surface. As noted earlier, Weyl gave for the first time a solid topological foundation for a Riemann surface. Then Rado showed that the triangulability of a Riemann surface is a consequence of other conditions. Thus, removing the triangulability assumption, he gave a simpler definition of a Riemann surface (1925). Following Rado's definition, we discuss analytic functions, differentials, integration, etc. on a Riemann surface. Then as the main goal of this chapter, we prove in §4 the existence of a differential and an analytic function which have given singularities. As mentioned in the historical note, this is what Riemann proved using the Dirichlet principle. We give the proof without using the Dirichlet principle following Weyl's method of using Hilbert spaces. Our motivation for this choice is mainly the success of Kodaira's work with the Hilbert space method in function theory in several complex variables. We also considered that, with Rado's definition, Weyl's method may be a better method for building Riemann surface theory. In the remainder of Chapter 3, we define a covering Riemann surface using the general theory of covering manifolds. Then, using the above existence theorem, we determined types of simply connected Riemann surfaces, from which we obtain the canonical types of Riemann surfaces. We follow closely Weyl's book for this discussion of the canonical types. The reader should notice how effective the group theoretic method is for such an analysis type problem.

In Chapter 4 we describe in detail the correspondence between algebraic function fields over the field of complex numbers and closed Riemann surfaces. In §1 analytic coordinates are induced in Dedekind-Weber's absolute Riemann surface, i.e., the set of prime divisors on a given algebraic function field K. Then we prove that the analytic function field on that closed Riemann surface is isomorphic to K. In §2 following Riemann, we prove that conversely, starting with a closed Riemann surface $\mathfrak{R}$, the analytic function field on $\mathfrak{R}$ is an algebraic function field over the field of complex numbers. Furthermore, we verify that the closed Riemann surface consisting of prime divisors of the above algebraic function field is isomorphic to $\mathfrak{R}$. Thus, up to isomorphisms, there is a one-to-one correspondence between algebraic function fields and closed Riemann surfaces. We also show that algebraic notions of an algebraic function field correspond to topological and analytical notions of a closed Riemann surface. In §3, by using several results in the theory of surfaces, we prove that the genus of an algebraic function field coincides with the genus of the corresponding closed Riemann surface, determining the topological properties of closed Riemann surfaces.

In Chapter 5, considering the correspondence obtained in Chapter 4, we obtain some of the main results in the classical algebraic function theory. In §1 we treat Abelian integrals, by which we prove that the additive group of differentials of the first kind in K and the first homology group of $\mathfrak{R}$ with real coefficients form a dual pair of topological Abelian groups in Pontrjagin's sense. This duality theorem refines the previous result that the algebraic genus of K coincides with the topological genus of $\mathfrak{R}$. A crucial part of the proof of the duality theorem is the computation of the Abelian integral along the circumference of a region in the simply connected Riemann surface over $\mathfrak{R}$. This method of proof is attributed to Riemann.

In §2 we consider a simply connected covering Riemann surface $\mathfrak{R}^*$ of $\mathfrak{R}$. Then one can regard an analytic function, i.e., an element of K, as a function on $\mathfrak{R}^*$ and conversely regard an analytic function on $\mathfrak{R}^*$ as a multivalued function on $\mathfrak{R}^*$. Among functions on $\mathfrak{R}^*$, we examine the simplest possible ones, i.e., additive functions and multiplicative functions. Additive functions are nothing but the Abelian integrals of the first kind and the second kind. Riemann introduced the notion of multiplicative functions in the strict sense. A multiplicative function is a generalization to an algebraic function of the well-known fact that any rational function can be factored into a product of linear functions. Using integral characters obtained by multiplicative functions, we prove our main theorem of §2 that, in Pontrjagin's sense, the divisor class group of degree 0 of K is dual to the first homology group of $\mathfrak{R}$ as topological Abelian groups. The reader should notice that the main results in this section and the previous section are obtained in the form of Pontrjagin's duality theorem for topological Abelian groups. Note that some proofs in §1 can be simplified if one uses $\mathfrak{R}^*$ in this section.

From the results in §2, we are able to obtain in §3 the classical Abel's theorem and Jacobi's theorem rather easily. The original forms of these theorems are slightly different from the way we state them here. As we mentioned in the historical note, Abel and Jacobi's works greatly influenced the development of algebraic function fields. The work of Riemann and Weierstrass to solve Jacobi's converse problem is considered a climax of classical mathematics.

We could make a short cut to the same results through modern mathematics. To appreciate the classical approach, we provide a short tour through Riemann's Abelian functions and ϑ-functions.

In §4 we consider extensions of an algebraic function field, in particular unramified extensions in terms of covering Riemann surfaces. The basic device for this study comes from §5 of Chapter 3 and §2 of Chapter 5. We show that unramified Abelian extensions of K are captured by the divisor class group of K. This proves that not only for an algebraic function field over a finite field, but also for such an extension over K as in the above, some analogous statements of the class field theory hold. Historically speaking, when Hilbert wrote the famous "Zahlbericht", outlining his grand idea of class field theory, he seems to have had in mind such results on unramified Abelian extensions over an algebraic function field. A similar theory can be developed for ramified extensions. But this material is not covered.

The last section, §5, is devoted to elliptic function theory. We treat elliptic function theory in the framework of algebraic function theory. That is, beginning with the $\wp$-function following Weierstrass, we derive the modular function from the invariant of an elliptic function field. Then we define $\zeta(u)$ and $\sigma(u)$ as an additive function and a multiplicative function, respectively. We end this book with a brief comment on unramified extensions and complex multiplication.

3. As the reader may have noticed in the above summary of this book, we emphasize the analytic theory more than the algebraic theory of algebraic function fields. We describe the algebraic theory sufficiently for applications to the analytic theory. We would like to explain why we have chosen classical theory. One of the primary reasons for this choice is that algebraic function theory in several variables, which is currently being developed rapidly, may benefit greatly from the single variable theory, and the second reason is that the literature on the algebraic theory is newer and more accessible to the reader. Our main intention is to show the process of modern mathematics being built through abstractions and generalizations of classical mathematics and how some of the classical results can be obtained clearly and rapidly from the point of modern mathematics.

The author first considered presenting algebraic function theory following the historical development. However, that would require profound knowledge in both classical and modern mathematics. In this book, we followed the local and logical order in the theory, disregarding the historical order. We used the methods of modern mathematics freely in order to obtain the main classical results as efficiently as possible. Consequently, the order of the presentation in the book reverses the order of historical developments.

Thus, we freely use fundamental results from algebra and topology which form a core of modern mathematics. For prerequisite knowledge, the reader can consult the books noted in various places throughout this book. The reference papers and books have been chosen for the interested reader from recent and easily accessible publications. For a classical reference, see W. Wirtinger, *Algebraishe Funktionen und ihre Integrale.*

We try to be consistent in our notation. We use the same notation for the corresponding concepts between the algebraic theory and the analytic one, e.g., a prime divisor on an algebraic function field and the point on the closed Riemann surface. However, the notations $\mathfrak{G}$, $\mathfrak{H}$, $\mathfrak{M}$ in Chapter 3 are used with different meanings in other chapters. We also follow common terminology usage with the exception of a few coined words.

It would be our great pleasure if the reader is influenced by nineteenth century mathematics through this book and finds something in this book that leads to future research.

Note. Most of the material in this book was written during the years 1947–1948. Recently, modern treatments of the theory have appeared in a book in Japanese by Inaba and in the book by C. C. Chevalley, *Introduction to the theory of algebraic functions of one variable*, Math. Surveys Monographs, vol. 6, Amer. Math. Soc., Providence, RI, 1951. In Chevalley's book the algebraic theory of algebraic functions over an arbitrary coefficient field is lucidly described. I had an opportunity to listen to E. Artin's lectures at Princeton University on algebraic number theory which included the algebraic theory of algebraic functions. I would have made several changes in this book, especially in Chapters 1 and 2, had I had the opportunity. I hope to have the opportunity to revise this book in the future. (Princeton, September 1951).

Kenkichi Iwasawa

CHAPTER 1

Preparation from Valuation Theory

§1. Valuations and prime divisors

The field of real numbers and the field of complex numbers have various properties other than being fields in the abstract sense. One of these outstanding properties is that for an arbitrary element a, i.e., a number, there is defined the absolute value of a, denoted $|a|$, satisfying

$$|a| \geq 0, \quad |a| + |b| \geq |a+b|, \quad \text{and} \quad |a|\,|b| = |ab|.$$

The notion of valuation for abstract fields is a generalization of absolute value. We will consider a special valuation, called a discrete or exponential valuation, rather than a general theory of valuations. (See Note 2 at the end of this section). We will begin with the definition.

DEFINITION 1.1. A function $v(a)$ defined for all a in a field K is said to be a *valuation* on K if v satisfies the following conditions:

(i) $v(0) = +\infty$ and $v(a)$ is a rational integer for $a \neq 0$,
(ii) $v(a) + v(b) = v(ab)$,
(iii) $v(1+a) \geq 0$ for $v(a) \geq 0$,
(iv) $v(a) \neq 0, \infty$ for some a.

Condition (ii) above means that v is a homomorphism from the multiplicative group $K^* = K - \{0\}$ to the additive group I of rational integers. Therefore, if we let

$$G = \{v(a)\,;\, a \in K^*\},$$

then G is a subgroup of I and is a called the *value group* of v. Since G contains a nonzero rational integer by (iv), G has a finite index e for I. Then let

$$v^*(a) = \frac{1}{e} v(a). \tag{1.1}$$

Then v^* also satisfies conditions (i) through (iv), and hence v^* is a valuation on K. The value group of v^* becomes the set of rational integers I. Generally, any valuation with value group I is said to be a *normalized valuation.* For an arbitrary valuation v, one can get a normalized valuation v^* by (1.1). We say that v^* belongs to v.

DEFINITION 1.2. If the normalized valuations v_1^* and v_2^* belonging to v_1 and v_2 are the same, then we say v_1 and v_2 are equivalent and write

$$v_1 \sim v_2.$$

This equivalence relation divides the valuations on K into equivalence classes called prime divisors of K. They are denoted by $P, Q, \ldots$.

Each prime divisor v of K contains a unique normalized valuation denoted by v_P.

Simple properties of valuations and prime divisors are given in the following lemmas.

LEMMA 1.1. *Let v be an arbitrary valuation on K, and let a and b be elements of K. Then we have*

$$v(1) = v(-1) = 0, \qquad v(-a) = v(a), \qquad v(1/a) = -v(a) \quad \textit{for } a \neq 0,$$

and

$$v(a+b) \geq \mathrm{Min}(v(a), v(b)), \tag{1.2}$$

where the above inequality becomes an equality for the case $v(a) = v(b)$.

PROOF. Let $a = b = 1$ or $a = b = -1$ in (ii) of Definition 1.1. Then we have $v(1) = v(-1) = 0$. Therefore we have $v(-a) = v(a)$ for $b = -1$ in (ii), and we also have $0 = v(1) = v(a \cdot 1/a) = v(a) + v(1/a)$. Hence $v(1/a) = -v(a)$. For $a = 0$ or $b = 0$, (1.2) obviously holds. So we may assume $a \neq 0$ and $b \neq 0$ and $v(b) \geq v(a)$, i.e., $v(b/a) \geq 0$. Then by (iii) we have

$$v(a+b) = v(a(1+b/a)) = v(a) + v(1+b/a) \geq v(a).$$

The case $v(a) \geq v(b)$ can be proved similarly. Therefore (1.2) has been proved. In general we have

$$v(a) = v(a+b-b) \geq \mathrm{Min}(v(a+b), v(-b)) = \mathrm{Min}(v(a+b), v(b)).$$

If $v(b) > v(a)$ holds, then the right side Min of the above inequality must be $v(a+b)$. Then we have

$$v(a) \geq v(a+b),$$

in which case equality holds in (1.2). For the other case $v(a) > v(b)$, the proof is the same.

LEMMA 1.2. *Let v be an arbitrary valuation on K. Let*

$$\mathfrak{o} = \{a : a \in K,\ v(a) \geq 0\} \quad \textit{and} \quad \mathfrak{p} = \{a' : a' \in K,\ v(a') > 0\}. \tag{1.3}$$

Then we have:

(i) *$\mathfrak{o}$ is a subring of K, and an arbitrary element c of K can be written as a quotient of elements of $\mathfrak{o}$, i.e., $c = a_1/a_2$, a_1 and $a_2 \in \mathfrak{o}$ and $a_2 \neq 0$.*

(ii) *The ideal* $\mathfrak{p}$ *is a prime ideal of* $\mathfrak{o}$, *and any nonzero ideal* J *of* $\mathfrak{o}$ *can be written uniquely as a power of* $\mathfrak{p}$.

(iii) *The ideal* (c) *generated by an element* $c \neq 0$ *of* K *may be written as* $\mathfrak{p}^n$ *from* (ii). *Then if you let*

$$v^*(c) = n,$$

v^* *is a normalized valuation on* K *belonging to* v.

PROOF. (i) Let a and b be elements of $\mathfrak{o}$, i.e., $v(a) \geq 0$ and $v(b) \geq 0$. Then we have

$$v(a \pm b) \geq \mathrm{Min}(v(a), v(b)) \geq 0 \quad \text{and} \quad v(ab) = v(a) + v(b) \geq 0.$$

Hence, $a + b$ and ab belong to $\mathfrak{o}$, and $\mathfrak{o}$ is a ring. If an element c of K does not belong to $\mathfrak{o}$, then let $c = 1/d$. We have $v(d) = -v(c) > 0$ and $d \in \mathfrak{o}$, i.e., $a_1 = 1$ and $a_2 = d$.

(ii) First $1 \in \mathfrak{o}$. Next, for a' and b' in $\mathfrak{p}$, $v(a') > 0$ and $v(b') > 0$ hold. Then $v(a' + b') \geq \mathrm{Min}(v(a'), v(b')) > 0$ implies $a' + b' \in \mathfrak{p}$. Let a be an arbitrary element of $\mathfrak{o}$. Since $v(a) \geq 0$, $v(aa') = v(a) + v(a') > 0$ implies $aa' \in \mathfrak{p}$. Suppose that $ab \in \mathfrak{p}$ for a and b in $\mathfrak{o}$. Then we have

$$v(a) \geq 0, \quad v(b) \geq 0, \quad \text{and} \quad v(ab) = v(a) + v(b) > 0.$$

Hence, $v(a) > 0$ or $v(b) > 0$ must hold, i.e., $a \in \mathfrak{p}$ or $b \in \mathfrak{p}$. The ideal $\mathfrak{p}$ is indeed a prime ideal of $\mathfrak{o}$.

Let $e = \mathrm{Min}(v(a'); a' \in \mathfrak{p})$, and let $v(t) = e$ for $t \in \mathfrak{p}$. Then e is the smallest positive value of $v(a)$, where a runs through the elements of K. Hence, for the value group G of v, we have

$$e = [I : G].$$

If J is any nonzero $\mathfrak{o}$-ideal J in K, one can find a nonzero element c in K such that $cJ \subseteq \mathfrak{o}$. Then

$$m = \mathrm{Min}(v(a'); a' \in J)$$

is finite. Hence, if we let

$$m = v(a_0), \qquad a_0 \in J, \quad m = ne,$$

we have

$$v(t^n a_0^{-1}) = nv(t) - v(a_0) = 0, \qquad b = t^n a_0^{-1} \in \mathfrak{o}.$$

Hence, $t^n = ba_0 \in J$ and $(t^n) \subseteq J$. Moreover, for an arbitrary element a' of J,

$$v(a't^{-n}) = v(a') - nv(t) = v(a') - m \geq 0, \qquad b' = a't^{-n} \in \mathfrak{o}$$

holds. Therefore, we have $a' = b't^n \in (t^n)$ and $J \subseteq (t^n)$. Then we have $J = (t^n) = (t)^n$. In particular, for $J = \mathfrak{p}$ and $a_0 = t$, $\mathfrak{p} = (t)$ holds. Hence, we obtain

$$J = \mathfrak{p}^n.$$

That is, J is a power of $\mathfrak{p}$. On the other hand, if one lets $J = \mathfrak{p}^{n'}$, then $J = (t^{n'})$. Consequently, we have

$$m = n'e \quad \text{and} \quad n = n'.$$

(iii) From the above proof we can let $(c) = \mathfrak{p}^n$. Then we have $v(c) = ne$. Hence

$$v^*(c) = (1/e)v(c) = n$$

is a normalized valuation on K belonging to v. Q.E.D.

If one replaces v with an equivalent valuation v' in the definition (1.3) then $\mathfrak{o}$ and $\mathfrak{p}$ will be unchanged. We have the following definition.

DEFINITION 1.3. Let P be a prime divisor on K and let v be an arbitrary valuation belonging to P. Then $\mathfrak{o}$ and $\mathfrak{p}$, as defined in (1.3), are said be the *valuation ring* and the *prime ideal* of P, and the field $\mathfrak{o}/\mathfrak{p}$ is called the *residue field* of P.

LEMMA 1.3. *Let v_1 and v_2 be valuations on K. If we always have $v_2(a) \geq 0$ whenever $v_1(a) \geq 0$, then v_1 and v_2 are equivalent.*

PROOF. Let $\mathfrak{o}_1$, $\mathfrak{p}_1$, $\mathfrak{o}_2$, and $\mathfrak{p}_2$ be corresponding valuation rings and prime ideals to v_1 and v_2, respectively. The hypothesis implies

$$\mathfrak{o}_1 \subseteq \mathfrak{o}_2.$$

Since $\mathfrak{o}_1 \cap \mathfrak{p}_2$ is a prime ideal of $\mathfrak{o}_1$ and the only prime ideal of $\mathfrak{o}_1$ is $\mathfrak{p}_1$, we obtain

$$\mathfrak{p}_1 = \mathfrak{o}_1 \cap \mathfrak{p}_2.$$

For $v_1(b) < 0$, we have $v_1(b^{-1}) > 0$, i.e., $b^{-1} \in \mathfrak{p}_1$. Then we have $b^{-1} \in \mathfrak{p}_2$, $v_2(b^{-1}) > 0$ and $v_2(b) < 0$. That is, $v_2(b) \geq 0$ implies $v_1(b) \geq 0$, i.e., $\mathfrak{o}_2 \subseteq \mathfrak{o}_1$. Hence, we have $\mathfrak{o}_1 = \mathfrak{o}_2$ and $\mathfrak{p}_1 = \mathfrak{p}_2$. Then by Lemma 1.2, the normalized valuations belonging to v_1 and v_2 are the same, i.e., $v_1 \sim v_2$.

LEMMA 1.4 (Ostrowski). *For nonequivalent valuations $v_1, v_2, \ldots, v_n$ $(n \geq 2)$, on K, there exists an element x in K such that*

$$v_1(1-x) > 0, \; v_2(x) > 0, \ldots, v_n(x) > 0 \tag{1.4}$$

holds.

PROOF. First we will prove the case where $n = 2$. Lemma 1.3 implies that one can choose a such that

$$v_1(a) \geq 0 \quad \text{and} \quad v_2(a) < 0$$

holds. Let

$$x_1 = 1/a \quad \text{for } v_1(a) = 0$$

and

$$x_1 = 1/(1+a) \quad \text{for } v_1(a) > 0.$$

Then by Lemma 1.1 we have $v_1(x_1) = 0$ and $v_2(x_1) > 0$. Similarly there exists an x_2 such that $v_1(x_2) > 0$ and $v_2(x_2) = 0$. Let $x = x_1/(x_1 + x_2)$. Since $v_1(x_1 + x_2) = v_2(x_1 + x_2) = 0$, we obtain

$$v_1(1 - x) = v_1(x_2/(x_1 + x_2)) = v_1(x_2) - v_1(x_1 + x_2) > 0$$

and

$$v_2(x) = v_2(x_1/(x_1 + x_2)) = v_2(x_1) - v_2(x_1 + x_2) > 0.$$

We have proved the case $n = 2$. We will prove the case $n \geq 3$ by mathematical induction. The hypothesis implies that there exist elements x_1 and x_2 in K such that

$$v_1(1 - x_1) > 0, \ v_3(x_1) > 0, \ v_4(x_1) > 0, \ldots, v_n(x_1) > 0$$

and

$$v_1(1 - x_2) > 0, \ v_2(x_2) > 0, \ v_4(x_2) > 0, \ldots, v_n(x_2) > 0$$

hold. Then we let

(1) $x = x_1 x_2$ if $v_2(x_1) \geq 0$ and $v_3(x_2) \geq 0$,
(2) $x = x_1/(1 + x_1(1 - x_1))$ if $v_2(x_1) < 0$, and
(3) $x = x_2/(1 + x_2(1 - x_2))$ if $v_2(x_1) \geq 0$ and $v_3(x_2) < 0$.

One can verify, by using Lemma 1.1, that each choice of x satisfies (1.4). The actual computation is left for the readers.

LEMMA 1.5. *Let $v_1, v_2, \ldots, v_n$ be nonequivalent normalized valuations on K. Then there exist elements $x_1, x_2, \ldots, x_n$ in K such that*

$$v_i(x_j) = \delta_{ij}$$

holds, where δ_{ij} is the Kronecker symbol.

PROOF. Since v_1 is a normalized valuation, there is an element a such that $v_1(a) = 1$. Let

$$x_1 = ax^m + (x - 1)^m,$$

where x is chosen so that (1.4) is satisfied, and where m is such a sufficiently large natural number as will be determined in the following computation. We have

$$x_1 = a + (x - 1)\left(a\frac{x^m - 1}{x - 1} + (x - 1)^{m-1}\right),$$

$$v_1\left(a\frac{x^m - 1}{x - 1} + (x - 1)^{m-1}\right) \geq 1.$$

Then (1.2) implies

$$v_1(x_1) = v_1(a) = 1.$$

Furthermore, for $i \geq 2$ and for a sufficiently large m, we have

$$v_i(ax^m) = v_i(a) + mv_i(x) > 0 \quad \text{and} \quad v_i((x-1)^m) = 0.$$

Once again by (1.2) we obtain

$$v_i(x_1) = 0.$$

One can choose $x_2, \ldots, x_n$ in a similar manner. Q.E.D.

Using these lemmas, we will prove the following theorem which will be needed later.

THEOREM 1.1 (approximation theorem). *Let $v_1, v_2, \ldots, v_n$ be mutually nonequivalent valuations on K and also let $y_1, y_2, \ldots, y_n$ be arbitrary elements of K. Then for any natural number m there exists an element y in K such that*

$$v_i(y - y_i) > m, \qquad i = 1, 2, \ldots, n.$$

PROOF. From Lemma 1.4 one can find $a_1, a_2, \ldots, a_n$ so that one may have

$$v_i(1 - a_i) > 0 \quad \text{and} \quad v_i(a_j) > 0 \quad \text{for } j \neq i.$$

For a sufficiently large N, choose y as follows:

$$y = \sum_{i=1}^{n} y_i(1 - (1 - a_i^N)^N).$$

Since we have

$$\begin{aligned} v_i(y - y_i) &\geq \operatorname{Min}(v_i(y_i(1 - a_i^N)^N),\ v_i(y_j(1 - (1 - a_j^N)^N)),\ (j \neq i)) \\ &\geq \operatorname{Min}(v_i(y_i) + Nv_i(1 - a_i),\ v_i(y_j) + Nv_i(a_j),\ (j \neq i)), \end{aligned}$$

for a sufficiently large N we can have the inequality

$$v_i(y - y_i) > m, \qquad i = 1, 2, \ldots, n.$$

THEOREM 1.2. *Let $P_1, P_2, \ldots, P_n$ be mutually distinct prime divisors on K, and let $v_i = v_{P_i}$ be the normalized valuation belonging to P_i. Also let*

$$\mathfrak{o} = \{a;\, a \in K,\ v_i(a) \geq 0,\ i = 1, 2, \ldots, n\},$$
$$\mathfrak{p}_i = \{a';\, a' \in \mathfrak{o},\ v_i(a') > 0\}, \qquad i = 1, 2, \ldots, n.$$

Then

(i) *$\mathfrak{o}$ is a subring of K, and an arbitrary element c of K can be expressed as a quotient of two elements in $\mathfrak{o}$.*

(ii) *$\mathfrak{p}_i$ $(i = 1, 2, \ldots, n)$ is a prime ideal of $\mathfrak{o}$, and any nonzero $\mathfrak{o}$-ideal J of K can be decomposed uniquely as*

$$J = \mathfrak{p}_1^{e_1}\mathfrak{p}_2^{e_2}\cdots\mathfrak{p}_n^{e_n}.$$

(iii) *For an element c in K such that*

$$(c) = \mathfrak{p}_1^{e_1}\mathfrak{p}_2^{e_2}\cdots\mathfrak{p}_n^{e_n},$$

we have

$$v_i(c) = e_i, \qquad i = 1, 2, \ldots, n. \tag{1.5}$$

PROOF. (i) Since $\mathfrak{o}$ is the intersection of the valuation rings of the prime divisors $P_1, P_2, \ldots, P_n$, it is clearly a subring of K. The presentation of an element of K as a quotient of elements of $\mathfrak{o}$ will be proved in (iii).

(ii) The proof for (ii) of Lemma 1.2 implies that $\mathfrak{p}_i$ is a prime ideal of $\mathfrak{o}$. Next, for an arbitrary $\mathfrak{o}$-ideal J, let $e_i = \mathrm{Min}(v_i(a); a \in J)$ and let $v_i(a_i) = e_i$ for some $a_i \in J$. Then e_i is finite as we saw earlier. Let $x_1, x_2, \ldots, x_n$ be as in Lemma 1.5 and let

$$a_0 = \sum_{i=1}^{n} a_i(x_1 \cdots x_{i-1}x_{i+1} \cdots x_n)^m,$$

where m is a sufficiently large natural number. Then a_0 is clearly in J, and a simple computation shows

$$v_i(a_0) = e_i, \qquad i = 1, 2, \ldots, n.$$

Hence, for an arbitrary element a of J we have

$$v_i(aa_0^{-1}) = v_i(a) - v_i(a_0) \geq 0, \quad \text{i.e., } b = aa_0^{-1} \in J.$$

This implies $a = ba_0 \in (a_0)$, i.e., $J \subseteq (a_0)$. On the other hand, $(a_0) \subseteq J$ is obvious. We obtain $J = (a_0)$. If we let

$$a' = x_1^{e_1}x_2^{e_2} \cdots x_n^{e_n}, \qquad v_i(a') = e_i,$$

then since $a_0a'^{-1}$ and $a'a_0^{-1}$ belong to $\mathfrak{o}$, we have

$$J = (a_0) = (a') = (x_1)^{e_1}(x_2)^{e_2} \cdots (x_n)^{e_n}.$$

In particular, for $J = \mathfrak{p}_i$, $\mathfrak{p}_i = (x_i)$ follows immediately since x_i is contained in $\mathfrak{p}_i$. Therefore, we generally have

$$J = \mathfrak{p}_1^{e_1}\mathfrak{p}_2^{e_2} \cdots \mathfrak{p}_n^{e_n}.$$

Suppose J has another decomposition

$$J = \mathfrak{p}_1^{e_1'}\mathfrak{p}_2^{e_2'} \cdots \mathfrak{p}_n^{e_n'}.$$

Then $J = (x_1^{e_1'}x_2^{e_2'} \cdots x_n^{e_n'})$ implies that $b = x_1^{e_1-e_1'}x_2^{e_2-e_2'} \cdots x_n^{e_n-e_n'}$ and b^{-1} must be in $\mathfrak{o}$. Hence, we have

$$e_1 = e_1', \; e_2 = e_2', \; \ldots, \; e_n = e_n'.$$

(iii) If $J = (c)$, we obviously have

$$e_i = \mathrm{Min}(v_i(a); a \in J) = v_i(c), \qquad i = 1, 2, \ldots, n.$$

Then the proof of (ii) plainly implies (1.5). If

$$(c) = \prod_{i=1}^{n} \mathfrak{p}_i^{e_i}, \qquad c = c'x_1^{e_1} \cdots x_n^{e_n},$$

then $c' \in \mathfrak{o}$. Let a_1 be the product of c' and $x_i^{e_i}$ for $e_i \geq 0$, and let a_2 be the product of $x_i^{e_i}$ for $e_i < 0$. Now we have

$$c = \frac{a_1}{a_2}, \qquad a_1, a_2 \in \mathfrak{o}.$$

This proves the second half of (i). Q.E.D.

In a way, the next theorem gives the converse of Theorem 1.2.

THEOREM 1.3. *Suppose $\mathfrak{o}$ is a subring of K such that*

(i) *an arbitrary element c of K can be written as a quotient of elements of $\mathfrak{o}$, and*

(ii) *any $\mathfrak{o}$-ideal of K can be decomposed uniquely as a product of prime ideals, and furthermore if $J \subseteq J'$, then we have $J = J'J''$ for some ideal $J'' \subseteq \mathfrak{o}$.*

For an arbitrary prime ideal $\mathfrak{p}$ of $\mathfrak{o}$, when an element a of K is divisible precisely by $\mathfrak{p}^n$, define

$$v_{\mathfrak{p}}(a) = n. \tag{1.6}$$

Then $v_{\mathfrak{p}}$ is a normalized valuation on K called the $\mathfrak{p}$-adic valuation. Its residue field (i.e., the residue field of the prime divisor belonging to $v_{\mathfrak{p}}$) is isomorphic to $\mathfrak{o}/\mathfrak{p}$. Furthermore, we certainly have $v_{\mathfrak{p}}(a) \geq 0$ for a in $\mathfrak{o}$. Conversely if we have a valuation v on K such that $v(a) \geq 0$ for all a in $\mathfrak{o}$, then v is equivalent to the $\mathfrak{p}$-adic valuation associated with some prime ideal $\mathfrak{p}$ as described above.

REMARK. The simplest example for K and $\mathfrak{o}$ is given by the field of rational numbers and the ring of rational integers. The general theory of ring $\mathfrak{o}$ may be found in the Chapter 14 of *Modern Algebra* II by B. L. van der Waerden, 2nd ed., Springer-Verlag, Berlin, 1940. As shown there, various properties in the arithmetic of algebraic number fields are valid for $\mathfrak{o}$. We will use some of those properties in the following proof. However, if the reader is not familiar with the general theory of the ring $\mathfrak{o}$, or with the arithmetic of algebraic number fields, then the reader may read the proof assuming that $K = k(x)$, the field of rational functions and $\mathfrak{o} = k[x]$, the ring of polynomials, since we only need this special case of K and $\mathfrak{o}$ for the following theory of algebraic function fields.

PROOF. It is easy to see that $v_{\mathfrak{p}}$ defined by (1.6) satisfies (i) through (iv) in Definition 1.1 from the definition of an ideal. Let $\mathfrak{o}_1$ and $\mathfrak{p}_1$ be the valuation ring and the prime ideal of $v_{\mathfrak{p}}$. Then we clearly have

$$\mathfrak{o}_1 \supseteq \mathfrak{o} \quad \text{and} \quad \mathfrak{p}_1 \cap \mathfrak{o} = \mathfrak{p}.$$

For an arbitrary element c of $\mathfrak{o}_1$ there exist a_1 and a_2 in $\mathfrak{o}$ such that $c = a_1/a_2$, where a_2 is not in $\mathfrak{p}$. Choose b in $\mathfrak{o}$ so that

$$a_2 b \equiv 1 \bmod \mathfrak{p}$$

may hold. Then $c' = c - a_1 b$ is in $\mathfrak{p}_1$, i.e., $\mathfrak{o}_1 = \mathfrak{o} + \mathfrak{p}_1$. Hence, the ring isomorphism theorem provides the following isomorphism

$$\mathfrak{o}_1/\mathfrak{p}_1 \cong \mathfrak{o}/\mathfrak{p}.$$

Conversely, if v is an arbitrary valuation on K such that $v(a) \geq 0$ for all a in $\mathfrak{o}$, then

$$\mathfrak{p} = \{a' \, ; \, a' \in \mathfrak{o}, \; v(a') > 0\}$$

is a prime ideal of $\mathfrak{o}$ (see the proof of Lemma 1.2(ii)). Let a_0 be an element in $\mathfrak{o}$ which is divisible precisely by $\mathfrak{p}$, and let c be an element of K which is precisely divisible by $\mathfrak{p}^n$. Then there exist a_1 and a_2 in $\mathfrak{o}$ such that

$$ca_0^{-n} = \frac{a_1}{a_2},$$

where $a_1, a_2 \notin \mathfrak{p}$. We obtain the following:

$$v(c) - nv(a_0) = v(ca_0^{-n}) = v(a_1 a_2^{-1}) = v(a_1) - v(a_2) = 0$$

and

$$v(c) = nv(a_0) = v(a_0)v_{\mathfrak{p}}(c).$$

Here a_0 is fixed and c is arbitrary in K. Hence, we have $v \sim v_{\mathfrak{p}}$. Q.E.D.

EXAMPLE 1. Let K be a finite algebra number field, and let $\mathfrak{o}$ be the collection of all algebraic integers. A fundamental theorem in the theory of algebraic number fields implies that conditions (i) and (ii) of Theorem 1.3 are satisfied. Therefore, each prime ideal of $\mathfrak{o}$ gives a valuation of K, and hence gives a prime divisor. Conversely, one can prove that all the prime divisors, in the sense of Definitions 1.1 and 1.2, come from prime ideals. However, we will not discuss this topic any further since there is no direct connection with our theory.

EXAMPLE 2. Let k be an arbitrary field, and let $K = k(x)$ be the field of rational functions of one variable x with coefficients in k. Let us find a normalized valuation v on K such that for any nonzero element α in k,

$$v(\alpha) = 0 \tag{1.7}$$

should hold. If we assume $v(x) \geq 0$, then we have

$$\begin{aligned} v(\alpha_0 + \alpha_1 x + \cdots + \alpha_n x^n) &\geq \mathrm{Min}(v(\alpha_i x^i); \; i = 0, 1, \ldots, n) \\ &= \mathrm{Min}(v(x^i)) \geq 0, \qquad \alpha_i \in k. \end{aligned}$$

That is, for any element a in the polynomial ring $\mathfrak{o} = k[x]$, we have

$$v(a) \geq 0.$$

For K the ring $\mathfrak{o}$ plainly satisfies the conditions (i) and (ii) of Theorem 1.3. Therefore the normalized valuation v is obtained from a prime ideal $(f(x))$ of $k[x]$, where $f(x)$ is an irreducible polynomial in $k[x]$. In the case $v(x) < 0$, we have $v(1/x) > 0$, and v is then obtained from a prime ideal of $k[1/x]$ as above. Furthermore, since this prime ideal contains $1/x$, it must be $(1/x)$. Conversely, the normalized valuations defined by the prime ideals $(f(x))$ in $k[x]$ and $(1/x)$ in $k[1/x]$ clearly satisfy (1.7).

In particular, if k is an algebraically closed field, irreducible polynomials in $k[x]$ are of the linear form $x-\alpha$, $\alpha \in k$. We will denote the normalized valuation and prime divisor obtained from this prime ideal as v_α and as P_α, respectively. The corresponding normalized valuation and prime divisor to the prime ideal $(1/x)$ in $k[1/x]$ are denoted as v_∞ and P_∞, respectively. Thus, there is a one-to-one correspondence between all the prime divisors satisfying (1.7) and the elements of k and ∞. Note that the notations P_α and P_∞ depend on the choice of the variable x, not on the field $K = k(x)$ alone.

NOTES 1. Readers are expected to have some background in abstract algebra in order to understand this chapter and Chapter 2. See, for example, B. L. van der Waerden, *Modern algebra* I, Springer, Berline, 1940 or A. Albert, *Modern higher algebra*, University of Chicago Press, Chicago, IL, 1956.

2. We did not describe the theory of general valuations since we will not need it in this treatise. However, we will give the definition here for the reader's reference.

A real-valued function $\varphi(a)$ on any field K is said to be a valuation on K if the following conditions are satisfied:

(i) $\varphi(0) = 0$, and $\varphi(a) > 0$ for $a \neq 0$,

(ii) $\varphi(ab) = \varphi(a)\varphi(b)$,

(iii) $\varphi(a+b) \leq \varphi(a) + \varphi(b)$.

For any field there is a least one function satisfying (i), (ii), (iii). That is, $\varphi(0) = 0$ and $\varphi(a) = 1$ for $a \neq 0$. Since this is a trivial valuation, we often exclude this valuation, and we add another condition:

(iv) There exists an element a such that $\varphi(a) \neq 0, 1$.

If a valuation $\varphi(a)$ on K satisfies the following condition stronger than (iii):

(iii′) $\varphi(a+b) \leq \mathrm{Max}(\varphi(a), \varphi(b))$,

then the valuation $\varphi(a)$ is said to be non-Archimedean.

For a non-Archimedean valuation $\varphi(a)$, let

$$v(a) = -\log \varphi(a).$$

Then $v(a)$ is not necessarily a rational integer valued function, yet $v(0) = +\infty$, and one can immediately verify (ii), (iii), (iv) of Definition 1.1. Conversely, if a real-valued function $v(a)$ satisfies these conditions, then the function

$$\varphi(a) = e^{-v(a)}$$

is a non-Archimedean valuation on K. Such a $v(a)$ as above is generally called an exponential valuation on K. In our treatise, $v(a)$ takes values in rational integers, i.e., the set $\{v(a); a \in K\}$ is discrete on the real line. We call such $v(a)$, "a discrete exponential valuation."

Note that some of the properties of $v(a)$ mentioned in this section are satisfied by a general valuation as well.

§2. The metric induced by a valuation

Let v be a valuation on the field K. For arbitrary elements a and b in K, define

$$\rho(a, b) = \exp(-v(a-b)) = e^{-v(a-b)}. \tag{2.1}$$

Then Definition 1.1(i) and Lemma 1.1 imply

(1) $\rho(a, a) = 0$ and $\rho(a, b) > 0$ for $a \neq b$,
(2) $\rho(a, b) = \rho(b, a)$,
(3) $\rho(a, b) + \rho(b, c) \geq \rho(a, c)$.

That is, ρ gives a metric on K. This $\rho = \rho_v$ is said to be the metric on K belonging to the valuation v. By using Definition 1.1(ii), and (1.2), a simple computation shows that the sum $a+b$ and the product ab are continuous for the above metric ρ:

$$\begin{aligned} \rho(a+b, a'+b') &\leq \operatorname{Max}(\rho(a, a'), \rho(b, b')), \\ \rho(ab, a'b') &\leq \operatorname{Max}(\rho(0, a)\rho(b, b'), \rho(0, b')\rho(a, a')). \end{aligned} \tag{2.2}$$

Moreover, since one can also show that a^{-1} is continuous for $a \neq 0$, K becomes a topological field such that $v(a)$ is a continuous function for the topology on K.

THEOREM 1.4. *A necessary and sufficient condition for valuations v_1 and v_2 to be equivalent is that the topologies induced by the metrics ρ_{v_1} and ρ_{v_2} belonging to v_1 and v_2 are equivalent.*

PROOF. First let $v_1 \sim v_2$. Then by definition there exists a positive rational number r such that

$$v_2(a) = rv_1(a) \quad \text{for } a \in K.$$

Then we have

$$\rho_{v_2}(a, b) = (\rho_{v_1}(a, b))^r.$$

It is now obvious that ρ_{v_1} and ρ_{v_2} give equivalent topologies on K.

Conversely, suppose that ρ_{v_1} and ρ_{v_2} give equivalent topologies in K. Let a be any element of K such that $v_1(a) < 0$ holds. We have

$$\rho_{v_1}(a^{-n}, 0) = \exp(nv_1(a)) \to 0 \quad \text{as } n \to \infty.$$

That is to say, the sequence: $a^{-1}, a^{-2}, a^{-3}, \dots$, converges to 0 with respect to ρ_{v_1}. Hence, this sequence must converge to 0 for ρ_{v_2} as well, i.e.,

$$\rho_{v_2}(a^{-n}, 0) = \exp(nv_2(a)) \to 0 \quad \text{as } n \to \infty.$$

Then $v_2(a) < 0$ clearly holds. Consequently, $v_2(b) \geq 0$ implies $v_1(b) \geq 0$. By Lemma 1.3 we obtain $v_1 \sim v_2$. Q.E.D.

Note that this theorem provides a substantial condition for valuations on K to be equivalent. Another implication is that each prime divisor of K corresponds to a distinct topology on K.

In a space S with a metric ρ, a sequence $\{a_n\}$ is said to be a fundamental sequence if it satisfies the Cauchy condition

$$\rho(a_m a_n) \to 0 \quad \text{as } m, n \to \infty.$$

For example, if a sequence $\{a_n\}$ converges to a point a in S, then it is clearly a fundamental sequence. The converse is also true for certain metric spaces, e.g., Euclidean space. That is, an arbitrary fundamental sequence $\{a_n\}$ converges to a point a in certain spaces. A metric space with this property is called a complete metric space. A metric space is not necessarily complete. However, there exists a complete metric space S', essentially unique, such that S' contains S as a dense subset, and such that the metric on S' is an extension of the metric on S. The space S' is said to be the completion of S. In particular, if S is complete, then $S' = S$ holds.

Let us apply the above general notion to the case where a metric $\rho = \rho_v$ is provided by a valuation v on K. Let K' be the completion of K with respect to the metric ρ. Since K is in K' as a dense subset, for arbitrary points a and b in K' there exist sequences $\{a_n\}$ and $\{b_n\}$ in K such that

$$a = \lim_{n\to\infty} a_n \quad \text{and} \quad b = \lim_{n\to\infty} b_n. \tag{2.3}$$

Then the sequences $\{a_n + b_n\}$ and $\{a_n b_n\}$ form fundamental sequences in K because of the continuity, for the metric ρ, of the operations of the sum and the product. Since K' is complete, there exist c and d in K such that

$$c = \lim_{n\to\infty} (a_n + b_n) \quad \text{and} \quad d = \lim_{n\to\infty} (a_n b_n) \tag{2.4}$$

hold. Instead of (2.3), choose other sequences $\{a'_n\}$ and $\{b'_n\}$ such that $a = \lim a'_n$ and $b = \lim b'_n$. However, (2.2) implies that we still obtain

$$c = \lim_{n\to\infty} (a'_n + b'_n) \quad \text{and} \quad d = \lim_{n\to\infty} (a'_n b'_n).$$

That is, the points c and d in K' are determined uniquely by a and b. Hence we can define

$$c = a + b \quad \text{and} \quad d = ab. \tag{2.5}$$

Notice that $\lim a_n^{-1}$ exists for $a \neq 0$ and it defines a^{-1}. It is clear from the definitions in (2.3) and (2.4) that K' is a field extension of K with the sum and the product. Let $\{a_n\}$ be a sequence satisfying the first part of (2.3) and let

$$v(a) = \lim v(a_n).$$

Then the right-hand side indeed converges, and the dependency on a alone can be proved similarly as before. Thus, we have a function $v(a)$ for $a \in K'$. One can observe that this function is an extension of the valuation v on K

and satisfies conditions (i) through (iv) of Definition 1.1. Together with the continuity of $v(a)$ on K, this extended v becomes a valuation on K'. The metric defined by this extended v is an extension of the metric ρ on K to K'. Then K' is said to be the completion of K with respect to the valuation v. The construction of K' from K is the imitation of G. Cantor's method of defining the real numbers from the rational numbers.

LEMMA 1.6. *Let L be a complete topological field with respect to a metric ρ'. Suppose that L contains K so that the restriction of ρ' on K is equivalent to the metric ρ_v determined by the valuation v on K. Then the closure $\overline{K}$ in L is a subfield of L which is isomorphic to K' over K, where K' is the completion of K with respect to v.*

PROOF. For an element a in K' choose a sequence $\{a_n\}$ in K such that

$$a = \lim_{n\to\infty} a_n \quad \text{in } K'. \tag{2.6}$$

Then the sequence $\{a_n\}$ is a fundamental sequence for ρ_v. By the hypothesis, it is also a fundamental sequence for ρ'. Hence the completeness of L implies that there exists a' in $\overline{K}$ such that

$$a' = \lim_{n\to\infty} a_n \quad \text{in } L \tag{2.7}$$

should hold. For another sequence $\{a_n'\}$ with $a = \lim a_n'$, we have $a' = \lim a_n'$ as well. Consequently, we obtain an injective map from K' into $\overline{K}$, i.e.,

$$a' = \varphi(a).$$

In particular, for a in K we have $\varphi(a) = a$. Conversely, for an arbitrary element a' in $\overline{K}$ there is a sequence $\{a_n\}$ in K such that (2.7) holds. Moreover there is a unique element a in K' satisfying (2.6). Therefore we conclude that φ is a one-to-one map from K' onto $\overline{K}$. In K' let

$$a = \lim a_n, \quad b = \lim b_n, \quad a + b = \lim(a_n + b_n), \quad \text{and} \quad ab = \lim a_n b_n.$$

Then we have the following in $\overline{K}$:

$$\varphi(a) = \lim a_n, \qquad \varphi(b) = \lim b_n,$$
$$\varphi(a+b) = \lim(a_n + b_n) \quad \text{and} \quad \varphi(ab) = \lim a_n b_n.$$

Hence, we obtain

$$\varphi(a+b) = \lim(a_n + b_n) = \lim a_n + \lim b_n = \varphi(a) + \varphi(b)$$

and

$$\varphi(ab) = \lim a_n b_n = \lim a_n \cdot \lim b_n = \varphi(a)\varphi(b).$$

That is, $\overline{K}$ is a subfield of L and φ is an isomorphism from K' onto $\overline{K}$ over K. Q.E.D.

Let v_1 and v_2 be equivalent valuations on K. Then ρ_{v_1} and ρ_{v_2} are equivalent topologies on K. Let K' and K'' be the completions of K for v_1 and v_2, respectively. By the above lemma, K' and K'' are isomorphic over K. That is, all the valuations belonging to a prime divisor P on K determine the same complete field K_P over K.

DEFINITION 1.4. We call such a complete extension field K_P of K determined by a prime divisor P as above the *completion* of K by P. Particularly, if $K_P = K$ holds, then K is said to be a *complete field* with respect to P.

All the valuations belonging to a prime divisor P are uniquely extended to K_P on which they all are equivalent valuations. Therefore, they determine a unique prime divisor on K_P, which we call the extension on K_P of P. When there is no fear of confusion, we will denote it by the same letter P.

Let $\mathfrak{o}$ be the valuation ring of P in K, and let $\bar{\mathfrak{o}}$ be the closure of $\mathfrak{o}$ in K_P. For an element a in $\bar{\mathfrak{o}}$, we have

$$a = \lim a_n \quad \text{for } a_n \in \mathfrak{o}, \ v(a_n) \geq 0, \tag{2.8}$$

where the valuation v belongs to P. The continuity of v implies that we have $v(a) \geq 0$. Conversely, if $a = \lim a_n$, $a_n \in K$ and $v(a) \geq 0$, then $v(a_n) \geq 0$ must hold for a sufficiently large n. This is because $v(a_n)$ is a rational integer such that $v(a) = \lim v(a_n)$. This implies that we have $a \in \bar{\mathfrak{o}}$. Therefore, $\bar{\mathfrak{o}}$ coincides with the valuation ring of P in K_P. Similarly one can prove that the closure $\bar{\mathfrak{p}}$ of a prime ideal $\mathfrak{p}$ in $\mathfrak{o}$ is a unique prime ideal of $\bar{\mathfrak{o}}$. Let a be an element of $\bar{\mathfrak{o}}$. Then (2.8) implies there is an element a_n in $\mathfrak{o}$ such that $v(a - a_n) > 0$, i.e., $a - a_n \in \bar{\mathfrak{p}}$. Hence, we have $\bar{\mathfrak{o}} = \mathfrak{o} + \bar{\mathfrak{p}}$. Since $\mathfrak{p} = \mathfrak{o} \cap \bar{\mathfrak{p}}$ holds, the ring isomorphism theorem provides us with

$$\mathfrak{o}/\mathfrak{p} \cong \bar{\mathfrak{o}}/\bar{\mathfrak{p}}.$$

That is, we have obtained the following theorem.

THEOREM 1.5. *Let K_P be the completion of K for a prime divisor P, let $\mathfrak{o}$ and $\mathfrak{p}$ be the valuation ring and the prime ideal of P in K. Then the closures $\bar{\mathfrak{o}}$ and $\bar{\mathfrak{p}}$ of $\mathfrak{o}$ and $\mathfrak{p}$ in K_P coincide with the valuation ring and the prime ideal of P in K_P. Furthermore, the residue field $\mathfrak{o}/\mathfrak{p}$ of P for K and the residue field $\bar{\mathfrak{o}}/\bar{\mathfrak{p}}$ for K_P are isomorphic.*

In various aspects complete fields have simpler properties than noncomplete fields. We will discuss some of these properties.

THEOREM 1.6 (expansion theorem). *Let K be a complete field with respect to a prime divisor P, i.e., $K_P = K$, and let $\mathfrak{o}$ and $\mathfrak{p}$ be the corresponding valuation ring and prime ideal of P. Let $M = \{u\}$ be a system of complete representatives of $\mathfrak{o}$* mod $\mathfrak{p}$, *i.e., the set of representatives from the residue classes of $\mathfrak{o}$ modulo the ideal $\mathfrak{p}$. We assume that the representative of the zero class in M is 0. Let t be an element of K such that $v_P(t) = 1$. For* *an arbitrary element a of K, if $v_P(a) = i_0$ holds, then a can be expanded uniquely as*

$$a = \sum_{i=i_0}^{\infty} u_i t^i, \qquad u_i \in M \text{ and such that } u_{i_0} \neq 0, \tag{2.9}$$

where the right-hand side means $\lim_{n\to\infty} \sum_{i=i_0}^{n} u_i t^i$. Conversely, for an arbitrary choice of u_i from M, the power series, $\sum_{i=i_0}^{\infty} u_i t^i$, $u_{i_0} \neq 0$, always converges and the series defines an element a in K such that $v_P(a) = i_0$. Therefore, the elements of K are exactly all the power series of the form (2.9) and 0.

PROOF. If a satisfies $v_P(a) = i_0$, then let

$$a' = at^{-i_0} \quad \text{so that } v_P(a') = 0.$$

Since we have $a' \in \mathfrak{o}$ and $a' \notin \mathfrak{p}$, there exists a unique element u_0 in M such that

$$a' \equiv u_0 \mod \mathfrak{p}, \qquad u_0 \neq 0$$

should hold. Suppose that one could choose $u_0, u_1, \dots, u_{n-1}$ from M so that one might have

$$a' \equiv u_0 + u_1 t + \cdots + u_{n-1} t^{n-1} \mod \mathfrak{p}^n.$$

Then the element

$$b = \frac{a' - (u_0 + u_1 t + \cdots + u_{n-1} t^{n-1})}{t^n}$$

belongs to $\mathfrak{o}$. Therefore, we can find a unique element u_n in M such that

$$b \equiv u_n \mod \mathfrak{p} \tag{2.10}$$

holds. Then (2.10) clearly implies

$$a' \equiv u_0 + u_1 t + \cdots + u_{n-1} t^{n-1} + u_n t^n \mod \mathfrak{p}^{n+1}.$$

That is, starting from u_0 one can determine $u_1, u_2, u_3, \dots$ recursively so that we have, by the note below,

$$a' = \lim_{n\to\infty} \sum_{i=0}^{n} u_i t^i = \sum_{i=0}^{\infty} u_i t^i.$$

Consequently,

$$a = \sum_{i=0}^{\infty} u_i t^{i+i_0}.$$

Hence, (2.9) can be obtained by rewriting u_i as u_{i+i_0}.

(Note: $a' - (u_0 + u_1 t + \cdots + u_n t^n)$ belongs to $\mathfrak{p}^{n+1}$, i.e.,

$$v_P(a' - (u_0 + u_1 t + \cdots + u_n t^n)) \geq n + 1.$$

Therefore, we obtain

$$\rho(a', u_0 + u_1 t + \cdots + u_n t^n) = \exp(-v_P(a' - (u_0 + u_1 t + \cdots + u_n t^n))) \le e^{-(n+1)}.$$

This implies that $\lim_{n\to\infty} \rho(a', u_0 + u_1 t + \cdots + u_n t^n) = 0$.)

We now prove uniqueness. Suppose we have another expansion

$$a = \sum_{i=i_0}^{\infty} u_i' t^i, \qquad u' \in M \text{ and } u_{j_0}' \neq 0.$$

Clearly, we have $j_0 = v_P(a) = i_0$. By comparing the two expansions of $a' = at^{-i_0}$

$$a' = \sum_{i=0}^{\infty} u_{i+i_0} t^i = \sum_{i=0}^{\infty} u_{i+i_0}' t^i,$$

we have

$$a' \equiv u_{i_0} \equiv u_{i_0}' \mod \mathfrak{p}.$$

Since u_{i_0} and u_{i_0}' are elements of M, i.e., the system of complete representatives mod $\mathfrak{p}$, we have $u_{i_0} = u_{i_0}'$. Hence,

$$\frac{a' - u_{i_0}}{t} = \sum_{i=1}^{\infty} u_{i+i_0} t^{i-1} = \sum_{i=1}^{\infty} u_{i+i_0}' t^{i-1}$$

holds, which implies

$$\frac{a' - u_{i_0}}{t} \equiv u_{i_0+1} \equiv u_{i_0+1}' \mod \mathfrak{p}.$$

Repeating the same argument as above we have

$$u_{i_0+1} = u_{i_0+1}'.$$

Successively we can establish $u_i = u_i'$.

For arbitrary u_i in M, let

$$s_n = \sum_{i=i_0}^{n} u_i t^i.$$

Then $v_P(s_n - s_m) \ge \mathrm{Min}(n, m)$ holds. That is, $\{s_n\}$ is a fundamental sequence in K. Since K is complete with respect to P by the hypothesis, the sequence converges to a in K. Then we certainly have $v_P(a) = i_0$ for $u_{i_0} \neq 0$. Q.E.D.

Remark. Instead of t^i one can use $\ldots, a_{-2}, a_{-1}, a_0, a_1, a_2, \ldots$ such that $v_P(a_i) = i$ in order to express an arbitrary element a of K as

$$a = \sum_{i=i_0}^{\infty} u_i a_i, \qquad u_i \in M.$$

DEFINITION 1.5. Let v_P be a normalized valuation belonging to a prime divisor P. Any element t of K is said to be a *prime element* of K for P (or at P) if we have

$$v_P(t) = 1.$$

EXAMPLE. Let $K = k(x)$ be the field of rational functions of a variable x with coefficients in k, and let $P = P_0$ be the prime divisor of K associated with the prime ideal (x). (See Example 2 in §1). From Theorem 1.3, the residue field of P is the residue field of $k[x]$ for (x), i.e., it is isomorphic to k. Therefore, by Theorem 1.5, the residue field of P in the completion K_P of K is also isomorphic to k. Hence, in this case, one can use all the elements of k for the system of complete representatives M in Theorem 1.6. Since x is obviously a prime element of K_P for P, by (2.9) an arbitrary element a $(a \neq 0)$ of K_P can be written uniquely as

$$a = \sum_{i=i_0}^{\infty} \alpha_i x^i, \qquad \alpha_i \in k, \ \alpha_{i_0} \neq 0 \text{ such that } v_P(a) = i_0.$$

That is, K_P coincides with the totality of the power series in x with coefficients in k. We denote this field of power series as $k((x))$, and we call the above P and v_P in this field $K_P = k((x))$ the *canonical divisor* and *canonical valuation* with respect to x.

Returning to the general theory, let K be a complete field for a prime divisor P, and let $\mathfrak{o}$ and $\mathfrak{p}$ be its valuation ring and the prime ideal. For an arbitrary polynomial with coefficients in $\mathfrak{o}$

$$f(x) = \sum a_i x^i, \qquad a_i \in \mathfrak{o}.$$

Replace each a_i in $f(x)$ by its residue class $\bar{a}_i$ of mod $\mathfrak{p}$. Then we obtain the polynomial with coefficients in the residue field $\boldsymbol{k} = \mathfrak{o}/\mathfrak{p}$

$$\sum \bar{a}_i x^i, \qquad \bar{a}_i \in \boldsymbol{k},$$

which we denote as $\bar{f}(x)$. Suppose that $f(x)$ and $g(x)$ are polynomials in $\mathfrak{o}[x]$ such that all the coefficients of $f(x) - g(x)$ are contained in $\mathfrak{p}^n$, $n = 1, 2, \ldots$. Then we write

$$f(x) \equiv g(x) \bmod \mathfrak{p}^n.$$

We have the following important theorem.

THEOREM 1.7 (Hensel). *Let K, P, $\mathfrak{o}$, $\mathfrak{p}$ and $\boldsymbol{k}$ be as above. Let $f(x)$ be a polynomial in $\mathfrak{o}[x]$ such that $\bar{f}(x)$ is decomposed into a product of relatively prime polynomials $g'(x)$ and $h'(x)$ in $\boldsymbol{k}[x]$, i.e., $\bar{f}(x) = g'(x)h'(x)$. Then there exist polynomials $g(x)$ and $h(x)$ in $\mathfrak{o}[x]$ such that*

$$f(x) = g(x)h(x), \qquad \bar{g}(x) = g'(x), \qquad \bar{h}(x) = h'(x),$$

and such that the degree of $g(x)$ equals the degree of $g'(x)$.

PROOF. Let the degrees of f, g', and h' be n, n', and n'', respectively. Since $n' + n''$ is the degree of $\overline{f}$, we have $n' + n'' \leq n$. First let g_1 and h_1 be polynomials in $\mathfrak{o}[x]$ so that we may have $\overline{g}_1 = g'$ and $\overline{h}_1 = h'$. Clearly the degrees of g_1 and h_1 can be assumed to be n' and n''. Suppose that $g_1, \dots, g_m$ are polynomials in $\mathfrak{o}[x]$ of degrees at most n' and that $h_1, \dots, h_m$ are polynomials in $\mathfrak{o}[x]$ of degrees at most $n - n'$ with the following congruence relations:

$$g_{i-1} \equiv g_i \quad \text{and} \quad h_{i-1} = h_i \bmod \mathfrak{p}^{i-1}, \qquad i = 2, \dots, m, \tag{2.11}$$

$$f \equiv g_i h_i \bmod \mathfrak{p}^i, \qquad i = 1, \dots, m. \tag{2.12}$$

We are going to show the existence of g_{m+1} and h_{m+1} with similar properties. Choose c such that $v_P(c) = m$. Then, for polynomials $u(x)$ and $v(x)$ in $\mathfrak{o}[x]$, let us try

$$g_{m+1}(x) = g_m(x) + cu(x) \quad \text{and} \quad h_{m+1}(x) = h_m(x) + cv(x).$$

Then $g_{m+1} \equiv g_m$ and $h_{m+1} \equiv h_m \bmod \mathfrak{p}^m$ clearly hold. Furthermore, in order to have $f \equiv g_{m+1}h_{m+1} \bmod \mathfrak{p}^{m+1}$, it suffices to have

$$f - g_m h_m \equiv c(g_m v + h_m u) \bmod \mathfrak{p}^{m+1}.$$

Since $c^{-1}(f - g_m h_m) = r(x)$ is a polynomial in $\mathfrak{o}[x]$, we only need to find u and v so that

$$g_m v + h_m u \equiv r \bmod \mathfrak{p}$$

may hold. Notice that $\overline{g}_m = \overline{g}_1 = g'$ and $\overline{h}_m = \overline{h}_1 = h'$ are relatively prime as polynomials in $\boldsymbol{k}[x]$. Therefore, there exist such u and v of degrees not greater than the degrees of g_m and h_m. We have proved the existence of g_{m+1} and h_{m+1}. That is, we have sequences of polynomials in $\mathfrak{o}[x]$

$$g_1, g_2, g_3, \dots; \qquad h_1, h_2, h_3, \dots,$$

whose degrees are not greater than n' and $n - n'$. If you let

$$g_i(x) = \sum_{k=0}^{n'} a_{ik}x^k \quad \text{and} \quad h_i(x) = \sum_{k=0}^{n-n'} b_{ik}x^k,$$

then by (2.11), $\{a_{1k}, a_{2k}, \dots\}$ and $b_{1k}, b_{2k}, \dots\}$ are fundamental sequences in K. Since K is complete, there exist a_k and b_k in K such that we have

$$a_k = \lim_{i \to \infty} a_{ik} \quad \text{and} \quad b_k = \lim_{i \to \infty} b_{ik}.$$

Moreover, since a_{ik} and b_{ik} are elements of $\mathfrak{o}$, a_k and b_k also belong to $\mathfrak{o}$. Then let

$$g(x) = \sum_{k=0}^{n'} a_k x^k \quad \text{and} \quad h(x) = \sum_{k=0}^{n-n'} bx^k.$$

We obtain $\bar{g} = \bar{g}_i = g'$ and $\bar{h} = \bar{h}_i = h'$, and also $f \equiv gh \mod \mathfrak{p}^m$ for all natural numbers m. Hence, $f = gh$ holds, and the degrees of g and h are at most n' and $n - n'$, respectively. Since the degree of the product is n, the degree of g must be precisely n'. This completes the proof of our theorem. Q.E.D.

Theorem 1.7 has many applications. We will give one of them.

LEMMA 1.7. *Let K be a complete field with respect to a prime divisor P and let $\mathfrak{o}$ be the valuation ring of P. If the last coefficient a_n of the irreducible polynomial*

$$f(x) = x^n + a_1 x^{n-1} + \cdots + a_n \in K[x]$$

belongs to $\mathfrak{o}$, then all other a_i also belong to $\mathfrak{o}$. That is to say, $v_P(a_n) \geq 0$ implies $v_P(a_i) \geq 0$ for $i = 1, \ldots, n-1$.

PROOF. Let $l = \mathrm{Min}(v_P(a_i),\ i = 1, \ldots, n)$. We will show that the assumption $l < 0$ leads to a contradiction. Choose b so that we may have $v_P(b) = -l$. Let $f_1(x)$ be the polynomial in $\mathfrak{o}[x]$ defined by

$$f_1(x) = bf(x) = b_0 x^n + b_1 x^{n-1} + \cdots + b_n, \qquad b_i = ba_i.$$

Then we have $v_P(b_n) > 0$ from the assumption. For some i, $0 < i < n$, $v_P(b_i) = 0$ must hold. Hence in $\boldsymbol{k}[x]$ we have the decomposition

$$\bar{f}_1(x) = x^k h'(x) \quad \text{such that } (x, h'(x)) = 1,\ 0 < k < n.$$

By the preceding theorem, in $K[x]$ we have a decomposition of $f_1(x)$ into the factors of degrees k and $n - k$, i.e., a decomposition of $f(x)$. This contradicts the irreducibility of $f(x)$. Q.E.D.

§3. Extension and projection of prime divisors

We will examine the effects on the valuation and the prime divisor when the ground field K is extended or contracted. Let L be an extension field of K, and let P and P' be prime divisors on K and L, respectively. One can restrict a valuation v' belonging to P' on K. If this restriction is a valuation belonging to P, then the restriction of any valuation v'' belonging to P' belongs to P as a function on K. In such a case let $\mathfrak{o}'$ and $\mathfrak{p}'$ be the valuation ring and the prime ideal of P' on L, and let $\mathfrak{o}$ and $\mathfrak{p}$ be the valuation ring and prime ideal of P on K. Then we clearly have

$$\mathfrak{p} = \mathfrak{o} \cap \mathfrak{p}'.$$

Since $(\mathfrak{o} + \mathfrak{p}')/\mathfrak{p}'$ is isomorphic to $\mathfrak{o}/\mathfrak{p}$ by the ring isomorphism theorem, the residue field $\boldsymbol{k} = \mathfrak{o}/\mathfrak{p}$ of P may be considered to be contained in the residue field $\boldsymbol{k}' = \mathfrak{o}'/\mathfrak{p}'$ of P'. Let

$$f = [\boldsymbol{k}' : \boldsymbol{k}]. \tag{3.1}$$

Note that for the normalized valuation $v_{P'}$ belonging to P' and the normalized valuation v_P belonging to P, we have the following: there exists a natural number e such that

$$v_{P'}(a) = ev_P(a) \quad \text{holds for any element } a \text{ of } K. \tag{3.2}$$

DEFINITION 1.6. A prime divisor P' on L is said to be the *extension* of a prime divisor P on K when P' and P satisfy the preceding conditions, and P is said to be the *projection* on K of P'. The natural number f in (3.1) is called the *relative degree* of P' with respect to P (or K), and e in (3.2) is called the *ramification index* of P' with respect to P (or K). When a valuation v on K is obtained by the restriction of a valuation v' on L, then v is called the projection of v' on K, and conversely v' is called an extension of v on L.

THEOREM 1.8. *Let L be a finite extension of K. Then an arbitrary prime divisor P' on L has a unique projection P on K, and its relative degree f is not greater than $[L:K]$.*

PROOF. Let v' be a valuation belonging to P'. Regarding v' as a function on K, we need to prove that it satisfies conditions (i) through (iv) of Definition 1.1. Conditions (i), (ii), and (iii) are plainly satisfied. Choose an element u in L such that $v'(u) \neq 0, \infty$. Replacing u by u^{-1} if necessary, we may assume $v'(u) < 0$. Let

$$u^n + a_1u^{n-1} + \cdots + a_n = 0, \qquad a_i \in K$$

be an equation satisfied by u over K. Suppose that $v'(a_i) = 0$ or ∞ for all i, $1 \le i \le n$. Then

$$nv'(u) = v'(u^n) = v'(a_1u^{n-1} + \cdots + a_n) \ge \mathrm{Min}(v'(a_iu^{n-i}),\ 1 \le i \le n),$$

where

$$v'(a_iu^{n-i}) = v'(a_i) + (n-i)v'(u) \ge (n-i)v'(u) > nv'(u).$$

This is a contradiction. Hence, there exists at least one a_i with $v'(a_i) \neq 0, \infty$, i.e., v' satisfies (iv) in K as well. Consequently P' has a unique projection P on K. Next let $\mathfrak{o}'$, $\mathfrak{p}'$, $\mathfrak{o}$, $\mathfrak{p}$ be valuation rings and prime ideals of P' and P, respectively. Let $x_1, x_2, \ldots, x_m$ be elements of $\mathfrak{o}'$ such that the residue classes $\overline{x}_i$ *modulo* $\mathfrak{p}'$ are linearly independent over $\boldsymbol{k} = \mathfrak{o}/\mathfrak{p}$. Then $x_1, x_2, \ldots, x_m$ are linearly independent over K. In fact assume that

$$b_1x_1 + \cdots + b_mx_m = 0, \qquad b_i \in K$$

holds with not every $b_i \in K$ being zero. Let

$$l = \mathrm{Min}(v_P(b_i),\ i = 1, 2, \ldots, m)$$

and then choose b such that $v_P(b) = -l$. Then we have

$$c_1x_1 + \cdots + c_mx_m = 0, \qquad c_i = bb_i,$$

where all the c_i belong to $\mathfrak{o}$ and at least one of them does not belong to $\mathfrak{p}$. Reducing this equation *modulo* $\mathfrak{p}'$, we will have the following

$$\overline{c}_1\overline{x}_1 + \cdots + \overline{c}_m\overline{x}_m = 0,$$

where $\overline{c}_i \neq 0$ for some i. This contradicts the assumption of the linearly independency of $\overline{x}_1, \overline{x}_2, \ldots, \overline{x}_m$ over $\boldsymbol{k}$. Therefore $x_1, \ldots, x_m$ are linearly independent over K. As a consequence, $f = [\boldsymbol{k}' : \boldsymbol{k}]$ is never greater than $[L : K]$. Q.E.D.

We will consider the converse of the above case, i.e., whether a prime divisor P on the ground field K can always be extended to L. First we will treat the case when K is complete.

THEOREM 1.9. *Let K be a complete field with respect to a prime divisor P. For a finite field extension L of K, there exists a unique extension P' on L of the prime divisor P such that L is complete with respect to P'. Then we have*

$$ef = [L : K],$$

where e is the ramification index and f is the relative degree.

NOTE. We will recall briefly the definitions of norm and trace for a finite extension field L over K. Let $w_1, w_2, \ldots, w_n$ be basis elements of L over K. For an arbitrary element a in L, let

$$aw_i = \sum_{i=1}^{n} \alpha_{ij} w_i, \qquad \alpha_{ij} \in K.$$

Denote the matrix with entries α_{ij} by $A = (\alpha_{ij})$. Then the map

$$a \to A$$

is a representation of the algebra L over K, the so-called regular representation of L/K with respect to the base $w_1, \ldots, w_n$. Then let

$$F(x; a) = |xE - A|, \quad \text{where } E \text{ is an identity matrix.}$$

$F(x; a)$ is a polynomial with the leading term x^n and depends upon only a, not the choice of a particular base $\{w_1, w_2, \ldots, w_n\}$. This polynomial is called the characteristic polynomial of a. Let

$$T(a) = \sum_{i=1}^{n} \alpha_{ii} \quad \text{and} \quad N(a) = |\alpha_{ij}|.$$

Then $-T(a)$ is the coefficient of the x^{n-1}-term of $F(x; a)$ and $(-1)^n N(a)$ is the constant term of $F(x; a)$. These are elements of K determined by a only, hence the above notation. We call $T(a)$ the trace of a for K and $N(a)$ the norm of a for K. Since the map $a \to A$ is a representation, we immediately obtain

$$T(a + b) = T(a) + T(b) \quad \text{and} \quad N(ab) = N(a)N(b)$$

for arbitrary elements a and b in K. Particularly, if a is an element of K, then we have

$$T(a) = na \quad \text{and} \quad N(a) = a^n.$$

PROOF OF THEOREM 1.9. Let $N(a)$ be the norm of an element a of L with respect to K. Consider a function v^* on L defined by

$$v^*(a) = v(N(a))$$

for the valuation $v = v_P$. Since $N(a) \neq 0$ for $a \neq 0$, $v^*(a)$ is a rational integer for $a \neq 0$. We clearly have $v^*(0) = +\infty$. Next $N(ab) = N(a)N(b)$ implies

$$v^*(ab) = v(N(ab)) = v(N(a)N(b)) = v(N(a)) + v(N(b)) = v^*(a) + v^*(b).$$

According to Definition 1.1, we need to show $v^*(1+c) \geq 0$ for $v^*(c) \geq 0$ to claim that v^* is a valuation on L. Let the irreducible polynomial of c with respect to K be given by

$$f(x) = x^m + a_1 x^{m-1} + \cdots + a_m, \qquad f(c) = 0, \ a_i \in K. \tag{3.3}$$

Since the characteristic polynomial $F(x)$ of c with respect to K equals $(f(x))^{n/m}$, by equating the constant terms of $F(x)$ and $(f(x))^{n/m}$, we obtain $N(c) = \pm(a_m)^{n/m}$, where $n = [L : K]$. Therefore, we have

$$v(a_m) = \frac{m}{n} v(N(c)) = \frac{m}{n} v^*(c) \geq 0.$$

Hence, we obtain $v(a_i) \geq 0$ for $i = 1, 2, \dots, m$ by Lemma 1.7. Since the characteristic polynomial for $1 + c$ with respect to K is $F(x-1) = (f(x-1))^{n/m}$, the constant term $N(1+c)$ of $F(x-1)$ is given by

$$N(1+c) = \pm(f(-1))^{n/m} = \pm((-1)^m + a_1(-1)^{m-1} + \cdots + a_m)^{n/m}.$$

Using $v(a_i) \geq 0$, we obtain

$$v^*(1+c) = v(N(1+c)) \geq 0.$$

Now let P' be the prime divisor of the valuation v^* on L, let $v' = v_{P'}$ be the normalized valuation belonging to P', and let

$$v^* = f'v' \quad \text{for a rational integer } f'.$$

For an arbitrary element a of K we have $N(a) = a^n$. Hence,

$$v^*(a) = nv(a)$$

holds. Therefore, we have

$$v'(a) = \frac{n}{f'} v(a), \qquad a \in K. \tag{3.4}$$

Since $v = v_P$ is a normalized valuation, the equation (3.4) indicates that P' is an extension of P on L and that $e = n/f'$ is the ramification index of P' with respect to K.

Next let $\mathfrak{o}'$ and $\mathfrak{p}'$ be the valuation ring and the prime ideal of P' on L, and let f be the relative degree of P' with respect to K. Choose $x_1, \dots, x_f$ in $\mathfrak{o}'$ so that their residue classes $\overline{x}_1, \dots, \overline{x}_f$ mod $\mathfrak{p}'$ may be linearly independent over the residue field $\boldsymbol{k} = \mathfrak{o}/\mathfrak{p}$ of P. Then we obtain a system of complete representatives, including 0, of $\mathfrak{o}'$ mod $\mathfrak{p}'$, $M' = \{u'\}$, where

$$u' = \sum_{i=1}^{f} u_i x_i,$$

and where u_i runs only once through a system of complete representatives M, including 0, of $\mathfrak{o}$ mod $\mathfrak{p}$. Let t be a prime element for P in K, and let t' be a prime element for P' in L, i.e., $v_P(t) = 1$ and $v_{P'}(t') = 1$. Let $a_0, a_1, \dots, a_{e-1}, a_e, a_{e+1}, \dots$ be

$$1, t', \dots, t'^{e-1}, t, t't, \dots, t'^{e-1}t, t^2, t't^2, \dots,$$

respectively. From (3.4), we have $v'(a_i) = i$. As in the proof of Theorem 1.6, for an arbitrary element a in $\mathfrak{o}'$ we can find u_{ijk} from M, $i = 0, 1, 2, \dots, e-1$, $j = 0, 1, \dots$, and $k = 1, 2, \dots, f$ such that

$$a \equiv \sum_{i=0}^{me-1} u_i' a_i = \sum_{i=0}^{e-1} \sum_{j=0}^{m-1} \left(\sum_{k=1}^{f} u_{ijk} x_k \right) t'^i t^j \bmod P'^{me} \tag{3.5}$$

for $m = 0, 1, 2, \dots$. Then let

$$b_{ikm} = \sum_{j=0}^{m-1} u_{ijk} t^j.$$

Since K is complete with respect to P, b_{ikm} converges to an element of $\mathfrak{o}$ as $m \to 0$, i.e.,

$$b_{ik} = \sum_{j=0}^{\infty} u_{ijk} t^j \in \mathfrak{o}.$$

Let

$$a' = \sum_{i=0}^{e-1} \sum_{k=1}^{f} b_{ik} x_k t'^i.$$

Then from (3.5) we have

$$a \equiv a' \bmod \mathfrak{p}'^{me}$$

for any m. Hence, $a = a'$ must hold. That is to say, an arbitrary element a of $\mathfrak{o}'$ may be written in the following form:

$$a = \sum_{i=0}^{e-1} \sum_{k=1}^{f} b_{ik} x_k t'^i, \qquad b_{ik} \in \mathfrak{o}. \tag{3.6}$$

Next assume $v'(a) \geq me$. Then from (3.5) we have $u'_0 \equiv u'_1 \equiv \cdots \equiv u'_{me-1} \equiv 0 \mod \mathfrak{p}'$. Since a representative of the class 0 is 0, all the u'_i must be 0:

$$\sum_{k=1}^{f} u_{ijk} x_k = 0 \quad \text{for } i = 0, 1, \ldots, e-1 \text{ and } j = 0, 1, \ldots, m-1.$$

Since the $\overline{x}_k$ are linearly independent over $\boldsymbol{k} = \mathfrak{o}/\mathfrak{p}$, the x_k are linearly independent over K (see the proof for Theorem 1.8). We obtain

$$u_{ijk} = 0 \quad \text{for } i = 0, 1, \ldots, e-1, \ j = 0, 1, \ldots, m-1, \ \text{and } k = 1, \ldots, f.$$

In other words, when a is expressed as in (3.6), $v(b_{ik}) \geq m$ must hold for $v'(a) \geq me$. In particular, if $a = 0$, i.e., $v'(a) = +\infty$, then we have $v(b_{ik}) = +\infty$, i.e., $b_{ik} = 0$ in (3.6). This immediately implies that there exists a unique way to express a in $\mathfrak{o}'$ in the form of (3.6). That is,

$$x_k t'^i \quad \text{for } i = 0, 1, \ldots, e-1 \text{ and } k = 1, \ldots, f$$

form a minimal base of $\mathfrak{o}'$ over $\mathfrak{o}$. Generally, any element b of L can be multiplied by some power of t so that the resulting element may be in $\mathfrak{o}'$. Therefore, the elements $x_k t'^i$ are basis elements of L over K as well. In particular, we have

$$ef = [L : K].$$

Since $v'(a) \geq me$ implies $v(b_{ik}) \geq m$, a necessary and sufficient condition for the sequence

$$a^{(s)} = \sum_{i=0}^{e-1} \sum_{k=1}^{f} b_{ik}^{(s)} x_k t'^i, \qquad s = 1, 2, \ldots$$

in L to be a fundamental sequence for P' is that each $\{b_{ik}^{(s)}\}$ is a fundamental sequence in K for P. Consequently, since K is a complete field, if $\{b_{ik}^{(s)}\}$ is a fundamental sequence, then there is an element b_{ik} in K such that

$$b_{ik} = \lim_{s \to \infty} b_{ik}^{(s)}.$$

Then let

$$a = \sum_{i=0}^{e-1} \sum_{k=1}^{f} b_{ik} x_k t'^i.$$

It is clear that $a = \lim_{s \to \infty} a^{(s)}$ holds. That is to say, L is complete with respect to P'.

Lastly we will prove the uniqueness of the extension P' of P. We need to prove the following useful lemma.

Lemma 1.8. *Let L be an extension field of K, and let v' be an arbitrary extension of a valuation v on K to L. If an element c of L satisfies the following*

$$c^m + a_1 c^{m-1} + \cdots + a_m = 0, \qquad a_i \in K \text{ and } v(a_i) \geq 0 \text{ for } i = 1, 2, \ldots, m, \tag{3.7}$$

then we have $v'(c) \geq 0$. Note that K need not be a complete field, and furthermore, L can be an arbitrary extension of K.

Proof. Since $v(a_i) = v(a_i)$ holds, we have

$$\begin{aligned} mv'(c) = v'(c^m) = v'(a_1 x^{m-1} + \cdots + a_m) &\geq \mathrm{Min}_{i=1,\ldots,m}(v'(a_i c^{m-i})) \\ &\geq \mathrm{Min}((m-i)v'(c);\ i = 1, \ldots, m). \end{aligned}$$

Therefore, $v'(c) \geq 0$ must hold. Q.E.D.

Now we return to the proof of Theorem 1.9. Let P'' be any extension of P to L, and let v'' be an extension belonging to P'' of v. Let equation (3.3) be the irreducible equation of an element c of L satisfying $v^*(c) \geq 0$. As we have already proved (following (3.3)) that $v(a_i) \geq 0$ for $i = 1, 2, \ldots, m$, we have $v''(c) \geq 0$ by Lemma 1.8 above. Thus, since $v^*(c) \geq 0$ always implies $v''(c) \geq 0$, we obtain $v^* \sim v''$ from Lemma 1.3, i.e., $P'' = P'$. In other words, an extension of P for L is none other than P'. We have completed the proof of Theorem 1.9. Q.E.D.

We can deduce the following theorem from the above proof and Lemma 1.8.

Theorem 1.10. *Let K, L, P, P' be as in Theorem 1.9. Also let $\mathfrak{o}$ and $\mathfrak{o}'$ be the corresponding valuation rings of P and P'. Then $\mathfrak{o}'$ coincides with the collection of integral elements in L over $\mathfrak{o}$. Note that an element c of L is said to be integral over $\mathfrak{o}$ if c is a root of a polynomial of the following form*:

$$f(x) = x^m + a_1 x^{m-1} + \cdots + a_m, \qquad a_i \in O \text{ for } i = 1, \ldots, m.$$

We will next consider an extension of P for the case where K is not necessarily complete for a prime divisor P. First we will treat the case when L is a finite simple extension of K, i.e., $L = K(a)$. Let $[L : K] = n$, and let the irreducible polynomial of α be given by

$$f(x) = x^n + a_1 x^{n-1} + \cdots + a_n, \qquad a_i \in K \tag{3.8}$$

such that $f(\alpha) = 0$. Let P' be an extension of P to L, and let $L_{P'}$ be the completion of L with respect to P'. Since $L_{P'}$ is a complete field containing K, from Lemma 1.6 the closure $\overline{K}$ of K in L coincides with the completion K_P of K with respect to P. Let $L' = K_P(\alpha)$, which is a subfield of L. Then, since L' is a finite extension of K_P, P can be extended uniquely to a prime divisor P'' on L' by Theorem 1.9. Moreover, L' is complete for P''.

On the other hand, L' contains $L = K(\alpha)$ and P'' is clearly an extension of P' in L. Hence, Lemma 1.6 implies that L' contains $L_{P'}$. Hence,

$$L_{P'} = L' = K_P(\alpha).$$

Thus, we have found a way to extend P to L. That is, we need to do the following. First, decompose $f(x)$ in (3.8) into irreducible polynomials in K_P, i.e.,

$$f(x) = \prod_{i=1}^{g} f_i(x)^{e_i}, \qquad f_i(x) \in K_P[x],$$

where $f_i(x)$ are distinct irreducible polynomials in $K_P[x]$. Let L' be the field obtained by adjoining the roots α_i of $f_i(x)$ to K_P:

$$L_i' = K_P(\alpha_i).$$

By Theorem 1.9, P on K_P can be extended uniquely to L_i'. Let us denote this extension as P_i^*. Since $f(\alpha_i) = 0$ holds, L_i' contains the subfield $K(\alpha_i)$, which is isomorphic to $L = K(\alpha)$ over K. Hence, identifying α_i with α, we may consider L' containing $L = K(\alpha)$. Let v_i^* be a valuation on L_i' belonging to P_i^*. Since v^*, as a mere function on K, is a valuation belonging to P, it is clear that it satisfies conditions (i) through (iv) of Definition 1.1 when it is considered as a function on L. As a consequence, P_i^* has the projection P_i' on L, and P_i' is an extension of P to L. Thus we obtain extensions $P_1', P_2', \dots, P_g'$ of P from $f_1(x), f_2(x), \dots, f_g(x)$, respectively. Then it is plain from the earlier consideration that P has none other than these extensions.

Next we will prove that $P_1', \dots, P_g'$ are all distinct prime divisors on L. Suppose $P_1' = P_2'$. Since L_1' and L_2' are the completions of $L = K(\alpha)$ $(= K(\alpha_1) = K(a_2))$ with respect to P_1' and P_2', respectively, by Lemma 1.6 there exists an isomorphism φ between L_1' and L_2' such that

$$\varphi(\alpha_1) = \alpha_2 \quad \text{and} \quad \varphi(b) = b \text{ for } b \in K.$$

Since the elements of K are invariant under the map φ, and φ does not change its distance, the elements of K_P are also invariant under φ. Hence the irreducible polynomials $f_1(x)$ and $f_2(x)$ of a_1 and a_2 must be the same, which is obviously a contradiction. As a consequence, we obtain $P_1' \neq P_2'$. We have proved the following theorem.

THEOREM 1.11. *Let $L = K(\alpha)$ be a finite extension of K, and let $f(x)$ be the irreducible polynomial in $K[x]$ satisfying $f(\alpha) = 0$. Also let P be an arbitrary prime divisor on K and let*

$$f(x) = \prod_{i=1}^{g} f_i(x)^{e_i}, \qquad f_i(x) \neq f_i(x) \text{ for } i \neq j$$

be the decomposition of $f(x)$ *into irreducible polynomials in* $K_P[x]$. *Then* P *has precisely* g *extensions* $P'_1, \dots, P'_g$.

Since a finite extension is obtained by the succession of simple extensions, we can deduce the next theorem from the above theorem.

THEOREM 1.12. *Let* L *be an arbitrary finite extension of* K. *Then any prime divisor* P *on* K *has at least one and at most* $[L : K]$ *many extensions on* L.

For any finite extension L of K, let $P'_1, \dots, P'_g$ be the totality of extensions of P to L, and let

$$\mathfrak{o}' = \{a; a \in L,\ v_i(a) \geq 0 \text{ for } i = 1, \dots, g\},$$

where $v_i = v_{P'_i}$ is the normalized valuation on L belonging to P'_i. From Theorem 1.2, $\mathfrak{o}'$ is a subring of L containing $\mathfrak{o}$. We will prove that $\mathfrak{o}'$ coincides with the collection of integral elements in L over $\mathfrak{o}$ (compare with Theorem 1.10).

Suppose c is an integral element in L over $\mathfrak{o}$. By definition and by Lemma 1.8, we have $v_i(c) \geq 0$ for $i = 1, \dots, g$, i.e., $c \in \mathfrak{o}'$. Conversely, let c be an arbitrary element of $\mathfrak{o}'$ and let

$$f(x) = x^m + a_1 x^{m-1} + \cdots + a_m, \qquad a_i \in K$$

be the irreducible polynomial of c in $K[x]$ satisfying $f(c) = 0$. Moreover, let the decomposition of $f(x)$ into irreducible polynomials over K_P be given by

$$f(x) = \prod_{i=1}^{k} f_i(x)^{e_i}, \qquad f_i = \sum a_{ij} x^j, \quad a_{ij} \in K_P. \tag{3.9}$$

Notice that, from Theorem 1.11, $f_i(x)$ determines an extension P''_i of P to $K(c)$. Since P''_i must be one of the projections of $P'_1, \dots, P'_g$ on $K(c)$, for a valuation v''_i belonging to P''_i, we have

$$v''_i(c) \geq 0.$$

Since P''_i is obtained through the method in the proof of Theorem 1.9 from the field $K_P(c) = (K(c))_{P''_i}$ adjoined by the root c of $f_i(x) = 0$ to the complete field $K_{P'}$, the inequality $v''_i(c) \geq 0$ implies $v(a_{ij}) \geq 0$ for all j. From (3.9) the coefficients a_k of $f(x)$ are polynomials in a_{ij}, hence, we have $v(a_k) \geq 0$, $k = 1, \dots, m$, i.e., c is integral over $\mathfrak{o}$.

Let $\mathfrak{p}$ be the prime ideal in the valuation ring $\mathfrak{o}$ of P, and also let

$$\mathfrak{p}_i = \{a; a \in \mathfrak{o}',\ v_i(a) > 0\}, \qquad i = 1, \dots, g.$$

By Theorem 1.2, $\mathfrak{p}_1, \dots, \mathfrak{p}_g$ are prime ideals of $\mathfrak{o}'$, and an arbitrary ideal of $\mathfrak{o}'$ can be decomposed uniquely into the product of $\mathfrak{p}_1, \dots, \mathfrak{p}_g$. Then let e_i

be the ramification index of P_i' with respect to K and let f_i be its relative degree. Since the residue field of P_i' is isomorphic to $\mathfrak{o}'/\mathfrak{o}\mathfrak{p}_i$ (see Theorem 1.3), one can choose elements y_{ij}, $(i = 1, \dots, g$ and $j = 1, \dots, f_i)$ of $\mathfrak{o}'$ so that the residue classes $\overline{y}_{ij}$ mod $\mathfrak{p}_i$ of y_{ij} form basis elements of $\mathfrak{o}'/\mathfrak{p}_i$ over $\mathfrak{o}/\mathfrak{p}$. By Theorem 1.1 we are allowed to assume

$$v_k(y_{ij}) \geq e_k \quad \text{for } i \neq k, \; j = 1, \dots, f_i. \tag{3.10}$$

From Lemma 1.5 we can choose $x_1, \dots, x_g$, so that

$$v_i(x_j) = \delta_{ij} \quad \text{for } i, j = 1, \dots, g$$

may hold. Then we will consider the following type of element in L:

$$a = \sum a_{ijk} y_{ij} x^k, \qquad a_{ijk} \in K, \tag{3.11}$$

where $1 \leq i \leq g$, $1 \leq j \leq f_i$, and $0 \leq k < e_i$. If $a_{ijk} \in \mathfrak{o}$, then we obviously have $a \in \mathfrak{o}'$. We will prove the converse: if $a \in \mathfrak{o}'$, then we have $a_{ijk} \in \mathfrak{o}$, i.e., $v(a_{ijk}) \geq 0$. Let

$$l = \operatorname{Min}(v(a_{ijk});\; 1 \leq i \leq g, \;\; 1 \leq j \leq f_i, \;\; 0 \leq k < e_i).$$

We need to show that the assumption $l > 0$ will lead to a contradiction. Let b be one of a_{ijk} such that $l = v(b)$ holds. Let $b = a_{1j_0k_0}$ and let $a_{ijk} b^{-1} = b_{ijk}$. Then we have $b_{ijk} \in \mathfrak{o}$. Since $l < 0$, we have

$$v_1(ab^{-1}) \geq v_1(b^{-1}) = e_1 v(b^{-1}) \geq (-l)e_1 \geq e_1.$$

Hence, when the element

$$ab^{-1} = \sum b_{ijk} y_{ij} x_i^k$$

is considered mod $\mathfrak{p}_1^{e_1}$, by (3.10) we obtain

$$\sum_j \sum_{k=0}^{e_1-1} b_{1jk} y_1 x_1^k \equiv 0 \bmod \mathfrak{p}_1^{e_1}. \tag{3.12}$$

Therefore,

$$\sum_j b_{1j0} y_{1j} \equiv 0 \bmod \mathfrak{p}_1$$

holds. Since $\overline{y}_{ij}$ are basis elements of $\mathfrak{o}'/\mathfrak{p}_1$ over $\mathfrak{o}/\mathfrak{p}$, from the above equation, we have $b_{1j0} \in \mathfrak{p}$, $\nu(b_{1/0}) > 0$, that is, $v_1(b_{1j0}) \geq e_1$, $b_{1j0} \in \mathfrak{p}_1^{e_1}$. Consequently, we obtain the following from (3.12):

$$\sum_j \sum_{k=1}^{e_1-1} b_{1jk} y_1 x_1^k \equiv 0 \bmod \mathfrak{p}_1^{e_1},$$

or

$$\sum_j \sum_{k=1}^{e_1-1} b_{1jk} y_1 x_1^{k-1} \equiv 0 \bmod \mathfrak{p}_1^{e_1-1}.$$

Just as before, we have $v(b_{1j1}) > 0$, $b_{1j1} \in \mathfrak{p}_1^{e_1}$. As a consequence, for every j and k such that $1 \leq j \leq f_1$ and $0 \leq k < e_1$, we have $v(b_{1jk}) > 0$. On the other hand, since we have $b_{1jk} = 1$ when $j = j_0$ and $k = k_0$, this is a contradiction.

If one has

$$\sum c_{ijk} y_{ij} x_i^k = 0, \qquad c_{ijk} \in K,$$

then $\sum (cc_{ijk}) y_{ij} x_i^k = 0$ holds for an arbitrary element c in K. Hence we have $v(cc_{ijk}) \geq 0$. Since c is arbitrary, $c_{ijk} = 0$ must hold. That is, $\sum_{i=1}^g e_i f_i$ many elements $y_{ij} x_i^k$ are linearly independent over K. Therefore we obtain

$$\sum_{i=1}^{g} e_i f_i \leq [L : K].$$

Since an arbitrary element of L can be written as in (3.11), if the equality holds in the above, then, as we saw, $y_{ij} x_i^k$ are minimal basis elements of $\mathfrak{o}'$ over $\mathfrak{o}$. Hence $\mathfrak{o}'$ is a finitely generated $\mathfrak{o}$-module. (See the note at the end of this chapter on finitely generated $\mathfrak{o}$-modules.)

We will next show that the converse is also true. First notice that for any element z in $\mathfrak{o}'$ there exists an element a of the form $a = \sum_{ijk} a_{ijk} y_{ij} x_i^k$, $a_{ijk} \in \mathfrak{o}$ such that

$$v_i(z - a) \geq e_i, \qquad i = 1, \ldots, g \tag{3.13}$$

may hold. From (3.10), the above (3.13) is equivalent to

$$z \equiv \sum_{j,k} a_{ijk} y_i x_i^k \mod P_i^{e_i}, \qquad i = 1, \ldots, g,$$

and the solvability of the above congruence equation follows easily from a similar method to that in the proof of Theorem 1.9. Let $\sum_{i=1}^g e_i f_i < [L : K]$, and let z be an element of $\mathfrak{o}'$ so that z and $y_{ij} x_i^k$ are linearly independent over K. For this z, let $a = a_1$ be an element which satisfies (3.13). Let t be a prime element for P in K. Then, since $v_i(t) = e_i$, we have $v_i(z_1) \geq 0$ for $z_1 = (z - a_1)t^{-1}$. That is, z_1 belongs to $\mathfrak{o}'$. Therefore, again by (3.13), there exists an element a_2 in the form of (3.11) such that

$$v_i(z_1 - a_2) \geq e_i, \qquad i = 1, \ldots, g$$

may hold. If we let $z_2 = (z_1 - a_2)t^{-1}$ and proceed with the method as above, then we will get elements $a_1, a_2, \ldots$ in the form of (3.11). For any natural number m, let

$$b_m = (a_1 + a_2 t + \cdots + a_m t^{m-1}) t^{-m}.$$

We have

$$v_k(zt^{-m} - b_m) \geq 0.$$

Since b_m is a linear combination of $y_{ij}x_i^k$ over K and $z/t^m - b_m$ belongs to $\mathfrak{o}'$, $\mathfrak{o}'$ cannot be finitely generated over $\mathfrak{o}$. (See the note below.) Summing up what has been said above, we have the following theorem:

THEOREM 1.13. *Let L be a finite extension of K, let $P_1', P_2', \dots, P_g'$ be all the extensions of a prime divisor P to L and let $v_i = v_{P_i}$ be the normalized valuation belonging to P_i'. Then the ring*

$$\mathfrak{o}' = \{a \in L: v_i(a) \geq 0,\ i = 1, \dots, g\}$$

coincides with the set of all the integral elements over the valuation ring $\mathfrak{o}$ of P in L. Furthermore, if we denote the index of ramification and the relative degree of P' with respect to K by e_i and f_i, respectively, we have

$$\sum_{i=1}^{g} e_i f_i \leq [L:K]. \tag{3.14}$$

The equality in (3.14) *holds if and only if $\mathfrak{o}'$ is finitely generated over $\mathfrak{o}$.*

According to E. Noether, for a separable extension L over K we have that $\mathfrak{o}'$ is a finitely generated $\mathfrak{o}$-module (see B. L. van der Waerden, *Modern algebra* II, Springer, Berlin, 1940, Chapter 14). As it will be proved in the next chapter, the equality in (3.14) always holds for an algebraic function field K. However, strict inequality can occur in the general case.

NOTE. We say that $\mathfrak{o}'$ is a finitely generated $\mathfrak{o}$-module if and only if there are finitely many elements $u_1, u_2, \dots, u_s$ in $\mathfrak{o}'$ such that an arbitrary element u of $\mathfrak{o}'$ can be written as:

$$u = d_1u_1 + d_2u_2 + \cdots + d_su_s, \qquad d_i \in \mathfrak{o}.$$

As in the above case of $y_{ij}x_i^k$, when those coefficients are uniquely determined for each choice of u, then $u_1, u_2, \dots, u_s$ are said to be minimal basis elements of $\mathfrak{o}'$ over $\mathfrak{o}$.

For a finitely generated $\mathfrak{o}$-module $\mathfrak{o}'$, let $z_1, z_2, \dots, z_l$ be linearly independent elements of L over K. Then for any element z of $\mathfrak{o}'$ of the form:

$$z = c_1z_1 + c_2z_2 + \cdots + c_lz_l, \qquad c_i \in K$$

the coefficients c_k must always satisfy

$$v(c_k) \geq m_0, \qquad k = 1, \dots, l,$$

where m_0 is an integer determined by $z_1, z_2, \dots, z_l$ alone. We will prove this statement since we will use it later in this treatise. First note that with a proper choice of $z_{l+1}, \dots, z_n$ we can make $z_1, z_2, \dots, z_l$ basis elements of L over K. Hence, one can assume that $l = n$ and $z_1, z_2, \dots, z_l\ (= z_n)$ are basis elements of L over K. Then let

$$u_i = \sum_{k=1}^{n} c_{ik}z_k, \qquad c_{ik} \in K.$$

Since z belongs to $\mathfrak{o}'$,

$$z = d_1u_1 + d_2u_2 + \cdots + d_su_s = \sum_{k=1}^{n}\left(\sum_{i=1}^{s} d_ic_{ik}\right) z_k, \qquad d_i \in \mathfrak{o}$$

holds. Hence,

$$c_k = \sum_{i=1}^{s} d_ic_{ik}, \qquad d_i \in \mathfrak{o} \text{ and } v(d_i) \geq 0.$$

For $m_0 = \mathrm{Min}(v(c_{ik});\ i = 1, \ldots, s, k = 1, \ldots, n)$, we have

$$v(c_k) \geq \mathop{\mathrm{Min}}_i(v(d_ic_{ik})) \geq \mathop{\mathrm{Min}}_i(v(c_{ik})) \geq m_0.$$

CHAPTER 2

Algebraic Theory of Algebraic Function Fields

§1. Algebraic function fields

The following is a survey of the algebraic theory of algebraic function fields. We will begin with a definition.

DEFINITION 2.1. An extension field K of a field k is said to be an *algebraic function field* with *coefficient field* k (or *constant field*), or simply, over k if the following two conditions are satisfied:

(i) for a certain element x in K that is transcendental over k, K is a finite extension of $k(x)$, and

(ii) k is algebraically closed in K, that is, all the algebraic elements in K for k belong to k.

REMARK 1. To be precise, the above K is the algebraic function field of one variable with the coefficient field k. However, since we seldom mention the algebraic function field of several variables in this treatise, we call it simply the algebraic function field.

REMARK 2. Condition (ii) above is for convenience and is not essential. In fact, if K is an extension field of k satisfying condition (i), and if k' is the collection of all the algebraic elements in K over k, then k' is a finite extension of k, and one can easily prove that conditions (i) and (ii) are satisfied for K and k'. Note also that condition (ii) is automatically satisfied for any algebraically closed field k.

Suppose that K is an algebraic function field over k as above. Let x' be an arbitrary element of K that is not in k. From condition (ii) x' is transcendental over k and also x is algebraic over $k(x')$. This is because when x is transcendental over $k(x')$, then $k(x, x')$ is the field of rational functions of two variables with coefficients in k. Hence, x' also has to be a transcendental element over $k(x)$, which contradicts condition (i). Therefore x is algebraic over $k(x')$, and $[k(x, x'); k(x')]$ is finite. From condition (i), $[K: k(x, x')]$ is finite, hence K is also finite over $k(x')$. That is, an arbitrary element x' in K, but not in k, satisfies condition (i) of Definition 2.1.

Thus, it is desirable to choose an element x so that the extension K over $k(x)$ may have properties as simple as possible to facilitate the study of the properties of the algebraic function field K. For example, if K is a separable extension over $k(x)$, then it is a simple extension. Hence, K can be obtained by adjoining two elements to k as $K = k(x, y)$. In general, an element x in K such that $K/k(x)$ is a finite separable extension is called a *separating element* of K. Then we have the following theorem.

THEOREM 2.1 (F. K. Schmidt). *If the coefficient field k is a perfect field, then there always exists a separating element for the algebraic function field K over k.*

PROOF. If the characteristic of k, therefore of K, is zero, any extension is separable. In this case there is nothing to prove. Let $p \neq 0$ be the characteristic of K, and let x be any element not in k. When x is a separating element, then there is nothing to be proved. Let us assume now that K over $k(x)$ is not separable. Let y be a nonseparating element (i.e., of the second kind) in K for $k(x)$. Let an irreducible polynomial in $k(x)[Z]$ for y be given as:

$$f(x, Z) = \sum_{i=1}^{m} \left(\sum \alpha_{ij} x^j\right) Z^{ip}, \qquad f(x, y) = 0, \alpha_{ij} \in k.$$

Let $\overline{K}$ be the algebraic closure of K, and let the subfield $K_1 = k(x^{1/p})$ of $\overline{K}$. Then the polynomial $f(x, Z)$ can be decomposed as follows:

$$f(x, Z) = \left(\sum_i \sum_j \alpha_{ij}^{1/p} x^{j/p} Z^i\right)^p = (f_1(x, Z))^p$$

as a polynomial in $K_1[Z]$. The perfectness of k implies that $\alpha_{ij}^{1/p}$ is contained in k. Hence $f_1(x, Z)$ is a polynomial in $K_1[Z]$. Since the degree of $f_1(x, Z)$ in Z is m, we have

$$[K_1(y) \colon K_1] \leq m.$$

On the other hand, since $[K_1 : k(x)] = p$ holds, we have $[K_1(y) : k(x)] \leq mp$. That is,

$$[k(x^{1/p}, y) : k(x)] \leq mp. \tag{1.1}$$

The degree of $f(x, Z)$ in Z is mp. Hence, we have

$$[k(x, y) : k(x)] = mp.$$

Therefore, from (1.1) we obtain $k(x^{1/p}, y) = k(x, y)$. In particular, $x^{1/p}$ is contained in K. If $x^{1/p}$ is a separating element of K, our proof is done. If it is not, then $(x^{1/p})^{1/p} = x^{1/p^2}$ is contained in K. Generally if x^{1/p^e} is contained in K, then

$$p^e = [k(x^{1/p^e}) : k(x)] \leq [K : k(x)]$$

must hold. By the repeated use of this process, the x^{1/p^e} for some c becomes a separating element of K. Q.E.D.

Hence, from the above proof we obtain the following corollary.

COROLLARY TO THEOREM 2.1. *If the coefficient field* k *is a perfect field of characteristic* $p \neq 0$ *and if* x *is not a separating element of* K, *then* $x^{1/p}$, *the* p*th root of* x, *is contained in* K.

For the case of a perfect field k we may take a separating element x so that $K = k(x, y)$ is as described earlier. Then one may take the separating element y^{1/p^e} of K so that $K = k(x, y^{1/p^e})$, instead of y. Hence, the algebraic function field over a perfect field k is obtained by adjoining two separating elements to k.

§2. Prime divisors on algebraic function fields

Let K be an algebraic function field over k. We will consider a prime divisor on K.

DEFINITION 2.2. A prime divisor P on K is said to be a *place*, or a *point* if for an arbitrary element $\alpha \neq 0$ in the coefficient field k

$$\nu_P(\alpha) = 0 \tag{2.1}$$

holds.

NOTE. Such a prime divisor P is called a place, or a point, as in the above definition since there is a one-to-one correspondence between P and a point on the Riemann surface of the algebraic function field K over the field of complex numbers.

In what follows we always consider such prime divisors on the algebraic function field as satisfy condition (2.1). That is, we consider only places.

THEOREM 2.2. *For an arbitrary element* x *in* K, *which does not belong to* k, *there exist at least two and at most finitely many places* P *such that* $\nu_P(x) \neq 0$. *In particular, for an element* a $(a \neq 0)$ *of* K *to be in* k *it is necessary and sufficient that* $\nu_P(a) = 0$ *for all the places* P *on* K.

PROOF. Let Q be the projection of a place P of K on $k(x)$ such that $\nu_P(x) \neq 0$. Then we obviously have $\nu_Q(x) \neq 0$. Hence, it is enough to find a prime divisor Q on $k(x)$ in order to find P as an extension to K as stated in the theorem. From Example 2, §1 of Chapter 1, there are precisely two prime divisors on $k(x)$ such that $\nu_Q(x) \neq 0$ and such that $\nu_Q(\alpha) = 0$ holds for $\alpha \in k$. That is, they are Q_0 and Q_∞ induced from the prime ideals (x) and $(1/x)$ in $k[x]$ and $k[1/x]$, respectively. Theorem 1.12 implies that each of Q_0 and Q_∞ has at least one and at most finitely many extensions on K, which proves the theorem.

DEFINITION 2.3. Let a be an arbitrary element of K. Then $\nu_P(a)$ is said to be the *order* of a at the place P. In particular, if $\nu_P(a) > 0$ holds, then a is said to be a *zero* of order $\nu_P(a)$, and if $\nu_P(a) < 0$ holds, then a is said to be a *pole* of order $-\nu_P(a)$.

THEOREM 2.3. *The residue class field* $\mathbf{k}$ *of a place* P *on* K *over* k *is a finite extension of* k (*or isomorphic to a finite extension of* k).

PROOF. Choose x in K such that $\nu_P(x) = 1$. As we saw in the proof of Theorem 2.2, P is an extension on K of the prime divisor Q_0 on $k(x)$ induced by the prime ideal (x) in $k[x]$. From Theorem 1.3 the residue class field of Q_0 is k. If f denotes the relative degree of P for $k(x)$, then $\boldsymbol{k}$ is an extension of k of degree f.

DEFINITION 2.4. Let P be a place of an algebraic function field K over k. The degree $[\boldsymbol{k}, k]$ of the residue class field $\boldsymbol{k}$ of P over k is called the (absolute) *degree* of P and is denoted by $n(P)$.

Next for a given algebraic function field K, we will consider a method to determine a place in K. With a little modification one could treat a more general case than the following, however, for the sake of simplicity we will consider the case where k is algebraically closed. First, assume a place P is given. Let K_P be the completion of K for P. From Theorem 1.5 the residue class fields of P in K and in K_P coincide. It is a finite extension of k by Theorem 2.3. Consequently, the residue class field of P in K_P is k itself since k is algebraically closed. As a system of complete representatives of the residue class field of P, one can take all the elements of k. Hence, by Theorem 1.6, for a prime element t of P each element of K_P can be written as a power series in t with coefficients in k, i.e., $K_P = k((t))$. Moreover, a prime divisor P on K_P is nothing but the canonical divisor P_1 for t on $k((t))$ (compare with the example in §2, Chapter 1). That is to say, whenever a place P is given, K admits a dense embedding in the power series field $k((t))$ and P coincides with the projection on K of the canonical divisor P_1 for t on $k((t))$.

Conversely, if K is mapped by a map σ into $k((t))$ so that the isomorphic image $K' = \sigma(K)$ may be a dense subset of $k((t))$, for the canonical divisor ν_1 on $k((t))$ for t, we can define a normalized valuation ν on K defined by

$$\nu(a) = \nu_1(\sigma(a)), \qquad a \in K.$$

Hence, we obtain a place P.

For such maps σ_1 and σ_2 from K into $k((t))$ as above, we will consider the condition that these maps define the same place P on K. Let K_P be the completion of K at P. Then from Lemma 1.6 the isomorphism σ_1 from K into $\sigma_1(K)$ can be extended uniquely to the isomorphism σ_1' from K_P into $k((t))$. Similarly σ_2 can be extended to σ_2' from K_P into $k((t))$. Therefore, the isomorphism $\sigma = \sigma_2\sigma_1^{-1}$ from $\sigma_1(K)$ to $\sigma_2(K)$ can be extended to the automorphism $\sigma' = \sigma_2'\sigma_1'^{-1}$ on $k((t))$ so that $\nu_1(\sigma'(b)) = \nu_1(b)$ may hold for $b \in k((t))$. Conversely, if $\sigma_2\sigma_1^{-1}$ is extended to such a σ' as above, it is clear that the valuations on K (hence the places) induced by σ_1 and σ_2 coincide. We have obtained the following theorem.

THEOREM 2.4. *Let K be an algebraic function field over an algebraically closed field k, let $k((t))$ be the field of power series with coefficients in k,* *and let* P_1 *and* ν_1 *be the canonical divisor and the canonical valuation for* t, *respectively. Let* σ *be an arbitrary isomorphism from* K *onto the dense subset* $\sigma(K)$ *in* $k((t))$ *such that* $\sigma(\alpha) = \alpha$ *for* $\alpha \in k$. *Define*

$$\nu(a) = \nu_1(\sigma(a)) \quad \text{for } a \in K.$$

Then ν *is a normalized valuation on* K, *and a place* P *is obtained. Moreover, every place* P *is obtained in this manner from a certain map* σ. *A necessary and sufficient condition for* σ_1 *and* σ_2 *to give the same place on* K *is the following: the isomorphism* $\sigma_2\sigma_1^{-1}$ *from* $\sigma_1(K)$ *to* $\sigma_2(K)$ *can be extended to an automorphism* σ' *on* $k((t))$ *so that* $\nu_1(\sigma'(b)) = \nu_1(b)$ *may hold for* $b \in k((t))$.

In the case where k is of characteristic zero, we are able to obtain more explicit results. We will first prove a lemma.

LEMMA 2.1. *Let* k *be an algebraically closed field of characteristic zero, and let* K *and* K' *be algebraic function fields over* k *such that* K' *is a subfield of* K. *For a place* P *on* K *we denote the projection on* K' *by* Q *and the ramification index by* e. *Then for an arbitrary prime element* u *for* Q *in the complete field* K'_Q *we may write*

$$u = t^e$$

for some prime element t *for* P *on* K_P.

PROOF. Take an arbitrary prime element t_1 for P on K_P. Then, as we saw in the proof of Theorem 2.4, K_P coincides with the power series field $k((t_1))$ in t_1. By the definition of ramification index we have

$$\nu_P(u) = e\nu_Q(u) = e.$$

Hence, u has an expansion in $K_P = k((t))$ as follows:

$$u = \alpha_e t_1^e + \alpha_{e+1} t_1^{e+1} + \cdots, \qquad \alpha_i \in k, \ \alpha_e \neq 0.$$

Therefore, since the characteristic of k is 0, for some power series

$$t = \beta_1 t_1 + \beta_2 t_1^2 + \cdots, \qquad \beta_i \in k, \ \beta_1 \neq 0, \tag{2.2}$$

we can write

$$u = t^e.$$

Then from (2.2) t is also a prime element for P on K_P. Q.E.D.

When K is an algebraic function field over an algebraically closed field k of characteristic zero, K is obtained by adjoining certain two elements x and y to k. Since y is algebraic over $k(x)$, for some irreducible polynomial $f(X, Y)$ in two variables X and Y with coefficients in k, we have

$$f(x, y) = 0.$$

The structure of K is uniquely determined by x, y, and the irreducible relation equation between them. We will express such a K as

$$K = k(x, y) \quad \text{such that } f(x, y) = 0.$$

Notice that the above remark holds for the nonzero characteristic case as well.

Let P be a place on such a field K and let Q be the projection of P on $k(x)$. Then we have either $Q = P_\alpha$, $\alpha \in k$, or $Q = P_\infty$ by Example 2, §1 of Chapter 1. Let us assume $Q = P_\alpha$. Since $u = x - \alpha$ is a prime element of Q in $k(x)$, from Lemma 2.1 there exists a prime element t for P in K_P such that

$$x - \alpha = t^e,$$

where e is the ramification index of P for $k(x)$. Since $K_P = k((t))$, let the power series expansion for y in t be:

$$\begin{cases} x = S_1(t) = \alpha + t^e, \\ y = S_2(t) = \beta_1 t^{e_1} + \beta_2 t^{e_2} + \cdots, \qquad e_1 < e_2 < \cdots, \ \beta_i \in k, \ \beta_i \neq 0. \end{cases} \tag{2.3}$$

$S_1(t)$ and $S_2(t)$ have the following two properties:

(1) For any t we have

$$f(S_1(t), S_2(t)) = 0.$$

This is clearly true.

(2) The greatest common divisor of $e, e_1, e_2, \ldots,$ is 1.

This is because any element a in K can be expressed as a rational function in x and y with coefficients in k, i.e.,

$$a = R(x, y).$$

Then, for the greatest common divisor d of $e, e_1, e_2, \ldots,$ the expansion of a for t in $a = R(S_1(t), S_2(t))$ is a power series in t^d. Since in K there exists a prime element a for P such that $\nu_P(a) = 1$, i.e.,

$$a = \gamma_1 t + \gamma_2 t^2 + \cdots, \qquad \gamma_i \in k, \ \gamma_1 \neq 0,$$

$d = 1$ must hold.

Similarly, for the case $Q = P_\infty$ we have the expansion

$$\begin{cases} x = S_1(t) = t^{-e}, \\ y = S_2(t) = \beta_1 t^{e_1} + \beta_2 t^{e_2} + \cdots, \qquad e_1 < e_2 < \cdots, \ \beta_i \in k, \ \beta_i \neq 0 \end{cases} \tag{2.4}$$

such that conditions (1) and (2) are satisfied. As before, e is the ramification index of P for $k(x)$.

Conversely, when power series $S_1(t)$ and $S_2(t)$ as in (2.3) or (2.4) satisfying the above conditions (1) and (2) are given, from (1) it follows that there is an embedding from $K = k(x, y)$ isomorphically into $k((t))$, the power series field in t, where $x = S_1(t)$ and $y = S_2(t)$. Let P_1 be the canonical divisor for t. Then, for $\nu_1 = \nu_{P_1}$ define

$$\nu(a) = \nu_1(a) \quad \text{for } a \in K.$$

It can be verified easily that (i) through (iv) in Definition 1.1 and (2.1) of Definition 2.2 are satisfied by this ν. That is, ν is a valuation on K. Let P be the place on K to which ν belongs and let e' be the ramification index of P for $k(x)$. Lemma 1.6 implies that the closure $\overline{K}$ of K in $k((t))$ coincides with K_P. There is a prime element t_1 for P in $\overline{K} = K_P$ by Lemma 2.1 so that we may have $x - \alpha = t_1^{e'}$, or $x = t_1^{-e'}$. Let ν_P be the normalized valuation belonging to P, and let $\nu = m\nu_P$. Since we have

$$e = \nu(t^e) = \nu(x-\alpha) = m\nu_P(x-\alpha) = m\nu_P(t_1^{e'}) = me', \qquad t^e = t_1^{e'},$$

we have

$$t_1 = \zeta t^m, \qquad \zeta^{e'} = 1.$$

As elements in $\overline{K} = K_P$, x and y can be expanded as power series in t_1. Hence, they are power series in t^m, however Property (2) implies $m = 1$. Consequently, $\overline{K} = K_P = k((t))$ in which K is dense so that $\nu = \nu_P$ is the normalized valuation on K.

Suppose that two pairs $\{S_1(t), S_2(t)\}$ and $\{S_1'(t), S_2'(t)\}$ of power series of the type in (2.3) or (2.4) satisfy (1) and (2) above, and that these two pairs give the same place P. Then Theorem 2.4 implies that the isomorphism σ from $k(S_1(t), S_2(t))$ to $k(S_1'(t), S_2'(t))$ determined by

$$\sigma(S_1(t)) = S_1'(t) \quad \text{and} \quad \sigma(S_2(t)) = S_2'(t)$$

can be extended to an automorphism σ' on $k((t))$ so that $\nu_1(\sigma'(b)) = \nu_1(b)$ holds for $b \in k((t))$. Since $\nu_1(\sigma'(t)) = \nu_1(t) = 1$ holds, we have

$$\sigma'(t) = \gamma_1 t + \gamma_2 t^2 + \cdots, \qquad \gamma_i \in k, \ \gamma_1 \neq 0.$$

Substituting this into

$$S_1'(t) = \sigma'(S_1(t)) = S_1(\sigma'(t)),$$

we find that the ramification indices e for $S_1(t)$ and $S_1'(t)$ coincide and that

$$\sigma'(t) = \zeta t$$

must hold. Note that ζ is an eth root of 1.

Conversely, for power series $S_1(t)$ and $S_2(t)$ of the type in (2.3) or (2.4) satisfying conditions (1) and (2) and for an eth root ζ of 1, define

$$S_1'(t) = S_1(\zeta t) \quad \text{and} \quad S_2'(t) = S_2(\zeta t).$$

Then it is clear that these power series are of the same type as $S_1(t)$ and $S_2(t)$, and they give the same place on K.

Let $x - \alpha = t^e = u$ in (2.3). We have

$$y = \beta_1 u^{e_1/e} + \beta_2 u^{e_2/e} + \cdots . \tag{2.5}$$

Consider a power series in a variable u of the following form:

$$\alpha_1 u^{r_1} + \alpha_2 u^{r_2} + \cdots , \qquad r_1 < r_2 < \cdots , \alpha_i \in k , \alpha_i \neq 0, \tag{2.6}$$

where $r_i = n_i/m_i$, m_i and n_i are relatively prime rational integers, and all the $m_i > 0$ and are bounded, and such that the number of negative r_i is finite. Such a power series is called a Puiseux series in u. The totality of Puiseux series in u forms a field with the formal series addition and multiplication. We will denote this field by $k\{u\}$. When $F(Y) = f(u+a, Y)$ is regarded as a polynomial in Y with coefficients in $k\{u\}$, then (2.5) is a root of $F(Y)$ in $k\{u\}$ because of condition (1) for $S_1(t)$ and $S_2(t)$. Conversely let the power series (2.6) be one of the roots of $F(Y)$ in $k\{u\}$, and let e be the least common multiple of the denominators m_i of r_i. Define

$$u = t^e , \quad S_1(t) = \alpha + t^e , \quad \text{and} \quad S_2(t) = \alpha_1 t^{er_1} + \alpha_2 t^{er_2} + \cdots .$$

Then we obtain power series $S_1(t)$ and $S_2(t)$ in integer powers of t satisfying conditions (1) and (2). The same results can be obtained when type (2.4) is considered, instead of (2.3). Consequently, in order to get all the places on K one needs to get all the roots of $F(u+\alpha, Y) = 0$, $\alpha \in k$, or $f(1/u, Y) = 0$ in $k\{u\}$. Then we have the following theorem.

THEOREM 2.5. *The field $k\{u\}$ of Puiseux series, in a variable u with coefficients in the algebraically closed field k of characteristic zero, is also algebraically closed.*

PROOF. Let $F(Y)$ be an irreducible polynomial with coefficients in $k\{u\}$ given as

$$F(Y) = Y^n + A_1 Y^{n-1} + \cdots + A_n , \qquad A_i \in k\{u\}.$$

We need to show that the assumption $n \geq 2$ leads to a contradiction. If one lets $Y' = Y - A_1/n$, then $F(Y)$ becomes an irreducible polynomial in Y' such that the coefficient of Y'^{n-1} is zero. Hence, we may assume $A_1 = 0$ in the above expression for $F(Y)$. Since $F(Y)$ is irreducible and $n \geq 2$, not all the coefficients $A_2, \ldots , A_n$ are zero. Let $\alpha_i u^{r_i}$, $\alpha_i \in k$, $\alpha_i \neq 0$ be the initial term of such a nonzero Puiseux series A_i, and let r be the minimum value of r_i/i for all of such $A_i \neq 0$. Then we have

$$r_i - ir \geq 0, \qquad A_i \neq 0, \tag{2.7}$$

and the equality holds for at least one i. Let m be the least common multiple of r and the denominators which appear in all the exponents of the expansions of $A_2, \ldots , A_n$. Let

$$t = u^{1/m} \quad \text{and} \quad Z = u^{-r} Y.$$

Then let

$$F(Y) = u^{nr}(Z^n + u^{-2r}A_2Z^{n-2} + \cdots + u^{-nr}A_n)$$
$$= u^{nr}(Z^n + B_2(t)Z^{n-2} + \cdots + B_n(t)),$$

and let

$$G(Z) = g(t, Z) = Z^n + B_2(t)Z^{n-2} + \cdots + B_n(t).$$

Notice that each $B_i(t)$ is a power series in t with integer exponents, and also $B_i(t)$ does not contain any terms with negative exponents by condition (2.7). Further, for certain i

$$B_i(t) = \alpha_i + \alpha_i't + \alpha_i''t^2 + \cdots, \qquad \alpha_i \neq 0$$

holds. Let P_1 be the canonical divisor $k(t)$ for t, let $\mathfrak{o}_1$ be the valuation ring and let $\mathfrak{p}_1$ be the prime ideal. Since $G(Z)$ is a polynomial in $\mathfrak{o}_1[Z]$, we obtain by reducing $\operatorname{mod} \mathfrak{p}_1$ the following polynomial with coefficients in $k = \mathfrak{o}_1/\mathfrak{p}_1$

$$\overline{G}(Z) = g(0, Z) = Z^n + \cdots + \alpha_i Z^{n-i} + \cdots, \qquad \alpha_i \neq 0 \ (i \geq 2).$$

Since k is an algebraically closed field, $\overline{G}(Z)$ can be decomposed as a product of linear factors in $k[Z]$. Then not all those linear equations can be the same since the coefficient of Z^{n-1} is zero and $\alpha_i \neq 0$ for $i \geq 2$ (remember that the characteristic of k is zero). Therefore, $\overline{G}(Z)$ may be written as a product of nonconstant relatively prime polynomials $g_1(Z)$ and $g_2(Z)$:

$$\overline{G}(Z) = g_1(Z)g_2(Z), \qquad (g_1, g_2) = 1.$$

Hence, by Theorem 1.7, $G(Z)$ is decomposed in $k((t))[Z]$, and therefore $F(Y)$ is also decomposed in $k\{u\}[Y]$. This contradicts our hypothesis on the irreducibility of $F(Y)$. Q.E.D.

From the above theorem we can summarize our results concerning the prime divisors in the following theorem.

THEOREM 2.6. *Let K be an algebraic function field over an algebraically closed field k of characteristic zero, and let*

$$K = k(x, y) \quad \textit{and} \quad f(x, y) = 0.$$

Let α be an arbitrary element of k. Then consider $F(Y) = f(u + \alpha, Y)$ as a polynomial in Y with coefficients in the field of Puiseux series $k\{u\}$. From Theorem 2.5, one can decompose $F(Y)$ into linear factors as

$$F(Y) = \prod_{i=1}^{n}(Y - Q_i(u)), \qquad Q_i(u) = \sum \alpha_{is}u^{r_{is}}, \qquad \alpha_{is} \in k. \tag{2.8}$$

For each i, let e^i be the least common multiple for the denominators of r_{is} in the expansion of $Q_i(u)$, and let

$$x = S_{1i}(t) = \alpha + t^{e_i}, \qquad y = S_{2i}(t) = \sum \alpha_{is}t^{e_i r_{is}},$$

then this gives an embedding of $k(x, y)$ *into the field* $k((t))$. *Hence, by Theorem* 2.4, *we obtain a place* P_i *on* K. *Notice that the projection of* P_i *on* $k(x)$ *is* P_α *which is determined by the prime ideal* $(x-\alpha)$ *and that* e_i *is its ramification index. Conversely, any extension of* P_α *is one of* P_i, $i = 1, 2, \dots, n$. *Furthermore,* P_i *and* P_j *coincide if and only if* $e_i = e_j$ *and* $Q_i(u, \zeta)$ *coincides with* $Q_j(u)$, *where* $Q_i(u, \zeta)$ *is the series obtained from* $Q_j(u)$ *by replacing* u^{1/e_i} *by* $\zeta u^{1/e_i}$, *and where* ζ *is an* e_i*th root of unity. When* $Q_i(u)$ *is given, from the definition of* e_i, *we have the following: as* ζ *runs through the* e_i*th roots of unity,* $Q_i(u, \zeta)$ *are distinct* e_i *roots of* $F(Y) = 0$. *Since* n, *the degree of* $F(Y)$, *is* $[K : k(x)]$, *we have the following from* (2.8): *for all the extensions* $P_1, \dots, P_g$ *on* K *of* P_α *and their ramification indices* $e_1, \dots, e_g$,

$$\sum_{i=1}^{g} e_i = [K : k(x)]. \tag{2.9}$$

Quite similarily, if we consider $F(Y) = f(1/u, Y)$ *instead of* $F(Y) = f(u+\alpha, Y)$, *we can obtain all the extensions on* K *of the place* P_∞ *on* $k(x)$. *Equation* (2.9) *holds for this case as well.*

Equation (2.9) may be generalized to a general coefficient field, which we discuss in the following section.

§3. Divisors on algebraic function fields

Let K be an arbitrary algebraic function field over a constant field k, and let P be a place on K. The totality of the formal powers of the symbol P becomes a multiplicative cyclic group $\mathfrak{D}_P$ with the rule $P^m P^{m'} = P^{m+m'}$. This multiplicative group is isomorphic to the additive group of rational integers. The direct sum $\mathfrak{D}$ of $\mathfrak{D}_P$ as P runs through all the places on K is said to be the *divisor group* of K, and each element of $\mathfrak{D}$ is called a *divisor* of K. That is to say, $\mathfrak{D}$ is a free abelian group generated by places P on K. Therefore, an arbitrary divisor D of K is uniquely, up to permutation, expressed as a product of places:

$$D = \prod_P P^{e_P}, \tag{3.1}$$

where rational integers e_P are all zero except finitely many e_P. The exponent e_P of P for D is denoted as

$$e_P = \nu_P(D).$$

The *degree* $n(D)$ of the divisor D in (3.1) is defined by

$$n(D) = \sum_P \nu_P(D) n(P),$$

where $n(P)$ is the degree of the place P as defined in Definition 2.4. Notice that $n(D)$ is a function of divisors D having their values in rational integers such that for any divisors D and D' we have

$$n(DD') = n(D) + n(D'). \tag{3.2}$$

One can rewrite D in (3.1) by taking only those $e_P \neq 0$ as a product of finitely many P_i:

$$D = \prod_{i=1}^{l} P_i^{e_i}, \qquad e_i = \pm 1, \pm 2, \dots . \tag{3.3}$$

For the identity element of $\mathfrak{D}$, i.e., the identity divisor E of K, the right-hand side of (3.3) becomes an empty product.

If $\nu_P(D) \geq 0$ for all the places P, the divisor D is said to be an *integral* (or *positive*) *divisor* of K. Any divisor D may be uniquely expressed as a quotient of two integral divisors

$$D = D_1 D_2^{-1} = \frac{D_1}{D_2}$$

such that D_1 and D_2 share no places in common, where D_1 is called the *numerator* of D and D_2 is called the *denominator* of D.

When, for divisors D and D', the quotient $DD'^{-1} = D/D'$ is an integral divisor, i.e., $\nu_P(D) \geq \nu_P(D')$ holds for every P, we say D is divisible by D' and write $D'|D$. We also say that D is a multiple of D', and that D' divides D. Just as numbers, the greatest common divisor (D, D') and the least common multiple $\{D, D'\}$ of D and D' can be defined. It is clear that the greatest common divisor and the least common multiple of divisors have the corresponding properties of these notions in numbers.

Let $x \neq 0$ be an arbitrary element of K. Then from Theorem 2.2, there are only finitely many places P such that $\nu_P(x) \neq 0$. Hence, there exists a unique divisor D of K such that

$$\nu_P(D) = \nu_P(x)$$

holds. This D is called the divisor of the element x in K, and is denoted by (x). Since $\nu_P(xy) = \nu_P(x) + \nu_P(y)$ holds for any x and y, we have

$$(xy) = (x)(y).$$

Therefore, we obtain a subgroup of $\mathfrak{D}$:

$$\mathfrak{D}_H = \{(x)\,;\, x \in K\,,\, x \neq 0\}.$$

This subgroup $\mathfrak{D}_H$ is said to be the principal divisor group of K. A divisor belonging to $\mathfrak{D}_H$ is called a principal divisor of K. By Theorem 2.2, it is necessary and sufficient for an element x of K to belong to the coefficient field k that (x) coincides with the unit divisor E. If $(x) = (y)$, i.e., $(x/y) = (x)/(y) = E$, there exists $\alpha \in k$ such that

$$y = \alpha x\,, \qquad a \neq 0.$$

The converse also holds. That is, a principal divisor determines an element of K up to multiplication by constant (i.e., multiplication by an element of k).

The quotient group

$$\overline{\mathfrak{D}} = \mathfrak{D}/\mathfrak{D}_H$$

is called the *divisor class group* of K, and each residue class is said to be a *divisor class*. Divisors D and D' belonging to the same class, denoted $D \sim D'$, are said to be equivalent.

Let L be an algebraic function field with the coefficient field k, and let L be an extension field of K. Suppose $P_1, \dots, P_g$ are extensions on L of a place P on K such that $e_1, \dots, e_g$ are the ramification indices. For an arbitrary divisor

$$D = \prod_P P^e, \qquad e = \nu_P(D)$$

on K, we have a divisor

$$\prod_P (P_1^{e_1} \cdots P_g^{e_g})^e$$

on L, which is called the extension of D to L, denoted as $\{D\}_L$. Since for the projection P on K of a place P' on L with the ramification index e, we have

$$\nu_{P'}(x) = e\nu_P(x)$$

for any element x of K, the extension $\{(x)\}_L$ to L of a principal divisor (x) on K coincides with the principal divisor (x) where x is considered as an element of L. Furthermore, if for divisors D_1 and D_2 on K

$$D_1 | D_2 \quad \text{or} \quad D_1 \sim D_2$$

holds, then on L we also have

$$\{D_1\}_L | \{D_2\}_L \quad \text{or} \quad \{D_1\}_L \sim \{D_2\}_L.$$

Hence, when there is no fear of confusion, we sometimes do not distinguish the divisor on K from its extension on L, and denote it simply by D rather than $\{D\}_L$.

We have the following important theorem.

THEOREM 2.7. *For an element x of K that is not in k, let D_0 be its numerator and D_∞ be its denominator of the principal divisor (x). Then we have*

$$n(D_0) = n(D_\infty) = [K : k(x)], \tag{3.4}$$

and in particular,

$$n((x)) = 0 \tag{3.5}$$

holds.

PROOF. Let P_0 and P_∞ be the numerator and the denominator of the principal divisor (x) considered on $k(x)$. Since we have $D_0 = \{P_0\}_K$ and

$D_\infty = \{P_\infty\}_K$, if $D_0 = \prod_{i=1}^{l} P_i^{e_i}$, $e_i > 0$, then $P_1, \dots, P_l$ are all the extensions of P_0 to K, and e_i is the ramification index of P_i. Since the residue field of P_0 coincides with k, the relative degree f_i of P_i with respect to $k(x)$ coincides with the (absolute) degree $n(P_i)$ of P_i. Hence, Theorem 1.13 implies that

$$n(D_0) = \sum_{i=1}^{l} e_i n(P_i) = \sum_{i=1}^{l} e_i f_i \leq [K : k(x)]. \tag{3.6}$$

Next we will show that the above inequality is actually an equality. Let

$$\mathfrak{o}_0 = \{a ; a \in k(x), \nu_0(a) \geq 0\}, \quad \text{where } \nu_0 = \nu_{P_0}$$

and let

$$\mathfrak{o}' = \{a' ; a' \in K, \nu_i(a') \geq 0, i = 1, \dots, l\}, \quad \text{where } \nu_i = \nu_{P_i}.$$

Again by Theorem 1.13, we need to show that the totality $\mathfrak{o}'$ of integral elements in K over $\mathfrak{o}$ is a finitely generated $\mathfrak{o}$-module.

Similarly, let $\mathfrak{o}''$ be the totality of integral elements in K over

$$\mathfrak{o}_\infty = \{b ; b \in k(x), \nu_\infty(b) \geq 0\}, \quad \text{where } \nu_\infty = \nu_{P_\infty}.$$

If one can prove that $\mathfrak{o}''$ is a finitely generated $\mathfrak{o}_\infty$-module, then we have $n(D_\infty) = [K : k(x)]$, which proves our theorem. Let us prove that $\mathfrak{o}'$ is a finitely generated $\mathfrak{o}_0$-module. One can also prove similarly that $\mathfrak{o}''$ is a finitely generated $\mathfrak{o}_\infty$-module.

Let $\mathfrak{o}^*$ be the intersection of valuation rings $\mathfrak{o}_P$ associated with all places P that are different from $P_1, \dots, P_l$, that is,

$$\mathfrak{o}^* = \bigcap \mathfrak{o}_P, \qquad P \neq P_i, \ i = 1, \dots, l.$$

First, $\mathfrak{o}^*$ is a subring of K. For an arbitrary element a of K, there is an element u in $k(x)$ such that $a' = ua$ belongs to $\mathfrak{o}^*$. Here is a proof: let a polynomial in $k(x)[Y]$, $f(Y) = Y^m + u_1 Y^{m-1} + \cdots + u_m$ be such that $f(a) = 0$ and $u_i \in k(x)$. Then we have

$$g(a') = 0, \quad \text{where } g(Y) = Y^m + uu_1 Y^{m-1} + u^2 u_2 Y^{m-1} + \cdots + u^m u_m.$$

We can choose the above u so that uu_1, $u^2 u_2, \dots, u^m u_m$ may belong to $k[1/x]$. Then for any valuation ν on $k(x)$ that does not belong to P_0,

$$\nu(u^i u_i) \geq 0, \qquad i = 1, 2, \dots, m$$

holds. Therefore, by Lemma 1.8, we have

$$\nu_P(a') \geq 0$$

for any place P that is not any of extensions $P_1, \dots, P_l$ of P_0. That is to say, $a' \in \mathfrak{o}_P$ for $P \neq P_i$, i.e., $a' \in \mathfrak{o}^*$.

Next, for any element z in K, let

$$\nu^*(z) = \mathrm{Min}([\nu_i(z)/e_i];\ i = 1, \ldots, l),$$

where $[r]$ indicates the greatest integer not larger than the rational number r. From the choices of ν_i and e_i, we have the following:

(1) $\nu^*(z_1 + z_2) \geq \mathrm{Min}(\nu^*(z_1), \nu^*(z_2))$,
(2) $\nu^*(z_1 z_2) \geq \nu^*(z_1) + \nu^*(z_2)$,
(3) In particular, for z_1 in $k(x)$ we have $\nu^*(z_1) = \nu_0(z_1)$ and $\nu^*(z_1 z_2) = \nu^*(z_1) + \nu^*(z_2)$.

Furthermore,

(4) for $y \in \mathfrak{O}^*$, $y \neq 0$, we have $\nu^*(y) \leq 0$.

While $\nu^*(y) = \nu_0(y) = 0$ for $y \in k$, if $y \notin k$, then we have

$$\nu_Q(y) < 0,$$

where Q is an extension on K of the place Q_∞ on $k(y)$ associated with the prime ideal $(1/y)$ in $K[1/y]$. The hypothesis that $\nu_P(y) \geq 0$ for $P \neq P_i$ implies that Q must be one of $P_1, \ldots, P_l$. Hence, there must be a negative $\nu_i(y)$, $i = 1, \ldots, l$, i.e., $\nu^*(y) < 0$ holds.

If $[K : k(x)] = n$, we choose $y_1, \ldots, y_n$ from $\mathfrak{o}^*$ as follows: from (4) there exists an element y_1 in $\mathfrak{o}^*$, $y_1 \neq 0$, such that $\nu^*(y_1)$ is the maximum. Next, suppose that $y_1, \ldots, y_{i-1}$ have been determined. Then choose an element y_i among those in $\mathfrak{o}^*$ that are linearly independant from $y_1, \ldots, y_{i-1}$ over $k(x)$ and that give the maximum value for ν^*. Since any element of K can be in $\mathfrak{o}^*$ after multiplying it by an element in $k(x)$, one can obtain n linearly independent elements $y_1, \ldots, y_n$ from the above method. Let

$$\nu^*(y_i) = r_i, \quad z_i = x^{-r_i} y_i, \qquad i = 1, \ldots, n,\ r_1 \geq r_2 \geq \cdots \geq r_n.$$

Then $z_1, \ldots, z_n$ are also linearly independent over $k(x)$, and by (3) we have

$$\nu^*(z_i) = \nu^*(x^{-r_i}) + \nu^*(y_i) = -r_i + r_i = 0.$$

Hence, all the z_i belong to $\mathfrak{o}'$, since $\nu_j(z_i) \geq 0$ for $j = 1, \ldots, l$. We will prove next that $z_1, \ldots, z_n$ are basis elements of $\mathfrak{o}'$ over $\mathfrak{o}_0$. I.e., for

$$z = \sum_{i=1}^{n} a_i z_i, \qquad a_i \in k(x),$$

if $z \in \mathfrak{o}'$ holds, then we need to show $\nu_0(a_i) \geq 0$, $i = 1, \ldots, n$. Let us assume: $z \in \mathfrak{o}'$ and

$$\mathrm{Min}(\nu_0(a_i);\ i = 1, \ldots, n) = -r < 0,$$
$$\nu_0(a_m) = -r,\ \nu_0(a_{m+1}) > -r, \ldots, \nu_0(a_n) > -r \qquad (m \leq n).$$

Then all the $a_i x^r$ belong to $\mathfrak{o}_0$. Therefore, for the prime ideal $\mathfrak{p} = (x)$ of $\mathfrak{o}_0$, there exists uniquely an element α_i in k satisfying the equation

$$a_i x^r \equiv \alpha_i \mod \mathfrak{p}.$$

By our assumption we have $\alpha_m \neq 0$ and $\alpha_{m+1} = \cdots = \alpha_n = 0$. Since zx^r is contained in the ideal $x\mathfrak{o}'$ of $\mathfrak{o}'$, we have

$$zx^r = \sum_{i=1}^{n} a_i z_i x^r \equiv 0 \mod x\mathfrak{o}'.$$

Hence, we obtain

$$w = \sum_{i=1}^{m} \alpha_i z_i \equiv 0 \mod x\mathfrak{o}', \qquad \alpha_m \neq 0.$$

Since $\nu_j(w) \geq e_j$ for $j = 1, \ldots, l$, $\nu^*(w) \geq 1$ holds. Furthermore, we have $r_m - r_i \leq 0$ for $i = 1, \ldots, m$ and $x^{r_m - r_i} \in \mathfrak{o}^*$. Hence,

$$wx^{r_m} = \sum_{i=1}^{m} \alpha_i y_i x^{r_m - r_i}, \qquad \alpha_m \neq 0,$$

which belongs to $\mathfrak{o}^*$ and is linearly independent from $y_1, \ldots, y_{m-1}$. On the other hand, we have

$$\nu^*(wx^{r_m}) = \nu^*(w) + \nu^*(x^{r_m}) \geq 1 + r_m > r_m = \nu^*(y_m),$$

contradicting the choice of y_m. Therefore, $\mathrm{Min}(\nu_0(a_i); i = 1, \ldots, n) \geq 0$ must hold, i.e., $\nu_0(a_i) \geq 0$. Consequently, all the a_i belong to $\mathfrak{o}_0$. We have proved that the z_i are minimal basis elements for $\mathfrak{o}_0$. Hence, the equality holds in (3.6). One can give a similar proof for the case $n(D_\infty) = [K : k(x)]$. The proof of Theorem 2.7 is complete.

REMARK 1. The above $\{y_1, \ldots, y_n\}$ is called a canonical basis for $K/k(x)$ with respect to P_0.

REMARK 2. Let L and K be algebraic function fields over the coefficient field k such that L is an extension field of K. For an arbitrary place P on K, let $Q_1, \ldots, Q_g$ be the extensions on L of P, and also let e_i and f_i be the ramification index and the relative degree of Q_i, respectively, with respect to K. Choose an element x of K such that $\nu_P(x) > 0$. We denote the numerator of the principal divisor (x) on $k(x)$ by P_0. Then let $P_1, \ldots, P_l$ be the extensions of P_0 to K, and let e_i' and f_i' be the ramification index and the relative degree of P_i with respect to $k(x)$. Moreover, for the extensions $Q_{i,1}, \ldots, Q_{i,g_i}$ on L of P_i, let $e_{i,j'}$, $f_{i,j'}$, and $e_{i,j}$, $f_{i,j}$ be the ramification indices and the relative degrees of $Q_{i,j}$ with respect to $k(x)$ and K, respectively. Since x has been chosen in such a way that P is an extension of P_0 to K, we can set, for example, $P = P_1$ and $Q_j = Q_{1,j}$. Then we can set $e_j = e_{1,j}$ and $f_j = f_{1,j}$. We also have

$$e_{i,j}' = e_i' e_{i,j} \quad \text{and} \quad f_{i,j}' = f_i' f_{i,j}. \tag{3.7}$$

From the proof of the above theorem applied to $K/k(x)$ and $L/k(x)$, we have

$$\sum_i e_i' f_i' = [K : k(x)], \qquad \sum_{i,j} e_{i,j}' f_{i,j}' = [L : k(x)].$$

Therefore, from (3.7) we obtain

$$[L : k(x)] = \sum_i \left(e_i' f_i' \sum_j e_{i,j} f_{i,j} \right). \tag{3.8}$$

However, in general we have

$$\sum_j e_{i,j} f_{i,j} \leq [L : K] \tag{3.9}$$

by Theorem 1.13. Substituting (3.9) into the right-hand side of (3.8), we have

$$[L : k(x)] \leq \sum_i e_i' f_i' [L : K] = [K : k(x)][L : K] = [L : k(x)].$$

Hence, we must always have equality in (3.9). In particular, for $i = 1$, we obtain

$$\sum_{j=1}^{g} e_j f_j = [L : K].$$

That is, for the case of the algebraic function field, we have shown that the equality in (3.14) in Theorem 1.13 always holds.

Theorem 2.7 and equation (3.2) imply that equivalent divisors have the same degree. Hence, one can define uniquely the degree of a divisor class containing D by $n(D)$.

Next we define the dimension of a divisor. Let A be any divisor on K. An element x in K is said to be a multiple of A if $A|(x)$. We denote the totality of the multiples of A^{-1} as

$$L(A) = \{x ; x \in K, A^{-1}|(x)\}.$$

Since $A^{-1}|(x)$ is equivalent to the statement that

$$\nu_P(x) \geq \nu_P(A^{-1}) = -\nu_P(A) \quad \text{holds for any } P, \tag{3.10}$$

for x and y in K and for α and β in k we have

$$\nu_P(\alpha x + \beta y) \geq \mathrm{Min}(\nu_P(\alpha x), \nu_P(\beta y)) = \mathrm{Min}(\nu_P(x), \nu_P(y)) \geq -\nu_P(A),$$

i.e., $\alpha x + \beta y \in L(A)$. Hence, $L(A)$ is a module over k.

DEFINITION 2.5. The dimension of $L(A)$ as a k-module is said to be the *dimension* of the divisor A, denoted by $l(A)$ or $\dim(A)$.

Before we prove the finite dimensionality of $l(A)$ for any A, we will mention a few examples. If $A = E$, then condition (3.10) becomes $\nu_P(x) \geq 0$. Theorem 2.7 implies that such an element x must be in k. Hence,

$L(A) = k$, i.e., $l(A) = 1$. If $A \neq E$ and A is an integral divisor, then for an element x in $L(A)$ we must always have $\nu_P(x) \geq 0$, and also for at least one P, $\nu_P(x) > 0$ must hold. However, for any α in k, $\alpha \neq 0$, $\nu_P(\alpha) = 0$ must hold. Therefore, $L(A)$ cannot contain any elements but 0. That is, $L(A) = 0$ and $l(A) = 0$. Suppose that $A \sim B$ such that $A = (y)B$ for some $y \in K$. Then we have

$$A^{-1} | (x) \Leftrightarrow B^{-1}(y)^{-1} | (x) \Leftrightarrow B^{-1} | (xy).$$

Hence, $L(B) = L(A)y \overset{\text{def}}{=} \{xy ; x \in L(A)\}$. Therefore, $L(A)$ and $L(B)$ are isomorphic to each other as k-modules. We have $l(A) = l(B)$. Consequently, the dimension of a divisor depends only upon the divisor class.

THEOREM 2.8. *A k-module $L(A)$ is finite dimensional*:

$$l(A) = \dim A < \infty.$$

PROOF. Let P be a place on K, and let $\boldsymbol{k}_P$ be the residue class field of P. Then $\boldsymbol{k}_P$ is a finite extension of k, and its extension degree is the degree $n(P)$ of the place P. Hence, there are elements $w_1^{(P)}, \ldots, w_n^{(P)}$ in K, where $n = n(P)$, so that the k-module

$$M_P = \{\alpha_1 w_1^{(P)} + \cdots + \alpha_n w_n^{(P)} ; \alpha_i \in k\}$$

is a system of complete representatives in K for $\boldsymbol{k}_P$. For any system of complete representatives as M_P above and a prime element $t = t_P$ for P, by Theorem 1.6, an arbitrary element a in the complete field K_P can be expressed uniquely as

$$a = \sum a_i t^i, \qquad a_i = \sum_{j=1}^{n} \alpha_{i,j}^{(P)} w_j^{(P)}, \qquad \alpha_{i,j}^{(P)} \in k.$$

Notice that $\alpha_{i,j}^{(P)}$ is determined uniquely by the element a. If we let

$$\alpha_{i,j}^{(P)} = f_{i,j}^{(P)}(a),$$

then $f_{i,j}^{(P)}(a)$ is a function on K_P with its values in k. Since M_P is a k-module, we have the following: let $a = \sum a_i t^i$ and let $b = \sum b_i t^i$, where a_i and b_i are elements of M_P. Then we have $\alpha a + \beta b = \sum (\alpha a_i + \beta b_i) t^i$ for $\alpha, \beta \in k$ such that $(\alpha a_i + \beta b_i) \in M_P$. That is, we have

$$f_{i,j}^{(P)}(\alpha a + \beta b) = \alpha f_{i,j}^{(P)}(a) + \beta f_{i,j}^{(P)}(b),$$

i.e., $f_{i,j}^{(P)}$ is a k-linear function on K_P.

Let B be an arbitrary divisor that divides A. We have $\nu_P(A) \geq \nu_P(B)$, and hence

$$L(B) \subseteq L(A).$$

Let r be an arbitrary natural number, and let M be a k-submodule of dimension r in $L(A)$. Suppose u is an element of M. Then u is in $L(B)$ if and only if

$$\nu_P(u) \geq -\nu_P(B) \quad \text{for any } P. \tag{3.11}$$

In terms of $f_{i,j}^{(P)}$, this can be rephrased as

$$f_{i,j}^{(P)}(u) = 0, \qquad i = 1, \ldots, n(P), \; j < -\nu_P(B). \tag{3.12}$$

Since u is already in $L(A)$, we have $\nu_P(u) \geq -\nu_P(A)$, and hence $f_{i,j}^{(P)}(u) = 0$ for $j < -\nu_P(A)$. We can restate (3.12) as

$$f_{i,j}^{(P)}(u) = 0, \qquad i = 1, \ldots, n(P), \; -\nu_P(A) \leq j < -\nu_P(B). \tag{3.13}$$

Let $m_1, \ldots, m_r$ be basis elements for M over k, and let

$$u = \xi_1 m_1 + \cdots + \xi_r m_r, \qquad \xi_i \in k.$$

Since $f_{i,j}^{(P)}$ is a linear function, from (3.13) we obtain that

$$\sum_{l=1}^{r} \xi_l f_{i,j}^{(P)}(m_l) = 0, \qquad i = 1, \ldots, n(P), \; -\nu_P(A) \leq j < -\nu_P(B). \tag{3.14}$$

Equation (3.14) is a system of linear equations in ξ_l, and the number of equations is

$$\begin{aligned}\sum_P n(P)(\nu_P(A) - \nu_P(B)) &= \sum_P n(P)\nu_P(A) - \sum_P n(P)\nu_P(B) \\ &= n(A) - n(B).\end{aligned}$$

The totality of solutions u is a k-submodule $M' = M \cap L(B)$ whose dimension is at least $r - (n(A) - n(B))$.

Since M' is contained in $L(B)$, we have

$$l(B) \geq r - (n(A) - n(B)). \tag{3.15}$$

In particular, take $B = A_1^{-1}$, where A_1 is the denominator of A. Then B divides A, and $A_1 = B^{-1}$ is an integral divisor. By the remark preceding this theorem, we have $l(B) = 0$ or 1. Hence, by (3.15) we have

$$n(A) - n(B) + 1 \geq r.$$

Recall that r was an arbitrary natural number not greater than $l(A)$. We have

$$n(A) - n(B) + 1 \geq l(A),$$

which proves the finite dimensionality of $l(A)$.

Since we have shown the finiteness of $l(A)$, we can let $r = l(A)$. Let $M = L(A)$. Then notice that the totality M' of the solutions u in (3.13) coincides with $L(B)$, where B is an arbitrary divisor that divides A. From (3.15) we have

$$n(A) - l(A) = n(B) - l(B) + r(A, B), \quad \text{where } r(A, B) \geq 0. \tag{3.16}$$

As one can see from the above proof, $r(A, B)$ is the number of independent relations that hold among the $f_{i,j}^{(P)}(u)$ in (3.14), $u \in L(A)$.

Next we show that $n(A) - l(A)$ is bounded above for any divisor A.

LEMMA 2.2. *Let x be an element of K that is not in k. Let D_∞ denote the denominator of (x). Then for an arbitrary divisor A there exists a divisor A' such that*

$$A|A' \quad \textit{and} \quad A' \sim D_\infty^m \quad \textit{for some } m = 0, \pm 1, \pm 2, \ldots.$$

PROOF. Clearly we can assume that A is an integral divisor. If one can give a proof for the case that A coincides with a place P, then the general case follows by considering the product of those A', each of which corresponds to a place appearing in A. Let Q be the projection of P on $k(x)$. If P_∞ denotes the denominator of the principal divisor (x) on $k(x)$, then $D_\infty = \{P_\infty\}_K$. Hence, if $Q = P_\infty$, then let A' be D_∞. Next, suppose $Q \neq P_\infty$. Since Q is induced by a prime ideal $(f(x))$ in $k[x]$, we have in $k(x)$ that

$$(f(x)) = \frac{Q}{P_\infty^m}, \qquad Q \sim P_\infty^m,$$

where m is the degree of the polynomial $f(x)$. Since also in K we have $\{Q\}_K \sim \{P_\infty^m\}_K = D_\infty^m$, one can let $A' = \{Q\}_K$.

THEOREM 2.9. *For an arbitrary divisor A on K, $n(A) - l(A)$ is bounded above.*

PROOF. For any A, choose A' and D_∞^m as in Lemma 2.2. Since we have $A|A'$, $n(A') - l(A') \geq n(A) - l(A)$ holds from (3.16). Moreover $A' \sim D_\infty^m$ implies $n(A') = n(D_\infty^m)$ and $l(A') = l(D_\infty^m)$. Therefore, it is sufficient to prove that $n(D_\infty^m) - l(D_\infty^m)$ is bounded for $m = 1, 2, \ldots$. Let $y_1, \ldots, y_n$ be a set of basis elements for K over $k(x)$, where $n = [K : k(x)]$, and let B be the least common multiple of the denominators of (y_i). By Lemma 2.2 we can find B' and D_∞^u such that $B|B'$ and $B' = (f(x))D_\infty^u$ for some rational function $f(x)$. Define

$$z_i = f(x)y_i, \qquad i = 1, \ldots, n.$$

Then $\{z_i\}$ is a basis for K over $k(x)$. For an arbitrary natural number ν, let $f_i(x)$ be any polynomial in $k[x]$ of degree at most ν. Let M be the set of elements a in K such that

$$a = \sum_{i=1}^{n} f_i(x) z_i.$$

Then the denominator of z_i is at most D_∞^u and the denominator of $f_i(x)$ is at most D_∞^ν. Hence,

$$D_\infty^{-(u+v)}|(a).$$

That is, $a \in L(D_{\infty}^{u+\nu})$, and therefore $M \subseteq L(D_{\infty}^{u+\nu})$. Since the dimension of M over k is clearly $n(\nu+1)$, we have

$$n(\nu+1) \leq l(D_{\infty}^{u+\nu}).$$

On the other hand, by Theorem 2.7,

$$n(D_{\infty}^{u+\nu}) = (u+\nu)n(D_{\infty}) = (u+\nu)n.$$

Therefore, we obtain

$$n(D_{\infty}^{u+\nu}) - l(D_{\infty}^{u+\nu}) \leq (u-1)n.$$

Since ν is arbitrary and the right-hand side is independent of ν, we conclude that $n(D_{\infty}^{m}) - l(D_{\infty}^{m})$ is indeed bounded above. Q.E.D.

This theorem tells us that, for all the divisors A on K, the maximum value g of $n(A) - l(A) + 1$ is finite.

DEFINITION 2.6. The above g is called the *genus* of the algebraic function field K.

Since $n(E) - n(E) = 0 - 1 + 1 = 0$ for a unit divisor E, we have $g \geq 0$, i.e., a genus is a nonnegative rational integer.

For an arbitrary A, let

$$r(A) = g - (n(A) - l(A) + 1).$$

Then, from the definition of g we have $r(A) \geq 0$. We call $r(A)$ the *index of speciality* of the divisor A. We call an A with $r(A) = 0$ a normal divisor. For a normal divisor A, multiplies of A are also normal divisors by (3.16).

THEOREM 2.10. *There exists a natural number m such that if $n(A) \geq m$, then A is a normal divisor.*

PROOF. Let D be a normal divisor, and let $m = n(D) + g$. Then $n(A) \geq m$ implies

$$l(AD^{-1}) \geq n(AD^{-1}) - g + 1 = n(A) - n(D) - g + 1 \geq 1.$$

Hence, $L(AD^{-1})$ must contain a nonzero element u. Then $AD^{-1} | (u)$ implies $(u) = DA^{-1}B$ for some integral divisor B, and therefore, we have $A \sim DB$. Hence,

$$n(A) - l(A) + 1 = n(DB) - l(DB) + 1.$$

Because of $D | DB$, the right-hand side of the above equation is not less than $n(D) - l(D) + 1 = g$. By the definition of g, we have

$$n(A) - l(A) + 1 = g.$$

That is, A is a normal divisor.

§4. Idele and differential

Let K be an algebraic function field over a coefficient field k, and let K_P be the completion of K at a place P. Denote the direct product of K_P for all P by K^*:

$$K^* = \prod_P K_P.$$

That is, K^* is the totality of $\overline{a} = (a_P)$, where a_P is taken arbitrarily from K_P. K^* becomes a ring with the operations:

$$\overline{a} + \overline{b} = (a_P) + (b_P) = (a_P + b_P), \qquad \overline{a}\overline{b} = (a_P)(b_P) = (a_P b_P).$$

For $\overline{a} = (a_P)$, let

$$\nu_P^*(\overline{a}) = \nu_P(a_P).$$

Then we have a function ν_P^* on K^*. When there is no possibility of confusion, we simply write ν_P for ν_P^*. As immediate consequences, we have

$$\nu_P(\overline{a} + \overline{b}) \geq \mathrm{Min}(\nu_P(\overline{a}), \nu_P(\overline{b})), \tag{4.1}$$

$$\nu_P(\overline{a}\overline{b}) = \nu_P(\overline{a}) + \nu_P(\overline{b}). \tag{4.2}$$

The function ν_P takes its values in rational integers; however, even for $\overline{a} \neq 0$, $\nu_P(\overline{a}) = +\infty$ can occur.

The totality $\widetilde{K}$ of those elements in K^* such that $\nu_P(\overline{a}) \geq 0$ for all but finitely many P forms a subring of K^* by (4.1) and (4.2).

DEFINITION 2.7. $\widetilde{K}$ is called the *idele ring* of the algebraic function field K, and each element $\overline{a}$ of $\widetilde{K}$ is called an idele of K.

NOTE. The notion of *idele* was introduced in an arithmetic proof for the class field theory by C. Chevalley as a substitute for ideal in number field. We defined the above idele for our algebraic function field modeled after that of Chevalley's. In Chevalley's theory, ideles form a multiplicative group, and the importance is on its multiplicative nature. However, the above $\widetilde{K}$ forms a ring, and as the reader will observe in what follows, our focus is on the additive nature. Though we will not discuss an arithmetic theory of algebraic function fields over finite fields, in which both Chevalley's and our idele are needed; it may be better to call them multiplicative idele and additive idele, respectively. (Additive ideles are now called adeles.)

For an arbitrary element x in K, let $a_P = x$ for all P. Then the $\nu_P(x)$ are all zero except for a finite number of P, i.e., Theorem 2.2. Therefore, (a_P) is an idele. Hence, for each element of K, we obtain an idele. Furthermore, this is clearly a one-to-one correspondence. We will identify the elements of K with the corresponding ideles. Thus, K becomes a subfield of the idele ring $\widetilde{K}$.

For any divisor A on K, define

$$\widetilde{L}(A) = \{\overline{a}; \in \widetilde{K}, \nu_P(\overline{a}) \geq -\nu_P(A)\}.$$

Then we have

$$L(A) = \widetilde{L}(A) \cap K.$$

Notice also that $\widetilde{L}(A)$ is a k-module just as $L(A)$ is. The k-module $\widetilde{L}(A)$ has the following properties:

(1) For the least common multiple C of A and B, we have

$$\widetilde{L}(A)+\widetilde{L}(B)=\widetilde{L}(C),$$

where the left-hand side is the totality of $\overline{a}+\overline{b}$, $\overline{a}\in\widetilde{L}(A)$, $\overline{b}\in\widetilde{L}(B)$.

(2) For the greatest common denominator D of A and B, we have

$$\widetilde{L}(A)\cap\widetilde{L}(B)=\widetilde{L}(D).$$

(3) $\widetilde{L}(A)\cdot\widetilde{L}(B)=\widetilde{L}(AB)$, where $\widetilde{L}(A)\cdot\widetilde{L}(B)=\{\overline{a}\overline{b}\,;\,\overline{a}\in\widetilde{L}(A),\ \overline{b}\in\widetilde{L}(B)\}$.

PROOF OF (1). We obviously have $\widetilde{L}(A)\subseteq\widetilde{L}(C)$ and $\widetilde{L}(B)\subseteq\widetilde{L}(C)$. Hence $\widetilde{L}(A)+\widetilde{L}(B)\subseteq\widetilde{L}(C)$ holds. Conversely, let $\overline{a}=(a_P)$ be an element in $\widetilde{L}(C)$. Then we have

$$\nu_P(a_P)\geq-\nu_P(C)=-\operatorname{Max}(\nu_P(A),\nu_P(B))=\operatorname{Min}(-\nu_P(A),-\nu_P(B)).$$

Let P_1 be those P such that $\nu_P(a_P)\geq-\nu_P(A)$, and let P_2 be the complement of those places. Then define

$$\overline{a}_1=(a_P^{(1)}),\qquad a_P^{(1)}=\begin{cases}a_P & \text{for } P=P_1,\\ 0 & \text{for } P=P_2,\end{cases}$$

$$\overline{a}_2=(a_P^{(2)}),\qquad a_P^{(2)}=\begin{cases}0 & \text{for } P=P_1,\\ a_P & \text{for } P=P_2.\end{cases}$$

Then we have $\overline{a}_1\in\widetilde{L}(A)$, $\overline{a}_2\in\widetilde{L}(B)$, and $\overline{a}=\overline{a}_1+\overline{a}_2$. Therefore, $\widetilde{L}(C)\subseteq\widetilde{L}(A)+\widetilde{L}(B)$ follows.

PROOF OF (2). In order for $\overline{a}=(a_P)$ to belong to $\widetilde{L}(D)$ we need to have $\nu_P(a_P)\geq-\nu_P(D)=-\operatorname{Min}(\nu_P(A),\nu_P(B))=\operatorname{Max}(-\nu_P(A),-\nu_P(B))$, which means that $\nu_P(a_P)\geq-\nu_P(A)$ and $\nu_P(a_P)\geq-\nu_P(B)$. That is, $\overline{a}\in\widetilde{L}(A)$ and $\overline{a}\in\widetilde{L}(B)$.

PROOF OF (3). For $\overline{a}\in\widetilde{L}(A)$ and $\overline{b}\in\widetilde{L}(B)$ we have $\overline{a}\overline{b}\in\widetilde{L}(AB)$ by (4.2), i.e., $\widetilde{L}(A)\widetilde{L}(B)\subseteq\widetilde{L}(AB)$. Conversely, let $\overline{a}=(a_P)\in\widetilde{L}(AB)$. Then we have $\nu_P(a_P)\geq-\nu_P(A)-\nu_P(B)$. Choose any $a_P^{(1)}\neq0$ from K_P so that $\nu_P(a_P^{(1)})=-\nu_P(A)$ holds and let $a_P=a_P^{(1)}a_P^{(2)}$, $\overline{a}_1=(a_P^{(1)})$, $\overline{a}_2=(a_P^{(2)})$. Then $\overline{a}_1=(a_P^{(1)})\in\widetilde{L}(A)$, $\overline{a}_2=(a_P^{(2)})\in\widetilde{L}(B)$ and $\overline{a}=\overline{a}_1\overline{a}_2$.

Next take the totality of $\overline{a}+\widetilde{L}(A)$ as neighborhoods of a point $\overline{a}$ in the idele ring $\widetilde{K}$, where A is an arbitrary divisor. Then, by (1) and (2) above, $\widetilde{K}$ becomes a Hausdorff topological space. Furthermore, (1) and (3) above imply that the addition and the multiplication in $\widetilde{K}$ are continuous for this topology. Hence we have

THEOREM 2.11. *The idele ring $\widetilde{K}$ is a topological ring.*

We will define a differential on the algebraic function field K according to A. Weil as follows.

DEFINITION 2.8. Let f be a map from the idele ring $\widetilde{K}$ of the algebraic function field K into the coefficient field k. Then f is said to be a *differential* in K if f satisfies the following three conditions:

(i) f is k-linear; that is,

$$f(\alpha\overline{a}+\beta\overline{b})=\alpha f(\overline{a})+\beta f(\overline{b}), \qquad \overline{a}, \overline{b}\in\widetilde{K} \text{ and } \alpha, \beta\in k.$$

(ii) When k is regarded as a subspace of the topological ring $\widetilde{K}$, then f is continuous.

If $n(A)<0$, then $\widetilde{L}(A)\cap k=0$. That is, the induced topology on k from $\widetilde{K}$ is discrete. Therefore, in light of (i), condition (ii) may be rephrased as

(ii′) one can choose a divisor A so that $f(\overline{a})=0$ holds for any $\overline{a}\in\widetilde{L}(A)$.

(iii) for an arbitrary element x of K, $f(x)=0$.

Let $\mathfrak{L}$ be the set of differentials on K, and let $\mathfrak{L}_B$ be those differentials f in $\mathfrak{L}$ with $f(\overline{b})=0$ for any idele $\overline{b}\in\widetilde{L}(B)$.

In order to study the properties of differentials, it is necessary to introduce more general maps from $\widetilde{L}(A)$ to k, i.e., in addition to linearity and continuity, the maps f from $\widetilde{L}(A)$ to k must satisfy $f(x)=0$ for an arbitrary x in $L(A)$. Let us denote the set of these differentials by $\mathfrak{L}^A$. Particularly, if A is a multiple of B, $\mathfrak{L}_B^A$ denotes the set of maps f in $\mathfrak{L}^A$ such that $f(\overline{b})=0$ for an arbitrary $\overline{b}\in\widetilde{L}(B)$.

It is clear that for differentials f and g, we have a differential $\alpha f+\beta g$ on K for arbitrary α and β. Therefore, the set $\mathfrak{L}$ of differentials is a k-module. Similarly, $\mathfrak{L}_B$, $\mathfrak{L}^A$, and $\mathfrak{L}_B^A$ are all k-modules. The dimension of $\mathfrak{L}_B$ over k is denoted by $\dim\mathfrak{L}_B$.

LEMMA 2.3. *For $B|A$, we have*

$$\dim\mathfrak{L}_B^A=r(A, B)=(n(A)-l(A))-(n(B)-l(B)).$$

PROOF. Using the same notation as in the proof of Theorem 2.8, an arbitrary element a_P in K_P can be expanded uniquely as

$$a_P=\sum_i\sum_{j=1}^{n(P)}f_{i,j}^{(P)}(a_P)w_j^{(P)}t_P^i, \qquad f_{i,j}^{(P)}(a_P)\in k.$$

For an idele $\overline{a}=(a_P)$ in $\widetilde{L}(A)$, the summation for i on the P-component a_P only ranges over $i\geq-\nu_P(A)$. Divide the sum into two partial sums over $-\nu_P(A)\leq i<-\nu_P(B)$ and $-\nu_P(B)\leq i$, respectively. That is,

$$a_P=a_P^{(1)}+a_P^{(2)}.$$

If we let $\overline{a}_1=(a_P^{(1)})$ and $\overline{a}_2=(a_P^{(2)})$, then we have $\overline{a}=\overline{a}_1+\overline{a}_2$. Since the idele $\overline{a}_2$ belongs to $\widetilde{L}(B)$, for a map f in $\mathfrak{L}_B^A$ we have

$$f(a_P)=f(a_P^{(1)})+f(a_P^{(2)})=f(a_P^{(1)}).$$

If we denote simply by a_P the idele whose P-component is a_P and the Q-components are all zero for $Q\neq P$, then we can write the above $\overline{a}_1$ as

$$\overline{a}_1=\sum_P\sum_{i=-\nu_P(A)}^{-\nu_P(B)-1}\sum_{j=1}^{n(P)}f_{i,j}^{(P)}(a_P)w_j^{(P)}t_P^i.$$

The summation on the right-hand side is over all the places P on K, but in fact the summation is over a finite sum of $\nu_P(A)-\nu_P(B)$ terms. Since f is k-linear, we have

$$f(\overline{a})=f(\overline{a}_1)=\sum_P\sum_i\sum_j \gamma^P_{i,j}f^{(P)}_{i,j}(a_P),$$

where $\gamma^P_{i,j}=f(w^{(P)}_j t^i_P)$. Notice that the constants $\gamma^P_{i,j}$ are independent from the choice of $\overline{a}$. Hence, the above equation shows that an arbitrary map $f(\overline{a})$ in $\mathfrak{L}^A_B$ can be written as a linear combination of $n(A)-n(B)$ linearly independent functions, i.e., for $\overline{a}=(a_P)$

$$f^{(P)}_{i,j}(\overline{a})=f^{(P)}_{i,j}(a_P),\quad -\nu_P(A)\le i\le -\nu_P(B),\quad \text{and}\quad j=1,\ldots,n(P).$$

Conversely, we will find a condition for the function $f(\overline{a})$ in $\widetilde{L}(A)$, which is a linear combination of $f^{(P)}_{i,j}(a_P)$, to belong to $\mathfrak{L}^A_B$:

$$f(\overline{a})=\sum_P\sum_{i=-\nu_P(A)}^{-\nu_P(B)-1}\sum_{j=1}^{n(P)}\gamma^P_{i,j}f^{(P)}_{i,j}(a_P)\quad \text{for } \overline{a}=(a_P),\ \text{where } \gamma^P_{i,j}\in k.$$

It is clear that such a function as above is linear for any $\gamma^P_{i,j}$, it is continuous, and $f(\overline{b})=0$ for an arbitrary $\overline{b}\in\widetilde{L}(B)$. Therefore, for $f(\overline{a})$ to belong to $\mathfrak{L}^A_B$, it is necessary and sufficient that

$$f(u)=\sum_P\sum_i\sum_j\gamma^P_{i,j}f^{(P)}_{i,j}(u)=0 \tag{4.3}$$

for an arbitrary element u in $\widetilde{L}(A)$. Hence, the dimension of $\mathfrak{L}^A_B$ is the dimension of $\{\gamma^P_{i,j}\}$ satisfying (4.3) for all u in $\widetilde{L}(A)$, that is, the number of independent linear relations among $f^{(P)}_{i,j}(u)$, $u\in L(A)$. By the remark at the end of the proof of Theorem 2.8, the number of relations is equal to $r(A,B)$. Q.E.D.

Let $B|A$ and $A|A'$, then $\widetilde{L}(A)\subseteq\widetilde{L}(A')$. Hence by restricting a map $f'(\overline{a})$ in $\mathfrak{L}^{A'}_B$ to $\widetilde{L}(A)$, one obtains a map f in $\mathfrak{L}^A_B$. The map f is said to be the projection on $\mathfrak{L}^A_B$, and conversely f' is said to be an extension of f in $\mathfrak{L}^{A'}_B$. One can also define the projection and the extension between $\mathfrak{L}^A_B$ and $\mathfrak{L}_B$.

LEMMA 2.4. *If $B|A$, $A|A'$, and if A is a normal divisor, then an arbitrary f in $\mathfrak{L}^A_B$ has a unique extension to f' in $\mathfrak{L}^{A'}_B$.*

PROOF. We will first prove the uniqueness. Suppose f' and f'' are two extensions of f to $\mathfrak{L}^{A'}_B$. Then, for $\overline{a}\in\widetilde{L}(A)$, we have

$$f'(\overline{a})=f''(\overline{a})=f(\overline{a}).$$

Hence, $f^* = f' - f''$ is clearly a map in $\mathfrak{L}_A^{A'}$. Since A and A' are normal divisors, by the preceeding lemma, we have

$$\begin{aligned}\dim \mathfrak{L}_A^{A'} &= r(A', A) = (n(A') - l(A') + 1) - (n(A) - l(A) + 1) \\ &= g - g = 0, \quad \text{where } g \text{ is the genus of } K.\end{aligned}$$

Therefore,

$$\mathfrak{L}_A^{A'} = 0, \quad f^* = 0, \quad \text{and} \quad f' = f''.$$

From the same lemma, we also have

$$\dim \mathfrak{L}_B^A = r(A, B) = g - 1 - (n(B) - l(B)).$$

Since the corresponding formula holds for $\mathfrak{L}_B^{A'}$, we have

$$\dim \mathfrak{L}_B^A = \dim \mathfrak{L}_B^{A'}.$$

Let r be the above dimension, and let $f_1, \ldots, f_r$ be the projections of $f_1', \ldots, f_r'$ on $\mathfrak{L}_B^A$, which are basis elements for $\mathfrak{L}_B^{A'}$ over k. Then $f_1, \ldots, f_r$ are linearly independent over k. This is because, for

$$\alpha_1 f_1 + \cdots + \alpha_r f_r = 0, \qquad a_i \in k,$$

we have a unique extension $\alpha_1 f_1' + \cdots + \alpha_r f_r'$ in $\mathfrak{L}_B^{A'}$ that must be zero. Then $\alpha_1 = \cdots = \alpha_r = 0$, since f_i', $i = 1, \ldots, r$, are basis elements for $\mathfrak{L}_B^{A'}$. Hence, $f_1, \ldots, f_r$ are linearly independent, and they are basis elements for $\mathfrak{L}_B^A$ over k. Therefore, any element f in $\mathfrak{L}_B^A$ can be written in the form $f = \beta_1 f_1 + \cdots + \beta_r f_r$, $\beta_i \in k$, having an extension $f' = \beta_1 f_1' + \cdots + \beta_r f_r'$ in $\mathfrak{L}_B^{A'}$.

LEMMA 2.5. *If $B|A$ and if A is a normal divisor, then an arbitrary element f in $\mathfrak{L}_B^A$ can be uniquely extended to a differential in $\mathfrak{L}_B$. Furthermore, we have*

$$\dim \mathfrak{L}_B = \dim \mathfrak{L}_B^A.$$

PROOF. For a given idele $\overline{a}$, one can always find A' such that $\overline{a} \in \widetilde{L}(A')$ and $A|A'$. From the preceeding lemma, there exists an extension f' in $\mathfrak{L}_B^{A'}$ of f. Then define

$$f^*(\overline{a}) = f'(\overline{a}). \tag{4.4}$$

Notice that the value of $f^*(\overline{a})$ is independent of the particular choice of A' in the above definition: choose another A'' such that $\overline{a} \in \widetilde{L}(A'')$ and $A|A''$, and let f'' be an extension of f to $\mathfrak{L}_B^{A''}$. Then let A''' be a divisor such that $A'|A'''$ and $A''|A'''$, and for this A''', let f_1 and f_2 be extensions of f' and f'', respectively. Since both f_1 and f_2 are extensions of f to $\mathfrak{L}_B^{A''}$, the uniqueness of this extension implies $f_1 = f_2$. Therefore, we have

$$f'(\overline{a}) = f_1(\overline{a}) = f_2(\overline{a}) = f''(\overline{a}),$$

i.e., $f^*(\overline{a})$ is uniquely determined. It is plain from definition (4.4) that the map f^* thus defined on $\widetilde{K}$ is a K-differential belonging to $\mathfrak{L}_B$. For instance, for arbitrary ideles $\overline{a}$ and $\overline{b}$, choose A' so that both $\overline{a}$ and $\overline{b}$ are in $\widetilde{L}(A')$, and $A|A'$. Let f' be the extension of f to $\mathfrak{L}_B^{A'}$. Then $f^*(\alpha\overline{a}+\beta\overline{b}) = f'(\alpha\overline{a}+\beta\overline{b}) = \alpha f'(\overline{a}) + \beta f'(\overline{b}) = \alpha f^*(\overline{a}) + \beta f^*(\overline{b})$ and so forth. Hence, f has an extension f^* in $\mathfrak{L}_B$.

Let f_1^* be an arbitrary extension of f to $\mathfrak{L}_B$. For a given $\overline{a}$, choose A' and f' as before. Since the projection f_1' of f_1^* on $\mathfrak{L}_B^{A'}$ is an extension of f to $\mathfrak{L}_B^{A'}$, f_1' coincides with f'. Therefore, we have

$$f_1^*(\overline{a}) = f_1'(\overline{a}) = f'(\overline{a}) = f^*(\overline{a}).$$

Hence, $f_1^* = f^*$.

Let $f_1, \dots, f_r$ be linearly independent maps in $\mathfrak{L}_B^A$ over k, and let f_i^* be an extension of f_i to $\mathfrak{L}_B$. Then $f_1^*, \dots, f_r^*$ are linearly independent in $\mathfrak{L}_B$. We have $\dim \mathfrak{L}_B^A \le \dim \mathfrak{L}_B$. Conversely, let $f_1^*, \dots, f_s^*$ be linearly independent maps in $\mathfrak{L}_B$ and let $f_1, \dots, f_s$ be the projections on $\mathfrak{L}_B^A$. Then, by the uniqueness of the extension as in the proof of Lemma 2.4, $f_1, \dots, f_s$ are also linearly independent in $\mathfrak{L}_B^A$. That is, $\dim \mathfrak{L}_B \le \dim \mathfrak{L}_B^A$. Therefore, we have $\dim \mathfrak{L}_B = \dim \mathfrak{L}_B^A$. Q.E.D.

For an arbitrary divisor B, there exists a normal divisor A such that $B|A$. Hence, from Lemma 2.3 and Lemma 2.5, we have

$$\begin{aligned}\dim \mathfrak{L}_B = \dim \mathfrak{L}_B^A &= r(A, B) = (n(A) - l(A)) - (n(B) - l(B)) \\ &= g - 1 - (n(B) - l(B)), \quad \text{where } g \text{ is the genus of } K.\end{aligned}$$

That is, we obtain

$$l(B) = n(B) - g + 1 + \dim \mathfrak{L}_B. \tag{4.5}$$

Next, let f be any differential in K such that f is not identically zero. Define

$$\mathfrak{M}_f = \{A;\, f \in \mathfrak{L}_A\}.$$

From condition (ii) on continuity of the differential, $\mathfrak{M}_f$ is not empty. Moreover, for a natural number m as in Theorem 2.10, consider a divisor D such that $n(D) \ge m$. Then we have $n(D) - l(D) = g$ since D is normal. Therefore, $\mathfrak{L}_D = 0$ by (4.5). Since $f \ne 0$, f is not contained in $\mathfrak{L}_D$. Hence, the degrees $n(A)$ of divisors A in $\mathfrak{M}_f$ are bounded above. Let A_0 be a divisor in $\mathfrak{M}_f$ that gives the maximum value of the degree $n(A)$. For an arbitrary divisor A in $\mathfrak{M}_f$, let C be the least common multiple of A and A_0. Since we have $f \in \mathfrak{L}_{A_0}$, $f \in \mathfrak{L}_A$, and $\widetilde{L}(C) = \widetilde{L}(A_0) + \widetilde{L}(A)$, we have $f \in \mathfrak{L}_C$. Therefore, C belongs to $\mathfrak{M}_f$. However, since $A_0|C$ implies $n(A_0) \le n(C)$, $C = A_0$, and $A|A_0$ must hold from the choice of A_0. Conversely, if $A|A_0$, A clearly belongs to $\mathfrak{M}_f$. Consequently, $\mathfrak{M}_f$ coincides with the set of all the divisors that divide A_0.

DEFINITION 2.9. The above divisor A_0 is called the divisor of the differential f, and is denoted by (f).

By the definition of A_0, we have $A|(f)$ if and only if $f \in \mathfrak{L}_A$.

We mentioned earlier that the totality $\mathfrak{L}$ of differentials in K is a k-module. We show that $\mathfrak{L}$ can be considered as a K-module as follows. Let f be an arbitrary differential in K, and let x be an arbitrary element in K. For an idele $\overline{a}$, define

$$f'(\overline{a}) = f(x\overline{a}).$$

Since $\widetilde{K}$ is a topological ring, one can observe that f' is also a differential in K. Write this f' as

$$f' = x \cdot f.$$

It should be plain that, with K acting on $\mathfrak{L}$ as above, $\mathfrak{L}$ becomes a K-module.

LEMMA 2.6. *For $x \neq 0$ and $f \neq 0$, we have $(x \cdot f) = (x)(f)$.*

PROOF. For any divisor A, $\widetilde{L}(x^{-1}A) = x\widetilde{L}(A)$. Hence, $f \in \mathfrak{L}_{(x)^{-1}A}$ and $x \cdot f \in \mathfrak{L}_A$ are equivalent. That is, $(x)^{-1}A \in \mathfrak{M}_f$ and $A \in \mathfrak{M}_{xf}$ are equivalent. Therefore, we have $\mathfrak{M}_{xf} = (x)\mathfrak{M}_f$. The least common multiple of divisors in $\mathfrak{M}_f$ is (f), and the least common multiple of divisors in $\mathfrak{M}_{xf}$ is (xf). Hence, we obtain $(xf) = (x)(f)$.

THEOREM 2.12. *As a K-module, the dimension of $\mathfrak{L}$ is one, i.e., choose a differential f_0, $f_0 \neq 0$, then an arbitrary differential f can be written uniquely as*

$$f = x \cdot f_0, \qquad x \in K.$$

PROOF. Suppose that f_1 and f_2 be linearly independent differentials over K. For any place P in K and for any natural number m, let $l(P^m(f_1)) = r$ and $l(P^m(f_2)) = s$. From (4.5), we have

$$r \geq n(P^m(f_1)) - g + 1 = m \cdot n(P) + n((f_1)) - g + 1, \tag{4.6}$$

$$s \geq n(P^m(f_2)) - g + 1 = m \cdot n(P) + n((f_2)) - g + 1. \tag{4.7}$$

Let $x_1, \dots, x_r$ be elements of $L(P^m(f_1))$ in K that are linearly independent over k, and similarly, let $y_1, \dots, y_s$ be elements in $L(P^m(f_2))$ that are linearly independent over k. Since f_1 and f_2 are linearly independent over K, $r+s$ differentials x_if_1 and y_jf_2 $(i = 1, \dots, r;\ j = 1, \dots, s)$ are linearly independent over k. Notice that $x_i \in L(P^m(f_1))$, i.e., $P^{-m}(f_1)^{-1}|(x_i)$, implies $P^{-m}|(x_if_1)$, which means $x_if_1 \in \mathfrak{L}_{P^{-m}}$. Similarly, we obtain $y_jf_2 \in \mathfrak{L}_{P^{-m}}$. Therefore, $\dim \mathfrak{L}_{P^{-m}} \geq r+s$. On the other hand, we have from (4.5)

$$\dim \mathfrak{L}_{P^{-m}} = l(P^{-m}) - n(P^{-m}) + g - 1 = m \cdot n(P) + g - 1.$$

Using (4.6) and (4.7), we have

$$m \cdot n(P) + g - 1 \geq 2m \cdot n(P) + n((f_1)) + n((f_2)) - 2g + 2.$$

That is,

$$3(g-1) \geq m \cdot n(P) + n((f_1)) + n((f_2)).$$

This inequality leads to a contradiction, since m was chosen to be arbitrary. Q.E.D.

By the virtue of Lemma 2.6 and Theorem 2.12, the totality of all the divisors (f) of differentials f, $f \neq 0$, which are called *differential divisors*, determines precisely one class in the divisor class group $\overline{\mathfrak{D}} = \mathfrak{D}/\mathfrak{D}_H$ of K.

DEFINITION 2.10. The divisor class consisting of all the differential divisors is said to be the *differential class*, or the *canonical class*, of the algebraic function field K.

THEOREM 2.13 (Riemann-Roch theorem). *Let W be an arbitrary differential divisor of the algebraic function field K of genus g. Then, for any divisor A on K, we have*

$$l(A) = n(A) - g + 1 + l(WA^{-1}). \tag{4.8}$$

NOTE. Riemann-Roch theorem is one of the fundamental theorems for the algebraic function field theory, and many proofs have been known for a long time. It was F. K. Schmidt who gave the first proof for the case of an arbitrary coefficient field in his paper *Zur arithmetischen Theorie der algebraischen Funktionen* I, Math. Z. **41** (1936). However, the approach that we have taken since the preceeding section is the idea of A. Weil, rather than of Schmidt. See the paper by A. Weil, *Zur algebraischen Theorie der algebraischen Funtionen*, Crelle's Jour. **179** (1938). We refer the reader to the preface.

PROOF. By the equation (4.5), it is enough to show $l(WA^{-1}) = \dim \mathfrak{L}_A$. Let f_0 be the differential in K such that $W = (f_0)$. Theorem 1.12 implies that any differential f can be expressed uniquely as $f = x \cdot f_0$ for some x in K. The differential $f = x \cdot f_0$ belongs to $\mathfrak{L}_A$ if and only if $A|(x \cdot f_0)$, i.e., $(f_0)^{-1}A|(x)$. That is, $x \in L(WA^{-1})$. Therefore, we have $\dim \mathfrak{L}_A = \dim L(WA^{-1}) = l(WA^{-1})$. Q.E.D.

Replacing A by WA^{-1} in (4.8), we obtain

$$l(WA^{-1}) = n(WA^{-1}) - g + 1 + l(A) = n(W) - n(A) - g + 1 + l(A).$$

Adding the left- and the right-hand sides of this equation to the corresponding sides of (4.8), we obtain

$$n(W) = 2g - 2.$$

Next let $A = W$ in (4.8). Then we have

$$l(W) = n(W) - g + 1 + l(E) = 2g - 2 - g + 1 + 1 = g.$$

Conversely, let X be a divisor of K such that $n(X) = 2g - 2$, $l(X) = g$. Then from (4.8) we have $l(WX^{-1}) = 1$. Therefore, there exists an element x in K such that $WX^{-1}|(x)$. Since we have $n(WX^{-1}) = n(x) =$

0, $WX^{-1} = (x)$ must hold. Hence, $X = (x)W$ belongs to the differential class. Therefore, we have obtained the following theorem.

THEOREM 2.14. *Let W be an arbitrary differential divisor of the algebraic function field K of genus g. Then we have*

$$n(W) = 2g - 2 \quad \textit{and} \quad l(W) = g.$$

Conversely, a divisor X of K such that $n(X) = 2g - 2$, $l(X) = g$, is a differential divisor.

DEFINITION 2.11. A differential f is said to be *of the first kind* if either $f = 0$, or (f) is an integral divisor for $f \neq 0$.

THEOREM 2.15. *The set of differentials of the first kind in the algebraic function field K of genus g forms a k-module of dimension g.*

PROOF. Fix $f_0 \neq 0$, and we will find a condition for $f = xf_0$ to be a differential of the first kind. For $x \neq 0$, the condition is $E|(xf_0)$, i.e., $(f_0)^{-1}|(x)$, i.e., $x \in L((f_0))$. Then $L((f_0)) = L(W)$ is a g-dimensional k-module by Theorem 2.14, completing the proof.

We will denote by $\overline{A}$ the class in $\overline{\mathfrak{D}}$ to which a divisor A belongs. In particular, for the differential class $\overline{W}$, the class $\overline{A}' = \overline{WA^{-1}}$ is said to be the complementary class of $\overline{A}$. And the complementary class of $\overline{A}'$ is $\overline{A}$ again. The functions $n(A)$ and $l(A)$ depend only upon the class $\overline{A}$, and furthermore, since we have

$$n(\overline{A}') = n(\overline{WA^{-1}}) = n(\overline{W}) - n(\overline{A}) = 2g - 2 - n(\overline{A}),$$

we can rewrite (4.8) as

$$l(\overline{A}) - \frac{n(\overline{A})}{2} = l(\overline{A}') - \frac{n(\overline{A}')}{2}. \tag{4.9}$$

The Riemann-Roch theorem in this symmetrical form can be applied to various situations.

§5. Hasse differential

The definitions of a differential and a differential divisor on an algebraic function field stated in the last section are from A. Weil. They are well suited for proving the Riemann-Roch theorem for a general coefficient field. However, Weil's definitions do not provide a clear linkage to the classical notion of a differential. On the other hand, H. Hasse established a theory of differentials for an arbitrary algebraic function field over a perfect field as an extension of the classical notion. See H. Hasse, *Theorie der Differentiale in algebraischen Functionenkörper mit vollkommenen Konstantenkörper*, Crelle's Jour. **172** (1935). In this section, we will describe the Hasse differential (referred to as the H-differential) and show that H-differentials essentially coincide with Weil's differentials (referred to as W-differentials).

We can give a definition of H-differential where the coefficient field is a perfect field k. However, for the sake of simplicity, we assume in what follows that k is an algebraically closed field.

Let P be a place on an algebraic function field K over k, and let K_P be the completion of K at P. Then, for a prime element $t = t_P$, an arbitrary element a in K_P can be uniquely expanded as

$$a = \sum \alpha_i t^i, \qquad i \geq i_0 = \nu_P(a), \alpha_{i_0} \neq 0, \alpha_i \in k,$$

as we saw in §2. Then we define the derivative da/dt of a as we do in calculus, i.e., by differentiation of the power series:

$$\frac{da}{dt} = \sum i\alpha_i t^{i-1}.$$

Clearly, this differentiation has the following properties:

$$\frac{d}{dt}(\alpha a + \beta b) = \alpha \frac{da}{dt} + \beta \frac{db}{dt}, \qquad a \text{ and } b \in K_P, \ \alpha \text{ and } \beta \in k$$

$$\frac{d}{dt}(ab) = a\frac{db}{dt} + b\frac{da}{dt}.$$

Since we have

$$\nu_P\left(\frac{da}{dt}\right) \geq \nu_P(a) - 1,$$

the operator da/dt is continuous with respect to a in K_P. Therefore, for example, the derivative of a convergent power series can be carried out term by term.

For another prime element t' for P on K_P, one can define da/dt' in the same way. That is, for $a = \sum \alpha_i' t'^i$ $(\alpha_i' \in k)$,

$$\frac{da}{dt'} = \sum i\alpha_i' t'^{i-1}.$$

On the other hand, by differentiating $a = \sum \alpha_i' t'^i$ term by term with respect to t, we have

$$\frac{da}{dt} = \sum \frac{d}{dt}(\alpha_i' t'^i) = \sum i\alpha_i' t'^{i-1} \frac{dt'}{dt} = \frac{da}{dt'}\frac{dt'}{dt}.$$

That is, the chain rule holds between the differentiation with respect to t and that with respect to t':

$$\frac{da}{dt} = \frac{da}{dt'}\frac{dt'}{dt}. \tag{5.1}$$

Let $\mathfrak{S}_P$ be the totality of pairs (b, a), where a and b are arbitrary elements in K_P. We denote $(b, a) \sim (b', a')$ when, for any prime element t for P on K_P,

$$b\frac{da}{dt} = b'\frac{da'}{dt}. \tag{5.2}$$

Then $\sim$ is an equivalence relation on $\mathfrak{S}_P$, and the elements in $\mathfrak{S}_P$ are classified by this equivalence relation.

DEFINITION 2.12. Each class in $\mathfrak{S}_P/\sim$ is said to be a *local Hasse differential* of K_P, and the class to which (b, a) belongs is denoted by $b\,da$.

If $(b, a) \sim (b', a')$, then $(cb, a) \sim (cb', a')$ holds for any c in K_P. Hence, one can define the product $c \cdot b\,da$ to be the unique class to which (cb, a) belongs. From this definition, any $b\,da$ may be regarded as the product of b and $da = 1 \cdot da$. For any prime elements t and t' for P, we have $\nu_P(dt'/dt) = 0$. Then (5.1) implies

$$\nu_P\left(b\frac{da}{dt}\right) = \nu_P\left(b\frac{da}{dt'}\frac{dt'}{dt}\right) = \nu_P\left(b\frac{da}{dt'}\right) + \nu_P\left(\frac{dt'}{dt}\right) = \nu_P\left(b\frac{da}{dt'}\right).$$

Hence, one may define the value $\nu_P(b\,da)$ by

$$\nu_P(b\,da) = \nu_P\left(b\frac{da}{dt}\right).$$

The value $\nu_P(b\,da)$ of $b\,da$ is said to be the *order* of $b\,da$ at P. The *zeros* and the *poles* can be defined as in function theory.

The differential $b\,da$ is said to be the zero differential, denoted by 0, if $b\,da/dt = 0$. From (5.1) we see that this definition does not depend upon the choice of a prime element.

We obtain the following important theorem.

THEOREM 2.16. *Let t and t' be arbitrary prime elements in K_P for P. Then, for the expansions*

$$b\frac{da}{dt} = \sum \beta_i t^i \quad \textit{and} \quad b\frac{da}{dt'} = \sum \beta'_i t'^i$$

of the local H-differential $b\,da$, the coefficients β_{-1} and β'_{-1} of $1/t$ and $1/t'$, respectively, coincide:

$$\beta_{-1} = \beta'_{-1}.$$

PROOF. Let

$$t' = \sum_{i=1}^{\infty} \gamma_i t^i, \qquad \gamma_1 \neq 0 \text{ and } \gamma_i \in k.$$

Then we have

$$b\frac{da}{dt} = b\frac{da}{dt'}\frac{dt'}{dt} = \sum_n \beta'_n \left(\sum_i \gamma_i t^i\right)^n \left(\sum_i i\gamma_i t^{i-1}\right).$$

When one expands the right-hand side as a power series in t, one should obtain $\sum \beta_i t^i$. For $n = -1$ we have

$$\beta'_{-1}\left(\sum_{i=1}^{\infty} \gamma_i t^i\right)^{-1} \left(\sum_{i=1}^{\infty} i\gamma_i t^{i-1}\right) = \frac{\beta'_{-1}}{t} + \cdots.$$

Therefore, if for $n \neq -1$, the expansion of

$$\left(\sum_{i=1}^{\infty} \gamma_i t^i\right)^n \left(\sum_{i=1}^{\infty} i\gamma_i t^{i-1}\right) \tag{5.3}$$

does not contain the term $1/t$, then one obtains $\beta_{-1} = \beta'_{-1}$.

Let us assume first that the characteristic of k is 0. Then from (5.3) we have

$$t'^n \frac{dt'}{dt} = \frac{1}{n+1}\frac{d}{dt}(t'^{n+1}) = \frac{1}{n+1}\frac{d}{dt}\left(\sum \varepsilon_i t^i\right) = \frac{1}{n+1}\sum i\varepsilon_i t^{i-1},$$

which does not contain the term $1/t$.

Next we will prove the case of nonzero characteristic $p \neq 0$ by reducing this case to the case of characteristic zero. Replacing $\gamma_1, \gamma_2, \ldots$ by variables $z_1, z_2, \ldots$, let $L = R_0(z_1, z_2, \ldots)$, the field obtained from the field R_0 of rational numbers by the adjunction of elements $z_1, z_2, \ldots$. The power series in t with coefficients in L

$$\left(\sum_{i=1}^{\infty} z_i t^i\right)^n \left(\sum_{i=1}^{\infty} i z_i t^{i-1}\right), \qquad n \neq -1, \tag{5.4}$$

does not have the term $1/t$ since the characteristic of L is 0. Let

$$\sum w_i t^i, \qquad w_i \in L \tag{5.5}$$

be the expansion of (5.4) as a power series in t. Then w_i is a rational function of z_i, $i = 1, 2, \ldots$, whose denominator is at most a power of z_1 and whose numerator is a polynomial in z_i with rational integer coefficients, as one can observe from

$$\left(\sum_{i=1}^{\infty} z_i t^i\right)^{-1} = \frac{1}{z_1 t} + \cdots.$$

When one replaces the integer coefficients by the residue classes $\bmod p$ of the polynomial in the numerator, one obtains a rational function $\overline{w}_i$ in z_i with its coefficients in the finite field $GF(p)$. The division to get (5.5) from (5.4) involves only z_1. Hence, regarding (5.4) as a power series in t with coefficients in $\overline{L} = GF(p)(z_1, z_2, \ldots)$, we have

$$\left(\sum_{i=1}^{\infty} z_i t^i\right)^n \left(\sum_{i=1}^{\infty} i z_i t^{i-1}\right) = \sum \overline{w}_i t^i. \tag{5.6}$$

Since k is a field of characteristic p, we can regard k as an extension field of $GF(p)$. Note that $\gamma_1 \neq 0$. Hence, if one replaces $z_1, z_2, \ldots$ in (5.6) with $\gamma_1, \gamma_2, \ldots$, the equality is still valid. That is, letting $\xi_i = \overline{w}_i(\gamma_1, \gamma_2, \ldots)$; then for

$$\left(\sum_{i=1}^{\infty} \gamma_i t^i\right)^n \left(\sum i\gamma_i t^{i-1}\right) = \sum \xi_i t^i,$$

$w_{-1} = 0$ and $\overline{w}_{-1} = 0$ imply $\xi_{-1} = 0$. This completes the proof.

DEFINITION 2.13. The coefficient β_{-1} of $1/t$ in the expansion in a prime element t

$$b\frac{da}{dt} = \sum \beta_i t^i \in K_P$$

of the local H-differential $b\,da$ is said to be the *residue* of $b\,da$. We write

$$\beta_{-1} = \mathrm{Res}_P(b\,da).$$

By Theorem 2.16 above, the residue is an element of k, uniquely determined by $b\,da$, and it is independent of the choice of a prime element t.

Next we will define an H-differential in an algebraic function field K using the notion of a local H-differential. As before, let $\mathfrak{S}$ be the totality of the pairs (y, x) where y and x are elements of K. For any place P on K, y and x are elements in K_P. Hence, (y, x) determines a local H-differential $(y\,dx)_P$ of K_P. We write $(y, x) \sim (y', x')$ for elements (y, x) and (y', x') in $\mathfrak{S}$ if

$$(y\,dx)_P = (y'\,dx')_P$$

for all P. Then $\mathfrak{S}$ is classified by this equivalence relation.

DEFINITION 2.14. Each class defined above is called an *H-differential* in K, and the class to which (y, x) belongs is denoted as $y\,dx$.

One can define a product between an H-differential and an element of K as we defined the product for the case of a local H-differential. In particular, $y\,dx$ is the product of y and dx. Furthermore, if $(y\,dx)_P = 0$ for all the P, then $y\,dx$ is called the zero-differential, denoted by 0.

THEOREM 2.17. *We have $dx \neq 0$ if and only if x is a separating element of K. Then, $(dx)_P \neq 0$ holds for any P.*

PROOF. Let P be an arbitrary place of K and let $t = t_P$ be a prime element for P. If the characteristic of k is 0, then we have $dx/dt = 0$ only for x in k. Hence, let the characteristic of k be $p \neq 0$. Suppose that x is not a separating element. Then $z = x^{1/p}$ belongs to K. We have

$$\frac{dx}{dt} = \frac{d}{dt}(z^p) = pZ^{p-1}\frac{dz}{dt} = 0.$$

Since P was taken arbitrarily, we obtain $dx = 0$. Next suppose x to be a separating element of K. Let

$$K = k(x, y) \quad \text{and} \quad f(x, y) = 0,$$

where $f(x, y) = 0$ is an irreducible equation between x and y with coefficients in k. By differentiation with respect to t in K_P we have

$$f_x(x, y)\frac{dx}{dt} + f_y(x, y)\frac{dy}{dt} = 0.$$

Since y is separable over $k(x)$, $f_y(x, y) \neq 0$. Hence, if we assume $dx/dt = 0$, then $dy/dt = 0$ must hold. That is, for

$$x = \sum \alpha_i t^i \quad \text{and} \quad y = \sum \beta_i t^i, \quad \alpha_i, \beta_i \in k,$$

$\alpha_i = \beta_i = 0$ must hold for all i that are not divisible by p. We may rephrase this by using the expansion of x and y as power series in t^p. Since all the elements in K are rational functions in x and y, they must be power series in t^p. However, there exists an element a in K such that $\nu_P(a) = 1$, contradicting the above. Hence, $dx/dt \neq 0$, i.e., $(dx)_P \neq 0$.

THEOREM 2.18. *Choose an H-differential $dx \neq 0$. Then any H-differential in K can be written uniquely as $y\,dx$ for some $y \in K$.*

NOTE. The existence of x such that $dx \neq 0$, i.e., the existence of a separating element x, i.e., Theorem 2.17, is guaranteed from Theorem 2.1.

PROOF OF THEOREM 2.18. We need to show that any dx' can be written as $y\,dx$. The uniqueness is clear. Since x' is algebraic over $k(x)$, let

$$g(x, x') = 0$$

be the irreducible relation equation between x and x'. Differentiate this equation with respect to a prime element $t = t_P$ for an arbitrary place P. Then we have

$$g_x(x, x')\frac{dx}{dt} + g_{x'}(x, x')\frac{dx'}{dt} = 0. \tag{5.7}$$

By the preceding theorem, x is a separating element, that is, x' is separable over $k(x)$, i.e., $g_{x'}(x, x') \neq 0$. Hence, if we let

$$y = -\frac{g_x(x, x')}{g_{x'}(x, x')}, \qquad y \in K,$$

then we obtain from (5.7) that

$$\frac{dx'}{dt} = y\frac{dx}{dt}$$

for any P. That is, $dx' = y\,dx$. Q.E.D.

Fixing $dx \neq 0$, one can define an addition for arbitrary H-differentials $y_1\,dx$ and $y_2\,dx$ by

$$y_1\,dx + y_2\,dx = (y_1 + y_2)\,dx.$$

Notice that the above definition of addition is independent of the choice of dx. Moreover, it is clear that the set of H-differentials forms a K-module of dimension one.

DEFINITION 2.15. Let $y\,dx$ be an arbitrary H-differential in K. The residue of the local H-differential $(y\,dx)_P$ in the completion K_P at a place P is said to be the *residue* of $y\,dx$ at P. We denote it by

$$\mathrm{Res}_P(y\,dx) \quad \text{or} \quad \int_P y\,dx.$$

NOTE. The notation $\int_P y\,dx$ for the residue of $y\,dx$ at P will be explained in Chapters 3 and 4.

We have the following important theorem.

THEOREM 2.19 (residue theorem). *The residues* $\mathrm{Res}_P(y\,dx)$ *at places* P *are all zero except at a finite number of* P. *Furthermore, the sum of all the residues over all* P *is zero, i.e.,*

$$\sum_P \mathrm{Res}_P(y\,dx) = 0. \tag{5.8}$$

PROOF. For P such that $\nu_P(x) \geq 0$ and $\nu_P(y) \geq 0$, the expansion $y\,dx/dt = \sum \alpha_i t^i$, $t = t_P$, does not contain terms with negative powers of t. Hence we have $\mathrm{Res}_P(y\,dx) = 0$. Therefore, the first part of the theorem follows from Theorem 2.2.

Next we will prove (5.8). If $dx = 0$, there is nothing to be proved. Assume $dx \neq 0$, i.e., x is a separating element in K. We will first give a proof for the case $K = k(x)$. Since the element y in K is a rational function of x, we have

$$y = \sum \alpha_i x^i + \sum_j \sum_i \frac{\gamma_{jl}}{(x-\beta_j)^l}, \qquad \alpha_i, \beta_j, \gamma_{jl} \in k.$$

Generally speaking, for any $y\,dx$ and $y'\,dx'$, we have the following

$$\mathrm{Res}_P(\alpha(y\,dx) + \beta(y'\,dx')) = \alpha \,\mathrm{Res}_P(y\,dx) + \beta\,\mathrm{Res}_P(y'\,dx'), \qquad \alpha, \beta \in k, \tag{5.9}$$

from the definition of addition of H-differentials and that of residues. Hence, we need to prove (5.8) only for

$$y = \begin{cases} x^i \\ \dfrac{1}{(x-\beta)^i}, \end{cases} \qquad i = 1, 2, \ldots$$

in the above expansion of y. Places on $K = k(x)$ are P_α, $\alpha \in k$, and P_∞ from Chapter 1, §1, Example 2. Take $t = x - \alpha$ and $t = 1/x$ as prime elements for P_α and P_∞, respectively. Expanding $y\,dx/dt$ in t, by a simple computation we have

$$\mathrm{Res}_P(x^i\,dx) = 0 \quad \text{for } P = P_\alpha \text{ and } P = P_\infty,$$

$$\mathrm{Res}_P\left(\frac{dx}{(x-\beta)^i}\right) = \begin{cases} 1 & \text{for } P = P_\beta \text{ and } i = 1, \\ -1 & \text{for } P = P_\infty \text{ and } i = 1, \\ 0 & \text{otherwise.} \end{cases}$$

Then (5.8) indeed holds.

For the case $K \neq k(x)$, we will prove the formula

$$\sum_{i=1}^{g} \mathrm{Res}_{P_i}(y\,dx) = \mathrm{Res}_Q(T(y)\,dx), \tag{5.10}$$

where Q is an arbitrary place on $k(x)$, and $P_1, P_2, \ldots, P_g$ are all the extensions of Q to K. Note also that $T(y)$ is the trace of y from K to $k(x)$ as discussed in the note following Theorem 1.9. Once (5.10) is proved, as Q runs through all the places in $k(x)$, we obtain

$$\sum_P \mathrm{Res}_P(y\,dx) = \sum_Q \mathrm{Res}_Q(T(y)\,dx).$$

Since $T(y)$ is an element of $k(x)$, the right-hand side is 0 as we saw, which proves (5.8) for this general case.

Since x is a separating element of K, there exists an element z such that

$$K = k(x, z) \quad \text{and} \quad f(x, z) = 0.$$

Decompose the polynomial $F(Z) = f(x, Z)$ in Z with coefficients in $k(x)$ into irreducible polynomials in the completion $(k(x))_Q$ of $k(x)$ for Q,

$$F(Z) = \prod_{i=1}^{g} f_i(Z).$$

Since z is a separable element for $k(x)$, $F(Z)$ has no multiple roots. Hence, each factor of the above decomposition has no multiple roots. Regarding K as an algebra over $k(x)$, extend the base field $k(x)$ to $(k(x))_Q$. Then we have the direct sum decomposition

$$K \times (k(x))_Q = K_1 + \cdots + K_g, \tag{5.11}$$

where K_i is the field obtained by adjoining the roots of $f_i(Z) = 0$ to $(k(x))_Q$. Here, the extensions $P_1, \ldots, P_g$ of Q to K correspond exactly to the extension fields $K_1, \ldots, K_g$ by Theorem 1.11. That is, K_i is nothing but the completion of K for P_i: $K_i = K_{P_i}$. If we denote a trace from K_i to $(k(x))_Q$ as T_i, then (5.11) implies

$$T(y) = \sum_{i=1}^{g} T_i(y) \quad \text{and} \quad \mathrm{Res}_Q(T(y)\,dx) = \sum_{i=1}^{g} \mathrm{Res}_Q(T_i(y)\,dx).$$

Therefore, for the proof of (5.10), it is sufficient to show that for a finite extension L of the complete field $(k(x))_Q$

$$\mathrm{Res}_P(y\,dx) = \mathrm{Res}_Q(T(y)\,dx), \tag{5.12}$$

where $y\,dx$ is a local H-differential and y is an arbitrary element of L, P is the unique extension of Q to L, and T is the trace from L to $(k(x))_Q$. If we let $u = u_Q$ be a prime element for Q, since dx/du is an element of $(k(x))_Q$, we obtain

$$T(y)\,dx = T(y)\frac{dx}{du}\,du = T\left(y\frac{dx}{du}\right)\,du.$$

Rewriting $y\,dx/du$ as y, instead of (5.12), we need to prove

$$\mathrm{Res}_P(y\,du) = \mathrm{Res}_Q(T(y)\,du) \tag{5.13}$$

for any element y in L. Since $T(y)$ is linear with respect to y over k and continuous, both sides of (5.13) are linear and continuous in y. Moreover, since an arbitrary element y in L has a unique expansion in a prime element $t = t_P$ of P as

$$y = \sum \alpha_i t^i, \qquad \alpha_i \in k;$$

it is sufficient to prove

$$\operatorname{Res}_P(t^n\,du) = \operatorname{Res}_Q(T(t^n)\,du), \qquad n = 0, \pm 1, \pm 2, \ldots \tag{5.14}$$

instead of (5.13). As a consequence, a proof of (5.8) is reduced to a proof of (5.14).

Since k is algebraically closed, the residue fields of both P and Q coincide with k, and furthermore, the relative degree of P over Q is 1. From Theorem 1.9, $e = [L : (k(x))_Q]$ coincides with the ramification index of P. Let us assume the characteristic of k is zero. Then Lemma 2.1 implies that $t^e = u$ may hold. Hence, L is the field obtained by adjoining t to $(k(x))_Q$ and the elements $1, t, t^2, \ldots, t^{e-1}$ form a basis for L over $(k(x))_Q$. Using these basis elements, $T(t^n)$ can be computed as

$$T(t^n) = \begin{cases} 0 & \text{for } n \text{ not divisible by } e, \\ eu^m & \text{for } n = me. \end{cases}$$

Therefore, we have

$$\operatorname{Res}_Q(T(t^n)\,du) = \begin{cases} 0 & \text{for } n \neq -e, \\ e & \text{for } n = -e. \end{cases}$$

On the other hand, since $t^n\,du/dt = et^{n+e-1}$,

$$\operatorname{Res}_P(t^n\,du) = \begin{cases} 0 & \text{for } n \neq -e, \\ e & \text{for } n = -e. \end{cases}$$

Hence, (5.14) holds.

Next we consider the case where the characteristic of k is $p \neq 0$. Also in this case, $[L : (k(x))_Q] = e$ coincides with the ramification index of P, and $L = (k(x))_Q(t)$. Let

$$\frac{t^e}{u} = a_0 + a_1 t + \cdots + a_{e-1} t^{e-1}, \qquad a_i \in (k(x))_Q.$$

Then it follows easily from $\nu_P(t^e/u) = 0$ that $\nu_Q(a_i) \geq 0$. In particular, we obtain $\nu_Q(a_0) = 0$. Since $a_0 u$ is a primitive element for Q in $(k(x))_Q$, we can rewrite $a_0 u$ as u. Since we have $(k(x))_Q = k((u))$,

$$t^e = u(1 + A_1(u)t + \cdots + A_{e-1}(u)t^{e-1}),$$

where

$$A_i(u) = \sum_{j=0}^{\infty} \alpha_{ij} u^j, \qquad \alpha_{ij} \in k$$

is a power series in $k((u))$.

Let $R_0(x_{ij})$ be the field obtained by adjoining the variables x_{ij}, $i = 1, 2, \dots, e-1$, $j = 1, 2, \dots$, corresponding to α_{ij}, to the field R_0 of rational numbers. Let k^* be an algebraically closed field containing $R_0(x_{ij})$. Let

$$A_i^*(u) = \sum_{j=0}^{\infty} x_{ij} u^j \in k^*((u))$$

be a power series in u with coefficients in k^*. Let L^* be the field obtained by adjoining t to $k^*((u))$ where t satisfies

$$t^e = u(1 + A_1^*(u)t + \cdots + A_{e-1}^*(u)t^{e-1}). \tag{5.15}$$

The canonical prime divisor Q^* for u in $k^*((u))$ is uniquely extended to a prime divisor P^* on L^*, i.e., Theorem 1.9. From (5.15), the ramification index for P^* is e, and then we have $[L^* : k^*((u))] = e$, and the equation (5.15) in t is irreducible. Since the characteristic of k^* is zero, by the previous argument, (5.14) holds for the extension $L^*/k^*((u))$. That is,

$$\operatorname{Res}_{P^*}(t^n\, du) = \operatorname{Res}_{Q^*}(T(t^n)\, du). \tag{5.16}$$

Note that we assumed $t^e = u$ earlier to prove (5.14). For this particular pair of t and u, (5.14) implies (5.12). Therefore, for an arbitrary prime element t for P and an arbitrary prime element u for Q, (5.14) holds. Hence, we obtain (5.16).

On the other hand, from (5.15) we have

$$u = t^e - u(A_1^*(u)t + \cdots + A_{e-1}^*(u)t^{e-1}). \tag{5.17}$$

Substitute the right-hand side of (5.17) into each u in the right-hand side of (5.17). This repeated substitution gives the following expansion of u in terms of t:

$$u = \sum_{l=e}^{\infty} B_l t^l, \qquad B_l \in k^*, \ B_e = 1.$$

By this method of computation, we obtain that the

$$B_l = B_l(x_{ij})$$

are all polynomials in x_{ij} with coefficients in rational integers. Therefore, $\operatorname{Res}_{P^*}(t^n\, du)$, the coefficient of $1/t$ in

$$t^n \frac{du}{dt} = \sum_{l=e}^{\infty} l B_l t^{l-1+n},$$

is also a polynomial in x_{ij} with rational integer coefficients.

On the other hand, we have from (5.15)

$$\frac{1}{t} = -A_1^*(u) - A_2^*(u)t - \cdots - A_{e-1}^*(u)t^{e-2} + \frac{t^{e-1}}{u}.$$

Let

$$T(t^n) = \sum C_{nl}(x_{ij})u^l, \qquad C_{nl}(x_{ij}) \in k^*.$$

Then all the $C_{nl}(x_{ij})$, in particular,

$$\mathrm{Res}_{Q^*}(T(t^n)\,du) = C_{n,-1}(x_{ij}),$$

are polynomials in x_{ij} with coefficients in rational integers. Therefore, the equation (5.16) is an identity equation between the two polynomials. Hence, by the same method as in the proof of Theorem 2.16, i.e., by regarding the coefficients in $GF(P)$ by reduction $\bmod p$, and substituting α_{ij} in k for x_{ij}, we obtain from (5.16) the equality (5.14) for the extension $L/(k(x))_Q$, which completes our proof.

NOTE. We explain the decomposition (5.11) as follows: Let $K_0 = k(x)$. Then K may be regarded as the residue ring of the polynomial ring $K_0[Z]$ modulo the prime ideal $(F(Z))$, i.e.,

$$K = K_0[Z]/(F(Z)).$$

Therefore,

$$\begin{aligned} K \times (K_0)_Q &= (K_0)_Q[Z]/(F(Z)) = (K_0)_Q[Z]/(\Pi f_i(Z)) \\ &= \sum_{i=1}^{g} (K_0)_Q[Z]/(f_i(Z)) \\ &= \sum_{i=1}^{g} K_i. \end{aligned}$$

Lastly in this section, we will study the relation between an H-differential and a W-differential. Let $y\,dx$ be an arbitrary H-differential in K. For an idele $\overline{a} = (a_P)$ of K, consider the function $f(\overline{a})$ given by

$$f(\overline{a}) = \sum_P \mathrm{Res}_P(a_P y\,dx). \tag{5.18}$$

Here, $\mathrm{Res}_P(a_P y\,dx)$ is the residue of the local H-differential $a_P y\,dx$ in K_P, and $\sum_P$ is considered over all the places P on K. From the definition of an idele, there are only finitely many P such that $\nu_P(a_P) < 0$. Hence, $\mathrm{Res}_P(a_P y\,dx)$ equals zero except for finitely many P. Therefore, the sum in (5.18) is a finite sum, determining the value $f(\overline{a})$ as an element in k. Next we will prove that $f(\overline{a})$ is a W-differential. It is clear that $f(\overline{a})$ is k-linear. If, for all the places P, we have

$$\nu_P(a_P) \geq -\nu_P(y\,dx), \tag{5.19}$$

then $\mathrm{Res}_P(a_P y\,dx) = 0$ by the definition of the residue, i.e., $f(\overline{a}) = 0$. However, for $\nu_P(x) \geq 0$ and $\nu_P(y) \geq 0$, we have $\nu_P(y\,dx) \geq 0$. Hence, there are only finitely many P such that $\nu_P(y\,dx) < 0$. Consider the product of these finite places

$$D = \prod P^e, \qquad e = \nu_P(y\,dx) \text{ and } \nu_P(y\,dx) < 0.$$

Then D is a divisor on K, and $f(\overline{a}) = 0$ holds for $\overline{a} \in \widetilde{L}(D)$. Consequently, $f(\overline{a})$ is continuous. Lastly, let z be an arbitrary element of K. Then Theorem 2.19 implies

$$f(z) = \sum_P \mathrm{Res}_P(zy\,dx) = 0.$$

Hence, $f(\overline{a})$ is a W-differential in K. For $y\,dx = 0$, the induced W-differential f is naturally 0. For $y\,dx \neq 0$, let $(y\,dx)_P \neq 0$. Then there exists an element a_P in K_P such that $\mathrm{Res}_P(a_P y\,dx) \neq 0$. Let a_P denote the idele whose Pth component is a_P, and all others are zero. Then we have

$$f(a_P) = \mathrm{Res}_P(a_P y\,dx) \neq 0,$$

i.e., $f \neq 0$. Furthermore, if $y_1\,dx \to f_1$ and $y_2\,dx \to f_1$, then we have the correspondences:

$$y_1\,dx + y_2\,dx \to f_1 + f_2 \quad \text{and} \quad z(y_1\,dx) \to z \cdot f_1 \quad \text{for } z \in K.$$

Since both H-differentials and W-differentials are one dimensional K-modules, the correspondence between the H-differential $y\,dx$ and the W-differential $f(\overline{a})$ in (5.18) provides a one-to-one correspondence between the sets of H-differentials and W-differentials, respectively. That is, we have obtained the following theorem.

THEOREM 2.20. *Let K be an algebraic function field over an algebraically closed field k. For an arbitrary H-differential $y\,dx$ in K, let*

$$f(\overline{a}) = \sum_P \mathrm{Res}_P(a_P y\,dx),$$

where $\overline{a} = (a_P)$ is an idele of K. Then $f(\overline{a})$ becomes a W-differential in K. The correspondence $y\,dx \leftrightarrow f$ defines an isomorphism between these K-modules.

Let f be the W-differential corresponding to $y\,dx \neq 0$. Let A be a divisor on K such that

$$\nu_P(A) \leq \nu_P(y\,dx) \tag{5.20}$$

holds for all P. Then, for $\overline{a} = (a_P)$ in $\widetilde{L}(A)$, we have

$$\nu_P(a_P) \geq -\nu_P(A) \geq -\nu_P(y\,dx).$$

Hence, $f(\overline{a}) = 0$. That is, $f \in \mathfrak{L}_A$, i.e., $A \in \mathfrak{M}_f$, i.e., $A|(f)$. Let B be a divisor such that, for a P,

$$\nu_P(B) > \nu_P(y\,dx).$$

Then there exists an idele a_P such that all of the components are zero except the P-component at which we have

$$\nu_P(a_P) \geq -\nu_P(B) \quad \text{and} \quad f(a_P) = \mathrm{Res}_P(a_P y\,dx) \neq 0.$$

Hence, $f \notin \mathfrak{L}_B$, i.e., $B \notin \mathfrak{M}_f$. Therefore, $\mathfrak{M}_f$ coincides with the set of divisors A satisfying (5.20). Since (f) is the least common multiple of all the divisors belonging to $\mathfrak{M}_f$, we conclude the following.

THEOREM 2.21. *Let $y\,dx$ be a nonzero H-differential in K. Then the divisor (f) of the corresponding W-differential f is determined by*

$$\nu_P((f)) = \nu_P(y\,dx).$$

Hence, in particular, all the $\nu_P(y\,dx)$ are zero except for finitely many P.

The relation between an H-differential and a W-differential is clear from Theorems 2.20 and 2.21. In what follows, we will write $f = y\,dx$ and $(f) = (y\,dx)$ by identifying $y\,dx$ and f. We will use f or $y\,dx$ as needed.

We will compute the divisor (dx) for the case $dx \neq 0$. For simplicity, we assume the characteristic of k is zero. Let P be an arbitrary place on K, and let Q be its projection on $k(x)$. From Chapter 1, §1, Example 2, either $Q = P_\alpha$ or $Q = P_\infty$. Hence, we can choose as a prime element $u = u_Q$, where

$$u = \begin{cases} x - \alpha & \text{for } Q = P_\alpha, \\ \dfrac{1}{x} & \text{for } Q = P_\infty. \end{cases}$$

Let e be the ramification index of P with respect to $k(x)$. Then, by Lemma 2.1, there exists a prime element $t = t_P$ for P such that $t^e = u$. Since we have

$$\frac{dx}{dt} = \frac{dx}{du}\frac{du}{dt} = et^{e-1}\frac{dx}{du} = \begin{cases} et^{e-1} & \text{for } Q = P_\alpha, \\ -\dfrac{et^{e-1}}{x^2} & \text{for } Q = P_\infty; \end{cases} \tag{5.21}$$

$$\nu_P(dx) = \begin{cases} e-1 & \text{for } Q = P_\alpha, \\ e-1-2e & \text{for } Q = P_\infty. \end{cases}$$

Notice that $Q = P_\infty$ when P appears precisely as P^e in the denominator $\{P_\infty\}_K = D_\infty$ of (x). Hence, from (2.21) we have the following theorem.

THEOREM 2.22. *Let x be an arbitrary element in K over k of characteristic zero such that $x \notin k$. Then we have*

$$(dx) = \frac{\prod P^{e-1}}{D_\infty^2}, \tag{5.22}$$

where P runs through all the places in K in the product, $e = e_P$ is the ramification index of P over $k(x)$, and D_∞ is the denominator of (x).

The divisor $\prod P^{e-1}$ in (5.22) is called the *different* of $K/k(x)$ and is denoted as $\mathfrak{D}(K/k(x))$. We gave a proof for Theorem 2.22 in the case of characteristic zero. However, one can give a definition of the notion of different so that one can obtain

$$(dx) = \frac{\mathfrak{D}(K/k(x))}{D_\infty^2},$$

for an arbitrary characteristic p. See the paper by H. Hasse cited at the beginning of this section for further study.

If $[K : k(x)] = n$, we have $n(D_\infty) = n$ by Theorem 2.7. Then (5.22) and Theorem 2.14 imply the following formula of Riemann for the algebraic function K of genus g in characteristic zero:

$$2g-2 = \sum_P (e_P - 1) - 2n \quad \text{or} \quad \frac{w}{2} = n+g-1, \quad \text{where } w = \sum_P (e_P - 1). \tag{5.23}$$

§6. Applications of the Riemann-Roch theorem

The Riemann-Roch theorem, proved in §4, is one of key theorems in the algebraic theory of algebraic functions, and it has many applications. We will mention some applications in this section. As before, we assume in this section that the coefficient field k is algebraically closed.

(i) Divisor Group. Let $\mathfrak{D}$ be the divisor group of the algebraic function field K over k. Let

$$\mathfrak{D}_0 = \{A_0 ;\ n(A_0) = 0\}.$$

Then, from (3.2) and Theorem 2.7, $\mathfrak{D}_0$ is a subgroup of $\mathfrak{D}$ containing the principal divisor group $\mathfrak{D}_H$:

$$\mathfrak{D}_H \subseteq \mathfrak{D}_0 \subseteq \mathfrak{D}.$$

Since k is algebraically closed, the degree of any place P is 1. Let D be an arbitrary divisor on K such that $n(D) = n$, and let

$$D_0 = DP^{-n}. \tag{6.1}$$

Then we have

$$n(D_0) = 0, \quad \text{i.e., } D_0 \in \mathfrak{D}_0.$$

Hence, for the cyclic subgroup generated by P,

$$\mathfrak{B} = \{P^m ;\ m = 0, \pm 1, \pm 2, \dots\},$$

we have $\mathfrak{B} \cap \mathfrak{D}_0 = E$, and $\mathfrak{D}$ is the direct product

$$\mathfrak{D} = \mathfrak{B} \times \mathfrak{D}_0.$$

Therefore, it is sufficient to examine the group structure of $\mathfrak{D}_0$ to know the group structure of $\mathfrak{D}$. Since the principal divisor group $\mathfrak{D}_H$ is isomorphic to the residue group K^*/k^*, where K^* and k^* are the multiplicative groups, it is essential to study the structure of $\mathfrak{D}_0/\mathfrak{D}_H$. In the case where k is the field of complex numbers, one can determine $\mathfrak{D}_0/\mathfrak{D}_H$. This will be treated in a later chapter. In this section we prove that each class of $\mathfrak{D}_0/\mathfrak{D}_H$ generally contains a certain type of divisor.

Let g be the genus of K, and let A be a divisor such that $n(A) = g$. For example, take $A = P^g$. Let D_0 be an arbitrary divisor in $\mathfrak{D}_0$. Then by the Riemann-Roch theorem, we have

$$\begin{aligned} l(AD_0^{-1}) &= n(AD_0^{-1}) - g + 1 + l(WA^{-1}D_0) \\ &\geq n(A) - n(D_0) - g + 1 \\ &= g - 0 - g + 1 = 1. \end{aligned}$$

Hence, $L(AD_0^{-1})$ contains a nonzero element u. Let $(u) = B/AD_0^{-1}$. Then by the definition of $L(AD_0^{-1})$, B is an integral divisor such that $n(B) = g$. If one lets $(u)^{-1} = H$, we obtain the following theorem.

THEOREM 2.23. *Let g be the genus of K and let A be a divisor of degree g. Then any divisor D_0 of degree 0 can be written as*

$$D_0 = \frac{B}{A}H,$$

where H is a principal divisor, and B is an integral divisor such that $n(B) = g$. That is, each class of $\mathfrak{D}_0/\mathfrak{D}_H$ has a representative of the type B/A.

(ii) Element of K having a pole at a given place. Any element of K that is not in k always has a pole at some place, i.e., Theorem 2.2. Conversely, for a given place P, let us find an element of K which has a pole at P.

THEOREM 2.24 (Weierstrass). *Let P be an arbitrary place on an algebraic function field over k. Then there exists an element of K which has a pole only at P. Furthermore, if the genus of K is g, for any natural number n, except for g natural numbers $n_1, n_2, \ldots, n_g$, such an element of K can be chosen so that the order at P is exactly n.*

PROOF. From the Riemann-Roch theorem, for an arbitrary natural number n, we have

$$l(P^n) = n(P^n) - g + 1 + l(WP^{-n}) = n - g + 1 + l(WP^{-n}).$$

Hence,

$$(l(P^n) - l(P^{n-1})) = 1 - (l(WP^{-n+1}) - l(WP^{-n})),$$

where both of the expressions in parentheses represent nonnegative integers (notice, e.g., $L(P^{n-1}) \subseteq L(P^n)$ implies $l(P^n) - l(P^{n-1}) \geq 0$). Therefore, $l(P^n) - l(P^{n-1})$ is either 0 or 1. In the case it is 1, there exists an element u that is in $L(P^n)$ but not in $L(P^{n-1})$. Then the element u of K has a pole only at P, and its order at P is exactly n. Consider

$$1 = l(P^0) \leq l(P^1) \leq \cdots \leq l(P^{n-1}) \leq l(P^n) \leq \cdots. \tag{6.2}$$

As observed, $l(P^i) - l(P^{i-1})$, $i = 1, 2, \ldots,$ is either 1 or 0, and for $n > 2g-2$, $n(WP^{-n}) < 0$ holds. Since $P^n W^{-1}|(x)$ implies $x = 0$, we have $l(WP^{-n}) = 0$. Therefore, we obtain

$$l(P^n) = n - g + 1.$$

Hence, for $n > 2g - 2$, the difference of consecutive dimensions must be 1. On the other hand, $l(P^{2g-1}) = g$ and $l(P^0) = 1$. Therefore, there are g natural numbers n_i such that $l(P^{n_i-1}) = l(P^{n_i})$ between $l(P^0)$ and $l(P^{2g-1})$, completing our proof.

The natural numbers $n_1, n_2, \dots, n_g$ are determined by the place P. In the case when the characteristic of k is zero, one can prove those natural numbers coincide with $1, 2, \dots, g$ except at a finite number of places. These finite places where $n_1, n_2, \dots, n_g$ differ from $1, 2, \dots, g$ are said to be the Weierstrass points of the algebraic function field K. It has been proved that, for an algebraic function field of genus $g \geq 2$ over a coefficient field of characteristic zero, there exist at least $2g+2$ Weierstrass points. For results on Weierstrass points the reader is referred to F. K. Schmidt, *Zur arithmetischen Theorie der algebraischen Funktionen.* II, Math. Z. **45** (1939). Weierstrass points play a significant role in the study of the group of automorphisms on the algebraic function field. See H. L. Schmid, *Über die Automorphismen eines algebraischen Funktionenkörpers von Primzahlcharakteristik*, Crelle's Jour. **179** (1939).

(iii) Classification of differentials. When $(f) = (y\,dx)$ is an integral divisor, that is, $y\,dx$ possesses no poles at any place on the algebraic function field K, $y\,dx$ is said to be of the first kind. See Definition 2.11.

DEFINITION 2.16. A differential is said to be *of the second kind* if it has a pole at only one place on K. A differential is said to be *of the third kind* if it has poles at only two places and the orders of these poles are one.

Compare the following theorem with Theorem 2.24.

THEOREM 2.25. *Let P be an arbitrary place on K, and let $m > 1$ be an arbitrary natural number. Then there exists a differential $y_m\,dx$ of the second kind having a pole of order exactly m at P. Any differential having at most a pole at P with order $\leq m$ may be written as a linear combination of $y_2\,dx, \dots, y_m\,dx$ and g linearly independent differentials of the first kind.*

PROOF. Choose dx such that $dx \neq 0$. Let $W = (dx)$. The differential $y\,dx$ has at most a pole at P of order m if and only if $P^{-m} | (y\,dx)$ holds. That is, $W^{-1}P^{-m} | (y)$, i.e., $y \in L(WP^m)$. Since we have

$$\begin{aligned} l(WP^m) &= n(WP^m) - g + 1 + l(P^{-m}) \\ &= 2g - 2 + m - g + 1 + 0 = m + g - 1, \end{aligned} \tag{6.3}$$

we get for $m \geq 2$, that

$$l(WP^m) = l(WP^{m-1}) + 1.$$

Then choose y_m in $L(WP^m)$ that is not in $L(WP^{m-1})$. Then it is clear that the differential $y_m\,dx$ has a unique pole at P exactly of order m. Clearly those $y_2\,dx, \dots, y_m\,dx$ and g differentials of the first kind are linearly independent. Hence, the $m+g-1$ dimensional set of differentials having at most a pole at P of order $\leq m$ are linear combinations of $y_2\,dx, \dots, y_m\,dx$ and the differentials of the first kind.

REMARK. Notice that one obtains $l(WP) = g$ for $m = 1$ in (6.3). That is, the dimension of the set of differentials having at most a pole at P of order ≤ 1 is g, i.e., the dimension of the set of differentials of the first kind. Therefore, there does not exist a differential that has a pole only at P precisely of order 1. One can also observe this fact from Theorem 2.19.

THEOREM 2.26. *Let P_1 and P_2 be arbitrary places on K such that $P_1 \neq P_2$. Then, there exists a differential $y_{12}\,dx$ of the third kind which has poles at P_1 and P_2 with residues 1 and -1, respectively. Any differential that has at most a pole at P_1 or P_2 of order 1 can be written as a linear combination of $y_{12}\,dx$ and g linearly independent differentials of the first kind.*

PROOF. Take $(dx) = W$ as before. The differential $y\,dx$ has at most a pole at P_1 and P_2 with order ≤ 1 if and only if $y \in L(WP_1P_2)$. As in (6.3), we have

$$l(WP_1P_2) = g + 1 \quad \text{and} \quad l(W) = g.$$

Hence, there exists y_{12} in $L(WP_1P_2)$ that is not in $L(W)$. Since $y_{12}\,dx$ is not a differential of the first kind, the denominator D is either P_1, P_2, or P_1P_2. As we noted, D can be neither P_1 nor P_2. Therefore, $y_{12}\,dx$ has indeed a pole of order 1 at P_1 and P_2. Let α_1 and α_2 be the residues at P_1 and P_2, respectively. Then Theorem 2.19 implies $\alpha_1 + \alpha_2 = 0$.

If we rewrite $(1/\alpha_1)y_{12}\,dx$ as $y_{12}\,dx$, then this differential is what is needed. The second half of the theorem can be proved similarly as the previous theorem. Q.E.D.

Let $y\,dx$ be an arbitrary differential in K, and let P be a place in the denominator of $y\,dx$. For a prime element $t = t_P$ for P, we have the t-expansion of $y\,dx$ defined by

$$y\frac{dx}{dt} = \alpha_{-m}t^{-m} + \alpha_{-m+1}t^{-m+1} + \cdots, \qquad \alpha_i \in k \text{ and } \alpha_{-m} \neq 0.$$

The differentials $y_2\,dx, \dots, y_m\,dx$ in Theorem 2.25 can be expanded at P as

$$y_2\frac{dx}{dt} = \beta_{-2}^{(2)}t^{-2} + \beta_{-1}^{(2)}t^{-1} + \cdots, \qquad \beta_i^{(2)} \in k,\ \beta_{-2}^{(2)} \neq 0,$$

$$\vdots$$

$$y_m\frac{dx}{dt} = \beta_{-m}^{(m)}t^{-m} + \beta_{-m+1}^{(m)}t^{-m+1} + \cdots, \qquad \beta_i^{(m)} \in k,\ \beta_{-m}^{(m)} \neq 0.$$

Hence, one can choose $\gamma_2, \dots, \gamma_m$ in k so that

$$y'\,dx = y\,dx - \sum_{i=2}^{m} \gamma_i y_i\,dx$$

can have a pole at P at most of order 1. Furthermore, at places that differ from P, $y'\,dx$ has poles of the same order as $y\,dx$. Therefore, by repeating the same construction for all the poles, one can choose differentials of the second kind $z_1\,dx, \dots, z_n\,dx$ such that

$$y^*\,dx = y\,dx - \sum_{i=1}^{n} z_i\,dx$$

has only poles of order 1 at $P_1, \ldots, P_s$. Let ρ_i be the residue of $y^* \, dx$ at P_i. Then, from Theorem 2.19, we have

$$\rho_1 + \cdots + \rho_s = 0.$$

On the other hand, Theorem 2.26 implies that there exists a differential of the third kind $y_{ij} \, dx$ having the residues 1 and -1 at P_i and P_i, $i \neq j$, respectively. Let

$$w \, dx = y^* \, dx - (\rho_1 y_{1s} + \rho_2 y_{2s} + \cdots + \rho_{s-1} y_{s-1,s}) \, dx.$$

Since the pole of $y^* \, dx$ at P_i, $i < s$, is cancelled by $\rho_i y_{is} \, dx$, the differential $w \, dx$ has a pole at most of order 1 at P_s. Therefore, by the remark following Theorem 2.25, $w \, dx$ must be a differential of the first kind. Then, from the equation

$$y \, dx = \sum_{i=1}^{n} z_i \, dx + \sum_{i=1}^{s-1} \rho_i y_{is} \, dx + w \, dx,$$

we obtain:

THEOREM 2.27. *Any differential in the algebraic function field K over an algebraically closed field k can be expressed as a sum of differentials of the first kind, the second kind, and the third kind.*

§7. Special function fields

We will apply our general theory to special algebraic function fields. We assume in this section that the coefficient field is always algebraically closed.

(i) Rational function field $K = k(x)$. We will find the divisor of the differential dx. As we proved the formula (5.22), we can also prove

$$(dx) = \frac{1}{D_\infty^2}, \tag{7.1}$$

where D_∞ denotes the denominator of x. In this case, the characteristic of k need not be 0. Since the degree of D_∞ is 1, we have, from (7.1),

$$n((dx)) = -2n(D_\infty) = -2. \tag{7.2}$$

If g is the genus of K, then $n((dx)) = 2g - 2$ by Theorem 2.14. Hence, from (7.2),

$$g = 0 \tag{7.3}$$

must hold. That is, the genus of the rational function field is 0. Therefore, Theorem 2.15 implies that 0 is the only differential of the first kind. An element w of K, i.e., a rational function, may be decomposed as a product of linear factors:

$$w = \frac{\prod_{i=1}^{m}(x - \alpha_i)}{\prod_{j=1}^{n}(x - \beta_j)}, \qquad \alpha_i, \beta_j \in k.$$

Then the divisor (w) of w is given ds

$$(w) = \frac{\prod_{i=1}^{m} P_{\alpha_i}}{\prod_{j=1}^{n} P_{\beta_j}} \cdot P_{\infty}^{n-m}, \tag{7.3}$$

where P_{α_i}, P_{β_j}, and P_{∞} are defined in Chapter I, §1, Example 2. From this, we see that any divisor D_0 is a principal divisor if $n(D_0) = 0$ holds. That is, $\mathfrak{D}_0 = \mathfrak{D}_H$. Also, by (7.1) and (7.3), one can write out the elements and the differentials in Theorems 2.24, 2.25, 2.26.

Next, let K be an arbitrary algebraic function field of genus 0 over k. For any place P, we have

$$l(P) = n(P) - g + 1 + l(WP^{-1}) \geq 1 - 0 + 1 = 2$$

from the Riemann-Roch theorem. Since the dimension of $L(P)$ is greater than 2, $L(P)$ contains not only all the elements of k but also an element x that is not in k. By the definition of $L(P)$, the denominator of x is at most P. But, since x does not belong to k, x has to be precisely P. Then, from Theorem 2.7, we have

$$[K : k(x)] = n(P) = 1, \quad \text{i.e., } K = k(x).$$

That is, K is the rational function field.

THEOREM 2.28. *The genus of an algebraic function field over an algebraically closed field k is 0 if and only if K is the rational function field over k.*

(ii) Elliptic function field. By Theorem 2.28, the rational function field $K = k(x)$ is the only algebraic function field of genus 0. Next, we will consider an algebraic function field of genus 1.

DEFINITION 2.17. An algebraic function field of genus 1 over an algebraically closed field k is called an *elliptic function field* over k.

We will study the structure of elliptic function fields. For the sake of simplicity, we will assume that the characteristic of k is neither 2 nor 3. Let P be an arbitrary place on K, and let W be an arbitrary differential divisor on K. For $m \geq 2$, we have

$$n(WP^{-m}) = 2g - 2 - m = 2 - 2 - m < 0 \quad \text{and} \quad l(WP^{-m}) = 0.$$

Hence, from the Riemann-Roch theorem

$$l(P^m) = n(P^m) - g + 1 + l(WP^{-m}) = m - 1 + 1 + 0 = m.$$

Therefore, in particular, the dimensions of $L(P^2)$ and $L(P^3)$ are 2 and 3, respectively. Then one can find an element x_1 in $L(P^2)$ such that $x_1 \notin k$ and an element y_1 in $L(P^3)$ such that $y_1 \notin L(P^2)$. Notice that the denominator of (x_1) is either P or P^2. From part (i), however, it cannot be P, because otherwise $K = k(x_1)$ would hold, contradicting our assumption that the genus of K is 1. Hence, the denominator of (x_1) is P^2. The denominator of (y_1) is certainly P^3. Therefore, the denominators of

$$1, x_1, x_1^2, x_1^3, y_1, x_1y_1 \tag{7.4}$$

are, respectively, P^0, P^2, P^4, P^6, P^3, P^5. Consequently, from the expansions at P, one sees that the elements in (7.4) are linearly independent over k. Since $l(P^6) = 6$, those elements in (7.4) are basis elements for $L(P^6)$ over k. Therefore, since y_1^2 belongs to $L(P^6)$, we obtain

$$y_1^2 = \alpha_0 + \alpha_1x_1 + \alpha_2x_1^2 + \alpha_3x_1^3 + \alpha_4y_1 + \alpha_5x_1y_1$$

for $\alpha_i \in k$. Let

$$y' = y_1 - \frac{\alpha_5}{2}x_1 - \frac{\alpha_4}{2}. \tag{7.5}$$

Then we have

$$y'^2 = \beta_0 + \beta_1x_1 + \beta_2x_1^2 + \beta_3x_1^3. \tag{7.6}$$

Since the denominator of (y') is P^3 by (7.5), we have $\beta_3 \neq 0$. Let

$$x = \frac{1}{\beta_3}x_1 + \frac{\beta_2}{3\beta_3^2} \quad \text{and} \quad y = \frac{2}{\beta_3^2}y'. \tag{7.7}$$

From (7.6) we obtain

$$y^2 = 4x^3 - \gamma_2x - \gamma_3, \quad \text{where } \gamma_2, \gamma_3 \in k. \tag{7.8}$$

From Theorem 2.7, we have

$$[K : k(x_1)] = n(P^2) = 2 \quad \text{and} \quad [K : k(y_1)] = n(P^3) = 3.$$

Since $[K : k(x_1, y_1)]$ is a common divisor of $[K : k(x_1)]$ and $[K : k(y_1)]$, we obtain

$$[K : k(x_1, y_1)] = 1 \quad \text{and} \quad K = k(x_1, y_1) = k(x, y).$$

Next we will show that the right-hand side of (7.8) does not possess a multiple root. Suppose that

$$y^2 = 4(x - \alpha)^2(x - \beta).$$

Let $z = y/(x - \alpha)$. Then we have $x = z^2/4 + \beta$ and $y = z(z^2/4 + \beta - \alpha)$. Hence, $K = k(z)$ holds, i.e., the genus of K is 0 by Theorem 2.28. This contradicts our assumption. Therefore, $4x^3 - \gamma_2x - \gamma_3$ does not have a multiple root, and the discriminant

$$\gamma_2^3 - 27\gamma_3^2 \neq 0. \tag{7.9}$$

Conversely, we will prove that $K = k(x, y)$ is an elliptic function field if (7.8) and (7.9) hold. First of all, we have from (7.8) that $[K : k(x)] = 2$.

Next, let ε_1, ε_2 and ε_3 be the roots of $4x^3 - \gamma_2x - \gamma_3 = 0$, and let

$$f(x, y) = y^2 - 4(x - \varepsilon_1)(x - \varepsilon_2)(x - \varepsilon_3). \tag{7.10}$$

The roots ε_1, ε_2, and ε_3 are distinct by (7.9). Let α be an element of k such that $\alpha \neq \varepsilon_i$. The reduction modulo the prime ideal $(x-\alpha)$ of $f(x, y)$ is

$$\overline{f}(x, y) = f(\alpha, y) = y^2 - 4(\alpha - \varepsilon_1)(\alpha - \varepsilon_2)(\alpha - \varepsilon_3).$$

Then $\overline{f}(x, y)$ is decomposed into a product of distinct linear factors in $k[y]$. Therefore, from Theorem 1.7, $f(x, y)$ is a product of two distinct linear factors in the completion of $k(x)$ at the place P_α. From Theorem 1.11, P_α has two extensions $P_{\alpha,1}$ and $P_{\alpha,2}$ on K. Let e_1 and e_2 be the ramification indices of $P_{\alpha,1}$ and $P_{\alpha,2}$, respectively, for $k(x)$. Then we have the following from Remark 2 following Theorem 2.7:

$$\sum_{i=1}^{2} e_i = \sum_{i=1}^{2} e_i f_i = [K : k(x)] = 2, \tag{7.11}$$

which implies $e_1 = e_2 = 1$. Therefore, for $P = P_{\alpha,1}$ or $P = P_{\alpha,2}$

$$\nu_P(x-\alpha) = \nu_{P_\alpha}(x-\alpha) = 1.$$

That is, $t = x - \alpha$ is a prime element in K for P. We have

$$\frac{dx}{dt} = 1 \quad \text{and} \quad \nu_P((dx)) = 0.$$

Let $\alpha = \varepsilon_1$, let P be an extension on K of the place $P_\alpha = P_{\varepsilon_1}$ on $k(x)$, and let e be the ramification index. Since $\varepsilon_1 \neq \varepsilon_2$, ε_3, we have

$$\begin{aligned} 2\nu_P(y) = \nu_P(y^2) &= \nu_P(4(x-\varepsilon_1)(x-\varepsilon_2)(x-\varepsilon_3)) \\ &= e\nu_{P_{\varepsilon_1}}(4(x-\varepsilon_1)(x-\varepsilon_2)(x-\varepsilon_3)) = e, \end{aligned} \tag{7.12}$$

so that $e \geq 2$. Hence, this equality similar to (7.11) implies that $e = 2$, and that P is the only extension on K of the place P_{ε_1}. Since $\nu_P(x - \varepsilon_1) = 2$, we have the following for a prime element $t = t_P$:

$$\begin{aligned} &x - \varepsilon_1 = \xi_2 t^2 + \xi_3 t^3 + \cdots, &&\xi_i \in k \text{ and } \xi_2 \neq 0, \\ &\frac{dx}{dt} = 2\xi_2 t + 3\xi_3 t^2 + \cdots, &&2\xi_2 \neq 0, \\ &\nu_P((dx)) = 1. \end{aligned}$$

One could replace ε_1 by either ε_2 or ε_3 to obtain similar results.

Finally, consider the place P_∞ on $k(x)$. Let P be an extension on K of P_∞, and let e be the ramification index of P. The equality similar to (7.12) implies

$$2\nu_P(y) = -3e.$$

Hence, e is a multiple of 2, and it follows that $e = 2$ and that P is the only extension of P_∞. Since we have $\nu_P(x) = 2\nu_{P_\infty}(x) = -2$, for $t = t_P$ we have the following:

$$x = \xi_{-2}t^{-2} + \xi_{-1}t^{-1} + \cdots, \qquad \xi_{-i} \in k \text{ and } \xi_{-2} \neq 0,$$
$$\frac{dx}{dt} = -2\xi_{-2}t^{-3} + \cdots, \qquad -2\xi_{-2} \neq 0,$$
$$\nu_P((dx)) = -3.$$

Thus, we have found the values of $\nu_P((dx))$ for all the places on K. In particular, we obtain

$$n((dx)) = \sum_P \nu_P((dx)) = 1 + 1 + 1 - 3 = 0.$$

Let g be the genus of K. Then $n((dx)) = 2g - 2$ implies $g = 1$.

THEOREM 2.29. *Let k be an algebraically closed field of characteristic $\neq 2, 3$. Let K be an elliptic function field over k. Then K is given as*

$$K = k(x, y), \quad y^2 = 4x^3 - \gamma_2 x - \gamma_3, \quad \gamma_2^3 - 27\gamma_3^2 \neq 0, \tag{7.13}$$

which is called the Weierstrass canonical form. Conversely, an algebraic function field given by (7.13) is always an elliptic function field over k.

Next, we will find a condition for K defined by (7.13) to be isomorphic to K_1 defined by

$$K_1 = k(x_1, y_1), \quad y_1^2 = 4x_1^3 - \gamma_2' x_1 - \gamma_3', \quad \gamma_2'^3 - 27\gamma_3'^2 \neq 0. \tag{7.14}$$

Suppose $K_1 \cong K$. Let φ be an isomorphism from K_1 to K. Then $K = k(\varphi(x_1), \varphi(y_1))$, and $\varphi(x_1)$ and $\varphi(y_1)$ satisfy the same equation as x_1 and y_1. Hence, we may assume $K_1 = K$. Then we will study the relationship between the two canonical forms (7.13) and (7.14) for K. From the proof of Theorem 2.29, the denominator of (x) of (7.13) is the second power of P, i.e., P^2. Note that the extension P on K of P_∞ on $k(x)$ satisfies $\nu_P((x)) = -2$. Hence, the denominator of (y) in $y^2 = 4x^3 - \gamma_2 x - \gamma_3$ is P^3. Similarly, the denominators of (x_1) and (y_1) are P_1^2 and P_1^3, respectively.

Consider first the case where $P = P_1$. Then x and x_1 both belong to $L(P^2)$ and $l(P^2) = 2$. Hence,

$$x_1 = \alpha x + \beta, \qquad \alpha, \beta \in k, \alpha \neq 0.$$

Moreover, y and y_1 belong to $L(P^3)$ and $l(P^3) = 3$. So we have

$$y_1 = \gamma y + \delta x + \varepsilon, \qquad \gamma, \delta, \varepsilon \in k, \gamma \neq 0.$$

Substitute the above x_1 and y_1 into $y_1^2 = 4x_1^3 - \gamma_2' x - \gamma_3'$. Since there exists a unique monic irreducible polynomial for y over $k(x)$, we obtain

$$\beta = \delta = \varepsilon = 0, \qquad \alpha^3 = \gamma^2,$$
$$\gamma_2' = \alpha^2 \gamma_2 \quad \text{and} \quad \gamma_3' = \alpha^3 \gamma_3,$$

noting that the characteristic of k is neither 2 nor 3. Consequently,

$$\frac{\gamma_2'^3}{\gamma_2'^3 - 27\gamma_3'^2} = \frac{\gamma_2^3}{\gamma_2^3 - 27\gamma_3^2}. \tag{7.16}$$

Next let $P \neq P_1$. Then by the Riemann-Roch theorem we have

$$l(PP_1) = n(PP_1) - g + 1 + l(WP^{-1}P_1^{-1}) = 2 - 1 + 1 + 0 = 2.$$

Hence, there is an element z in $L(PP_1)$ that is not in k. By the definition of $L(PP_1)$, the nominator of (z) is at most PP_1. As before, one can show that it is actually PP_1. (For example, if P were the denominator, then $K = k(z)$ would hold.) Hence,

$$[K : k(z)] = 2.$$

Since the characteristic of k is not 2, $K/k(z)$ is a Galois extension. Therefore, besides the identity, there exists an automorphism σ of K such that $\sigma(a) = a$ for all $a \in k(z)$. Note that $\sigma^2 = 1$ holds. Let

$$x_2 = \sigma(x_1) \quad \text{and} \quad y_2 = \sigma(y_1).$$

Then we have

$$K = k(x_2, y_2), \qquad y_2^2 = 4x_2^3 - \gamma_2' x_2 - \gamma_3'. \tag{7.17}$$

For an arbitrary place Q on K, let

$$\nu'(a) = \nu_Q(\sigma^{-1}(a)), \qquad a \in K.$$

Then $\nu'(a)$ is a normal valuation on K. Hence, ν' determines a place Q' on K. Let us write $\sigma(Q)$ for Q'. That is,

$$\nu_{\sigma(Q)}(\sigma(a)) = \nu_Q(a), \qquad a \in K.$$

In particular, for $a = z$ and $Q = P_1$, $\sigma(z) = z$ implies

$$\nu_{\sigma(P_1)}(z) = \nu_{P_1}(z) = -1.$$

Since $\sigma(P_1)$ is also contained in the denominator of (z),

$$\sigma(P_1) = P_1 \quad \text{or} \quad \sigma(P_1) = P$$

must hold. In this case $\sigma(P_1) = P_1$ is impossible. Suppose that $\sigma(P_1) = P_1$. Then the denominators of x_2 and x_3 would be respectively P_1^2 and P_1^3. As shown previously,

$$\sigma(x_1) = x_2 = \alpha_1^2 x_1, \quad \sigma(y_1) = y_2 = \alpha_1^3 y_1, \qquad \alpha_1 \in k, \alpha_1 \neq 0.$$

Then $\sigma^2 = 1$ implies

$$\sigma^2(x_1) = \alpha_1^4 x_1 = x_1, \qquad \sigma^2(y_1) = \alpha_1^6 y_1 = y_1,$$

i.e., $\alpha_1^2 = 1$ and $\sigma(x_1) = x_1$. Since the ramification index of P_1 with respect to $k(x_1)$ is 2 and $\nu_{P_1}(z) = -1$, z does not belong to $k(x_1)$. Hence, $\sigma(z) \neq$

z would follow, which is a contradiction. Consequently, $\sigma_1(P_1) = P$ must hold, and the denominators of (x_2) and (y_2) are P^2 and P^3, respectively. Therefore, (7.17), together with what we have already proved implies that (7.16) also holds for this case.

Conversely, let (7.13) be a canonical form of K, and let γ_2' and γ_3' be arbitrary elements of k such that (7.16) is satisfied and such that $\gamma_2'^3 - 27\gamma_3'^2 \neq 0$. If $\gamma_2 = 0$ or $\gamma_3 = 0$ holds, then from (7.16) the corresponding γ_2' or γ_3' is zero. In this case, there is an element α in k which satisfies (7.15). Next, let $\gamma_2 \neq 0$ and $\gamma_3 \neq 0$; therefore, $\gamma_2' \neq 0$ and $\gamma_3' \neq 0$. From (7.16) we have

$$\left(\frac{\gamma_2'}{\gamma_2}\right)^3 = \left(\frac{\gamma_3'}{\gamma_3}\right)^2.$$

Then (7.15) holds for

$$\alpha = \left(\frac{\gamma_2'}{\gamma_2}\right)^{-1}\left(\frac{\gamma_3'}{\gamma_3}\right).$$

For $\gamma \in k$ such that $\alpha^3 = \gamma^2$, let

$$x_1 = \alpha x \quad \text{and} \quad y_1 = \gamma y.$$

Then we obtain

$$K = k(x_1, y_1), \qquad y_1^2 = 4x_1^3 - \gamma_2' x_1 - \gamma_3'.$$

Hence, we have the following theorem.

THEOREM 2.30. *Let k be an algebraically closed field of characteristic $\neq 2, 3$ and let K be an elliptic function field over k such that*

$$K = k(x, y), \quad y^2 = 4x^3 - \gamma_2 x - \gamma_3, \quad \gamma_2^3 - 27\gamma_3^2 \neq 0$$

is a canonical form of K. Then

$$\delta = \frac{\gamma_2^3}{\gamma_2^3 - 27\gamma_3^2} \tag{7.18}$$

is an invariant *determined by K only and is independent of the particular choice of a canonical form. Two elliptic function fields K_1 and K_2 over k are isomorphic if and only if the invariant δ_1 of K_1 coincides with the invariant δ_2 of K_2.*

For an arbitrary element δ in k, one can always find elements γ_2 and γ_3 in k so that (7.18) is satisfied. Therefore, there are as many nonisomorphic elliptic function fields as the number of elements in k, which is infinite. Notice that this was not the case for the genus-zero function field.

From the proof of Theorem 2.30, there is an automorphism σ of K over k such that $\sigma(P_1) = P$ for any given places P_1 and P on K. It is clear that in the case of the rational function field $k(x)$, one can find a similar automorphism on $k(x)$ of the form

$$\sigma(x) = \frac{\alpha x + \beta}{\gamma x + \delta}, \qquad \alpha, \beta, \gamma, \delta \in k \text{ such that } \alpha\delta - \beta\gamma \neq 0.$$

Hence, we have the following theorem.

THEOREM 2.31. *Let K be an algebraic function field of genus either 0 or 1 over an algebraically closed field k. For any two places P and P_1 on K, there exists an automorphism σ on K such that*

$$\sigma(P_1) = P.$$

In particular, the group of automorphisms on K over k is an infinite group.

For the earlier proof, we needed the condition that the characteristic of k should not be 2 or 3. However, Theorem 2.31 holds in general without the restriction on the characteristic. (See the paper: M. Deuring, *Invariante und Normalformen elliptischer Funktionenkörper*, Math. Z. **47** (1940).)

Theorem 2.31 is interesting because of the following fact: the group of automorphisms on an arbitrary algebraic function field K of genus $g \geq 2$ over an algebraically closed field k is always a finite group. For the proof of this remarkable fact, the reader is referred to the earlier cited paper: H. L. Schmid, *Über die Automorphismen eines algebraischen Funktionenkörpers von Primzahlcharakteristik*, Crelle's Jour. **179** (1938), or K. Iwasawa, T. Tamagawa, *Automorphisms of algebraic function field*, Sugaku, Vol. 1, No. 4, Math. Soc. Japan, Tokyo, 1948. (Japanese)

Let the characteristic of $k \neq 2, 3$ be as before, and let K be an elliptic function field whose canonical form is given as in (7.13). Keeping the notation as in the proof of Theorem 2.29, let Q_1, Q_2, Q_3, Q_∞ be extensions on K of $P_{\varepsilon_1}, P_{\varepsilon_2}, P_{\varepsilon_3}, P_\infty$ on $k(x)$. From that proof, we have

$$(dx) = (y) = \frac{Q_1 Q_2 Q_3}{Q_\infty^3}.$$

Therefore, the differential dx/y is of the first kind. Since the genus of K is 1, Theorem 2.15 implies that any differential in K is given by the form

$$\alpha \frac{dx}{y}, \qquad \alpha \in k.$$

Furthermore, from Theorem 2.23, for a fixed place, e.g., Q_∞, each class of $\overline{\mathfrak{D}}_0 = \mathfrak{D}_0/\mathfrak{D}_H$ can be represented by a divisor in the form

$$\frac{P}{Q_\infty}. \tag{7.19}$$

Suppose that for $P/Q_\infty \sim P'/Q_\infty$, i.e., $P \sim P'$. Then there exists an element u such that $(u) = P'/P$. If $P' \neq P$, then $[K : k(u)] = n(P) = 1$, i.e., $K = k(u)$, which is a contradiction. Hence, $P = P'$. Therefore, the correspondence between a place P and the class in $\overline{\mathfrak{D}}_0$ containing the divisor of the form (7.19) is one-to-one.

Those facts mentioned above make the study of an elliptic function field much simpler than that of an algebraic function field of genus $g \geq 2$. For example, the addition theorem in the classical theory of elliptic functions can be generalized to an elliptic function field over a general coefficient field. See H. Hasse, *Zur Theorie der abstrakten elliptischen Funktionenköper*, I, II, III, Crelle's Jour. **175** (1936).

(iii) Hyperelliptic function field. As a generalization of the canonical form (7.13), consider an algebraic function field K over k given by

$$K = k(x, y), \quad y^2 = (x - \alpha_1)(x - \alpha_2)\cdots(x - \alpha_{2n+1}), \qquad n \geq 2, \quad (7.20)$$

where $\alpha_1, \alpha_2, \ldots, \alpha_{2n+1}$ are mutually distinct elements in k. Such an algebraic function field K as in (7.20) is said to be a hyperelliptic function field over k. If the characteristic of k is not 2, then the places P_{α_i} and P_∞ on $k(x)$ have unique extensions Q_i and Q_∞ on K, respectively, and their ramification indices are both 2. One can also prove, as in (ii), that

$$(dx) = \frac{Q_1 \cdots Q_{2n+1}}{Q_\infty^3} \quad \text{and} \quad (y) = \frac{Q_1 \cdots Q_{2n+1}}{Q_\infty^{2n+1}}.$$

Therefore, the genus g of K can be computed as

$$2g - 2 = n((dx)) = 2n + 1 - 3 = 2n - 2,$$

i.e., $g = n$. Furthermore, all the differentials of the first kind in K can be written as

$$\frac{\beta_0 + \beta_1 x + \cdots + \beta_{n-1}x^{n-1}}{y}\, dx, \qquad \beta_i \in k.$$

Thus, given any integer $g \geq 0$, there exists an algebraic function field of genus g as long as the characteristic of k is not 2.

In the case where the characteristic of k is 2, one can construct K as follows. Let $\alpha_1, \alpha_2, \ldots, \alpha_{n+1}$ be mutually distinct elements of k. Then let

$$K = k(x, y) \quad \text{such that } y^3 = (x - \alpha_1)(x - \alpha_2)\cdots(x - \alpha_{n+1}).$$

Then the places P_{α_i} and P_∞ on k are extended to Q_i and Q_∞ on K such that their ramification indices are 3. Moreover, since we have

$$(dx) = \frac{(Q_1 Q_2 \cdots Q_{n+1})^2}{Q_\infty^4},$$

$2g - 2 = n((dx)) = 2(n + 1) - 4 = 2n - 2$ holds, i.e.,

$$g = n.$$

THEOREM 2.32. *Let* k *be any algebraically closed field and let* g *be any nonnegative rational integer. Then there exists an algebraic function field of genus* g *over* k.

We have so far discussed some fundamental results in the algebraic theory of algebraic functions over a general coefficient field k. However, in order to obtain more delicate results than those obtained from the general theory, we need to specify the coefficient field. For example, for a finite field k, algebraic function fields over k share many properties analogous to those of algebraic number fields. For instance, it is known that fundamental theorems in class field theory can be carried out for such function fields. See M. Moriya, *Rein arithematisch-algebraishen Aufbrau der Klassenkörpertheorie über algebraischen Fünktionenkörper einer Unbestimmten mit endlichen Konstantenkörper*, Japan J. Math. **14** (1938). (In this article, the works of F. K. Schmidt and H. Hasse are listed. The reader is referred to the preface for more comments.) Since such a study as mentioned belongs to number theory, it will not be discussed here. From the next chapter on, the coefficient field k will always be taken to be the field of complex numbers, and we will study the classical theory of algebraic function fields over the complex field by using not only algebraic methods, but also by freely using analytic methods.

CHAPTER 3

Riemann Surfaces

§1. Riemann surfaces and analytic mappings

We will present a general theory of Riemann surfaces as preparation for the study of the classical theory of algebraic functions.

DEFINITION 3.1. Let S be a connected topological space. Suppose that for each point P on S, there exists a function φ_P such that

(i) the function φ_P is defined in a neighborhood U_P, and φ_P maps U_P homeomorphically to the interior of the unit disc, i.e., $|z| < 1$, so that $\varphi_P(P) = 0$;

(ii) for P and Q on S, if $V = U_P \cap U_Q$ is not an empty set, then the map $\varphi_Q \varphi_P^{-1}$ from $\varphi_P(V)$ onto $\varphi_Q(V)$ is a one-to-one conformal map, namely, $\varphi_Q \varphi_P^{-1}$ is a complex regular function.

Then $\{\varphi_P, P \in S\}$ is said to be an R-function system on S and is denoted by $\mathfrak{F} = \{\varphi_P, U_P\}$.

DEFINITION 3.2. Let $\mathfrak{F} = \{\varphi_P, U_P\}$ and $\mathfrak{F}' = \{\varphi'_P, U'_P\}$ be R-function systems on S. For an arbitrary point P on S, if the function $\varphi'_P \varphi_P^{-1}$ defined in $\varphi_P(U_P \cap U'_P)$ is regular at the origin, then we say that the R-function systems $\mathfrak{F}$ and $\mathfrak{F}'$ are equivalent and write $\mathfrak{F} \sim \mathfrak{F}'$.

Notice that $\mathfrak{F} \sim \mathfrak{F}'$ is an equivalence relation. Hence, one can classify all the R-function systems on S in terms of this equivalence relation. We will define a Riemann surface via these classes as follows.

DEFINITION 3.3. The pair

$$\mathfrak{R} = \{S, \{\mathfrak{F}\}\} \tag{1.1}$$

of a connected topological space S and an equivalence class $\{\mathfrak{F}\}$ is said to be a *Riemann surface.* The space S is called the *underlying space,* and an R-function system $\mathfrak{F}$ belonging to the class $\{\mathfrak{F}\}$ is said to be an R-function system defining $\mathfrak{R}$, or an R-function system on R. We sometimes abbreviate *Riemann surface* by *R-surface.*

A point on the underlying space S, an open subset of S, a closed subset of S, etc., are said to be a point, an open subset, a closed subset, etc. of the R-surface $\mathfrak{R}$, respectively. Similar topological terminology for S is used for $\mathfrak{R}$.

The class $\{\mathfrak{F}\}$ is determined by an R-function system. Hence, one may write

$$\mathfrak{R} = \{S, \mathfrak{F}\}$$

instead of (1.1) to describe the R-surface. However, note that strictly speaking, a Riemann surface $\mathfrak{R}$ is determined by the class $\{\mathfrak{F}\}$, not by a representative $\mathfrak{F}$.

Definition 3.1 requires conditions for the underlying space S to possess an R-function system. That is, S must be a two-dimensional manifold to be the underlying space of an R-surface. Moreover, (ii) of the definition implies that S is an oriented manifold. For an orientation on a manifold, see H. Weyl, *Die Idee der Riemannschen Fläche*, Teubner, Berlin, 1913. Later we will show that S satisfies the second countability axiom. On the other hand, the converse is also true; that is, it has been proved that any connected two-dimensional manifold with the second countability axiom is indeed the underlying space of a Riemann surface. A general proof of the above assertion requires more preparation than we will be able to provide here. The reader is referred to Stoilow, *Leçons sur les principes topologiques de la théorie des fonctions analytiques*, Gauthier-Villars, Paris, 1913. In the following chapter, we will provide a detailed discussion of the case where S is a compact space.

Let

$$\mathfrak{R} = \{S, \{\mathfrak{F}\}\}, \quad \text{where } \mathfrak{F} = \{\varphi_P, U_P\}$$

be a Riemann surface. Let C_P be a small circle in $\varphi_P(U_P)$ with center at 0. Let ψ_P be a linear transformation that maps C_P on the unit circle $\varphi_P(U_P)$. Then

$$\mathfrak{F}' = \{\varphi'_P = \psi_P \varphi_P, U'_P = \overset{-1}{\varphi_P}(C_P)\}$$

is an R-function system on S equivalent to $\mathfrak{F}$. Therefore, when any neighborhood W_P of each point P is given beforehand, there exists an R-function system $\mathfrak{F}' = \{\varphi'_P, U'_P\}$ defining $\mathfrak{R}$ such that

$$U'_P \subset W_P.$$

Particularly, if W_P is taken so that $\overline{W}_P$ is compact, one can choose U'_P so that $\overline{U}_P$ is compact.

Let S_1 be an arbitrary domain in $\mathfrak{R}$, i.e., S_1 is a connected open subset of the underlying space S. By what we have described above, we can find an $\mathfrak{R}$-function system $\mathfrak{F}' = \{\varphi'_P, U'_P\}$ on $\mathfrak{R}$ satisfying

$$U'_P \subset S_1 \quad \text{for } P \in S_1. \tag{1.2}$$

Then let

$$\mathfrak{F}'_1 = \{\varphi'_P, U'_P; P \in S_1\}. \tag{1.3}$$

It is clear that $\mathfrak{F}'_1$ is an $\mathfrak{R}$-function system on S_1. Notice also that if $\mathfrak{F}''$, which is equivalent to $\mathfrak{F}'$, satisfies the condition corresponding to (1.2), then $\mathfrak{F}''_1$ constructed as in (1.3) is an $\mathfrak{R}$-function on S_1 equivalent to $\mathfrak{F}'_1$.

DEFINITION 3.4. For a domain S_1 in $\mathfrak{R}$, there is induced a unique R-surface

$$\mathfrak{R}_1 = \{S_1, \{\mathfrak{F}_1'\}\},$$

which is called a Riemann subsurface.

The Riemann sphere is the simplest but an important example of a Riemann surface. As is well known, all the points on the extended complex plane with the point at infinity correspond to the points on the unit sphere S_0^* in a one-to-one fashion via a stereographic projection. One can express each point on S_0^* in terms of complex numbers including ∞. Consider S_0^* as an underlying space and let
for $P = z_0 \neq \infty$,

$$\varphi_P(z) = z - z_0, \qquad z \in U_P = \{z\,;\, |z - z_0| < 1\};$$

for $P = \infty$,

$$\varphi_P(z) = \frac{1}{z}, \qquad z \in U_P = \{z\,;\, |z| > 1\}.$$

Then $\mathfrak{F} = \{\varphi_P, U_P\}$ is an R-function system on S_0^*, i.e.,

$$\mathfrak{R}_0^* = \{S_0^*, \mathfrak{F}\}$$

is an R-surface, i.e., the Riemann sphere. From an arbitrary domain in $\mathfrak{R}_0^*$, we can construct an R-surface as in Definition 3.4. For example, the interior of the unit disc $\mathfrak{R}_1^* = \{z\,;\, |z| < 1\}$ is an R-surface.

DEFINITION 3.5. Let $\mathfrak{R} = \{S, \{\mathfrak{F}\}\}$ and $\mathfrak{R}' = \{S, \{\mathfrak{F}'\}\}$ be any R-surfaces, let $\mathfrak{F} = \{\varphi_P, U_P\}$ and $\mathfrak{F}' = \{\varphi_P', U_P'\}$ be their R-function systems, respectively. Let f be a map from a domain U in $\mathfrak{R}$ into $\mathfrak{R}'$, and let $P' = f(P)$ for P in U. If the complex valued function

$$\varphi_{P'} f \varphi_P^{-1}$$

defined in $\varphi_P(U_P)$ is regular at the origin, then the map f is called *analytic* at P. When f is analytic at every point in U, then f is said to be an *analytic map* from U into $\mathfrak{R}'$.

Note that the definition of analyticity of f at a point P does not depend upon the choice of representatives $\mathfrak{F}$ and $\mathfrak{F}'$ from the classes $\{\mathfrak{F}\}$ and $\{\mathfrak{F}'\}$.

DEFINITION 3.6. When there exists an analytic map f from a domain U in an R-surface $\mathfrak{R}$ onto a domain U' in the Riemann sphere $\mathfrak{R}_0^*$ such that U' does not contain ∞ and such that f is one-to-one, then the domain U is called an *analytical domain* in $\mathfrak{R}$. Let

$$f(P) = t = x + iy \quad \text{for } P \in U.$$

Then $t = f(P)$ is called an *analytical variable* in U, and $(x(P), y(P))$ are called *analytical coordinates*. If U is a neighborhood of a point P_0 such that

$$f(P_0) = 0,$$

then $t = f(P)$ is called a *locally uniformizing variable* at P_0.

Since f in the above definition is one-to-one, a point in U is uniquely determined by $t = t(P)$, or by $(x(P), y(P))$. We therefore speak of a point t in U, or a point (x, y) in U.

Let P_0 be an arbitrary point on $\mathfrak{R}$. Then there is a locally uniformizing variable at P_0. For example, for an R-function system $\mathfrak{F} = \{\varphi_P, U_P\}$ defining $\mathfrak{R}$, $t = \varphi_{P_0}(P)$ is a locally uniformizing variable. From a locally uniformizing variable we can construct as many locally uniformizing variables as desired.

Theorem 3.1. *For two locally uniformizing variables $t = t(P)$ and $t' = t'(P)$ at a point P_0 on an R-surface $\mathfrak{R}$, one can write near P_0,*

$$t' = a_1 t + a_2 t^2 + \cdots, \qquad a_1 \neq 0, \tag{1.4}$$

a power series in t. Conversely, for a locally uniformizing variable t, a function $t' = \varphi'(P)$ expressed as a power series as in (1.4) *is also a locally uniformizing variable.*

Proof. Let U and U' be domains where $t = \varphi(P)$ and $t' = \varphi'(P)$ are defined, and let $V = U \cap U'$. Then the map $t' = F(t) = \varphi'\varphi^{-1}(t)$ is a one-to-one and conformal map from $\varphi(V)$ onto $\varphi'(V)$. Since $F(0) = 0$ holds, we can obtain a power series as in (1.4) in a neighborhood of $t = 0$, i.e., P_0. The converse is clear.

If one uses locally uniformizing variables, the notion of an analytic map in Definition 3.5 becomes clear. That is, for a map f from $\mathfrak{R}$ to $\mathfrak{R}'$, let $P_0' = f(P_0)$. Suppose t and t' are locally uniformizing variables at P_0 and P_0', respectively. Then f can be considered to be a map from the t-plane to the t'-plane such that $t' = f(t)$ in a neighborhood of $t = 0$, i.e., $P = P_0$. The analyticity of f in the sense of Definition 3.5 is nothing but the regularity of the complex-valued function $f(t)$ at $t = 0$. Since $f(0) = 0$, we have

$$t' = f(t) = a_1 t + a_2 t^2 + \cdots \tag{1.5}$$

in a neighborhood of P_0. If $f(t)$ is not identically zero, $a_1 = a_2 = \cdots = a_{n-1} = 0$, and $a_n \neq 0$, then choose a locally uniformizing variable t_1 at P_0, i.e.,

$$t_1 = b_1 t + b_2 t^2 + \cdots, \qquad b_1 \neq 0,$$

such that

$$t' = t_1^n.$$

Hence, we obtain the following theorem.

Theorem 3.2. *If a map f from an R-surface to an R'-surface is analytic at P_0, then one can find locally uniformizing variables t and t' at P_0 and $P_0' = f(P_0)$, respectively, with the following property*:

$$t' = f(t) \equiv 0 \quad \text{or} \quad t' = f(t) = t^n, \qquad n \geq 1.$$

In the case $t' = t^n$, a neighborhood of P_0 is mapped by f n-fold onto a neighborhood of P_0'. Therefore, the integer n is determined independently from the choice of locally uniformizing variables, and n is called the *ramification exponent* of the map f at P_0. In particular, P_0 is said to be a branch point if $n > 1$, and an unramified point if $n = 1$. Since $t' = t^n$, P_0 is the only branch point in a sufficiently small neighborhood of P_0. Hence the set of branch points of f does not have an accumulation point in $\mathfrak{R}$. In the case where $\mathfrak{R}$ and $\mathfrak{R}'$ are compact R-surfaces, the properties mentioned above become equivalent to a certain property of algebraic function fields. We will come back to this topic in Chapter 4.

From Theorem 3.2 we obtain

THEOREM 3.3. *Let f be an analytic map from a domain U of an R-surface to an R-surface $\mathfrak{R}'$. Then $f(U)$ is either a point or a domain in $\mathfrak{R}'$.*

PROOF. Let us call a point P_0 in U a point of the first kind if f has the same value in a neighborhood of P_0, i.e., $t' = f(t) \equiv 0$. Otherwise we call P_0 a point of the second kind. Let U_1 be the set of points of the first kind, and let U_2 be the set of points of the second kind. Then, by the definition, both U_1 and U_2 are open sets and we have

$$U_1 \cup U_2 = U \quad \text{and} \quad U_1 \cap U_2 = \emptyset.$$

Since U is a connected set, $U_1 = U$ or $U_2 = U$ must hold. If $U_1 = U$ holds, then $f^{-1}(P_0')$ is open and closed for any point P_0' in $f(U)$. Hence, $U = f^{-1}(P_0')$, i.e., $P_0' = f(U)$, since U is connected. If $U_2 = U$ holds, then for the open map f, $f(U)$ is open. Furthermore, since f is continuous, $f(U)$ is connected. That is, $f(U)$ is a domain.

Note that the above proof implies the following corollary.

THEOREM 3.3 (corollary). *If an analytic map f from a domain U in an R-surface $\mathfrak{R}$ to an R-surface $\mathfrak{R}'$ takes the same value P_0' in a neighborhood of a point P_0, then $f(P) = P_0'$ for all the points P in U.*

REMARK. The assumption in the above corollary may be weakened. That is, it is enough to assume that $f(P_i) = P_0'$, $i = 1, 2, \dots$, for a convergent sequence of points $\{P_i\}$.

DEFINITION 3.7. A map f from an R-surface $\mathfrak{R}$ onto an R-surface $\mathfrak{R}'$ is called an *isomorphism* if f is one-to-one and analytic. When such an isomorphism f exists, $\mathfrak{R}$ and $\mathfrak{R}'$ are said to be *isomorphic* Riemann surfaces, denoted $\mathfrak{R} \cong \mathfrak{R}'$. We also say that $\mathfrak{R}$ and $\mathfrak{R}'$ are conformally equivalent.

Any R-surface $\mathfrak{R}$ is isomorphic to $\mathfrak{R}$ itself, i.e., for f take the identity map. Let $\mathfrak{R} \cong \mathfrak{R}'$ with an isomorphism f, and for an arbitrary point P on $\mathfrak{R}$ let $P' = f(P)$. Since f is one-to-one, one can find locally uniformizing variables t and t' at P and P', respectively, such that

$$t' = f(t) = t$$

by Theorem 3.2. Hence, its inverse f^{-1} is also analytic, i.e., $\mathfrak{R}' \cong \mathfrak{R}$. Let f and f' be isomorphisms so that $\mathfrak{R} \cong \mathfrak{R}'$ and $\mathfrak{R}' \cong \mathfrak{R}''$ hold. Then $f'f$ is an isomorphism, i.e., $\mathfrak{R} \cong \mathfrak{R}''$. Therefore, we can divide R-surfaces into equivalence classes of R-surfaces.

The totality of automorphisms, i.e., isomorphisms from $\mathfrak{R}$ to $\mathfrak{R}$, forms a group, as one can observe by setting $\mathfrak{R} = \mathfrak{R}' = \mathfrak{R}''$ in the above. This group is called the *group of automorphisms* of $\mathfrak{R}$, and is denoted by $A(\mathfrak{R})$.

Let $\mathfrak{R} \cong \mathfrak{R}'$ and let $\mathfrak{F} = \{\varphi_P, U_P\}$ be an R-function system determining the R-surface $\mathfrak{R}$. For an isomorphism f from $\mathfrak{R}$ to $\mathfrak{R}'$,

$$f(\mathfrak{F}) = \{\varphi_P f^{-1}, f(U_P)\}$$

is an R-function system defining $\mathfrak{R}'$. Similarly, for an arbitrary $\mathfrak{F}'$ defining $\mathfrak{R}'$, one can obtain $f^{-1}(\mathfrak{F}')$ defining $\mathfrak{R}$. The map f is a homeomorphism between the underlying spaces S and S' of $\mathfrak{R}$ and $\mathfrak{R}'$, respectively. Hence, if $\mathfrak{R} \cong \mathfrak{R}'$ holds, $\mathfrak{R}$ and $\mathfrak{R}'$ have the same structure as abstract Riemann surfaces. The theory of Riemann surfaces is the study of the properties of such abstract Riemann surfaces, as well as problems such as determining when two concretely given Riemann surfaces are isomorphic.

§2. Functions on a Riemann surface

We will study complex-valued functions $f(P)$ defined on an R-surface $\mathfrak{R}$ or on a domain U in $\mathfrak{R}$. We allow functions to take value ∞, but we exclude functions taking the value ∞ identically in an open set.

DEFINITION 3.8. Let $f = f(P)$ be as stated above, and let P_0 be a point on $\mathfrak{R}$. Then f is said to be *analytic* at P_0 if it is analytic at P_0 when regarded as a function from $\mathfrak{R}$ to the Riemann sphere $\mathfrak{R}_0^*$. Moreover, f is said to be *regular* at P_0 if $f(P_0) \neq \infty$ holds. If f is analytic at every point on $\mathfrak{R}$, then f is said to be an *analytic function* on $\mathfrak{R}$. Similar definitions are valid for a function defined in a domain U.

Let $z = f(P)$ be a function which is analytic at P_0, and let $f(P_0) = a_0 \neq \infty$. Since $z - a_0$ is a locally uniformizing variable at a_0 on $\mathfrak{R}_0^*$, for an arbitrary locally uniformizing variable t at P_0, we have from (1.5) that in a neighborhood of P_0

$$z - a_0 = a_1 t + a_2 t^2 + \cdots .$$

That is,

$$z = f(P) = a_0 + a_1 t + a_2 t^2 + \cdots . \tag{2.1}$$

Let us denote the least exponent of t that has a nonzero coefficient by

$$v_{P_0}(f).$$

This is called the *order* of f at P_0. If the right-hand side of (2.1) is identically zero, then $v_{P_0}(f) = +\infty$. In particular, in the case where $a_0 = 0$, i.e., $v_{P_0}(f) > 0$, f is said to have a *zero* at P_0 of order $v_{P_0}(f)$. Notice that the value $v_{P_0}(f)$ is invariant when t is replaced by another locally uniformizing variable t'. This is because t and t' are related as in equation (1.4) of Theorem 3.1.

NOTE. The reason we use the same notation v_{P_0} as in Chapter 1 is that v_{P_0} is in fact a normalized valuation on the analytic function field $K(\mathfrak{R})$ on $\mathfrak{R}$. In particular, when $\mathfrak{R}$ is a compact Riemann surface and $K(\mathfrak{R})$ is an algebraic function field, this relationship is decisive. See Chapter 4 for details.

For $f(P_0) = \infty$, $1/z$ is a locally uniformizing variable at ∞ on $\mathfrak{R}_0^*$. As before, we have

$$\frac{1}{z} = a_1 t + a_2 t^2 + \cdots ,$$

where the right-hand side is not identically zero in a neighborhood of P_0 by the hypothesis on f. If $a_1 = \cdots = a_{n-1} = 0$ and $a_n \neq 0$, then in a neighborhood of P_0, we have

$$z = f(P) = a'_{-n} t^{-n} + a'_{-n+1} t^{-n+1} + \cdots , \qquad a'_{-n} \neq 0. \tag{2.2}$$

We call

$$-n = v_{P_0}(f)$$

the order of f at P_0, and P_0 is called a *pole* of f of order n. The integer $v_{P_0}(f)$ is determined independently of the choice of t.

Conversely, if a complex-valued function $f(P)$ has an expansion as in (2.1) or (2.2) in a neighborhood of P_0, then f is clearly analytic at P_0 (regular in the case of (2.2)).

Let $f(P)$ and $g(P)$ be analytic functions on $\mathfrak{R}$. Then

$$f(P) + g(P) \quad \text{and} \quad f(P)g(P)$$

are also analytic functions on $\mathfrak{R}$, which can be observed by expanding $f(P)$ and $g(P)$ at each point on $\mathfrak{R}$. In particular, if $f(P)$ is not identically zero on $\mathfrak{R}$, then from Theorem 3.3, $f(P)$ is not identically zero in a neighborhood of P_0. Hence, $1/f(P)$ can be expanded as in (2.1) or (2.2), in a neighborhood of each point. That is, $1/f(P)$ is an analytic function on $\mathfrak{R}$. Therefore, the totality $K(\mathfrak{R})$ of analytic functions on $\mathfrak{R}$ forms a field. In particular, $K(\mathfrak{R})$ contains $f(P) \equiv a$, i.e., $K(\mathfrak{R})$ is an extension field of the field k of complex numbers. Hence, we have the following theorem.

THEOREM 3.4. *The set $K(\mathfrak{R})$ of analytic functions on a Riemann surface $\mathfrak{R}$ is an extension field of the field k of complex numbers.*

The field $K(\mathfrak{R})$ is called the *analytic function field* of $\mathfrak{R}$, or belonging to $\mathfrak{R}$. It will become clear that the field $K(\mathfrak{R})$ profoundly reflects the structures of $\mathfrak{R}$. First, we prove

THEOREM 3.5. *If R-surfaces* $\mathfrak{R}$ *and* $\mathfrak{R}'$ *are isomorphic, then* $K(\mathfrak{R})$ *and* $K(\mathfrak{R}')$ *are isomorphic over* k.

PROOF. Let f be an isomorphism from $\mathfrak{R}$ to $\mathfrak{R}'$, and let $g' = g'(P')$ be an arbitrary function in $K(\mathfrak{R}')$. Then,

$$g(P) = g'(f(P)), \qquad P \in \mathfrak{R}$$

is an analytic function on $\mathfrak{R}$. Conversely, an analytic function $g'(P')$ can be defined by

$$g'(P') = g(f^{-1}(P')), \qquad P' \in \mathfrak{R}'.$$

Therefore, we obtain a one-to-one correspondence

$$g \leftrightarrow g'$$

between $K(\mathfrak{R})$ and $K(\mathfrak{R}')$. It is clear from the definition of this correspondence that this correspondence is an isomorphism over k.

Next let us consider real-valued functions on $\mathfrak{R}$. Let f be a real-valued function on $\mathfrak{R}$, where f is allowed to take values $\pm\infty$. Let $t = x + iy$ be a locally uniformizing variable in a neighborhood U of P_0. Then in a neighborhood of P_0

$$f(P) = f(t) = f(x + iy) = F(x, y)$$

defines a real-valued function $F(x, y)$.

DEFINITION 3.9. When the function $F(x, y)$ is n-times continuously differentiable at the origin $P_0 = (0, 0)$, $f(P)$ is called an n-times continuously differentiable function at P_0. We write $C_n(U)$ for the totality of n-times continuously differentiable functions at each point in U.

If we replace t by another locally uniformizing variable t',

$$t' = x' + iy',$$

then Theorem 3.1 implies that

$$x' = x'(x, y) \quad \text{and} \quad y' = y'(x, y)$$

have derivatives of all order with respect to (x, y). Similarly, we have the same for $x = x(x', y')$ and $y = y(x', y')$. Hence, the property that $f(P)$ is n-times continuously differentiable at P_0 is defined independently of the choice of a locally uniformizing variable.

Furthermore, if U is an analytical domain and (x, y) are analytical coordinates in U, then the statement $f(P) \in C_n(U)$ means that $f(P) = F(x, y)$ is n-times continuously differentiable in U with respect to (x, y).

For an arbitrary domain U in $\mathfrak{R}$, we have

$$C_1(U) \supseteq C_2(U) \supseteq \cdots.$$

Furthermore, we also have the set $C_0(U)$, the collection of all continuous functions on U, which contains $C_1(U)$. However, we will need an even wider class of functions, that of measurable functions on U.

DEFINITION 3.10. Let $f(P)$ be a function (either real valued or complex valued) in a domain U in an R-surface $\mathfrak{R}$. Let U_1 be any analytical domain contained in U and let (x, y) be analytical coordinates in U_1. If for any such U_1, the function

$$f(P) = F(x, y)$$

is a measurable function in the sense of Lebesgue with respect to (x, y) on U_1, then f is said to be measurable in U. Moreover, if the characteristic function of a subset M in U,

$$\chi_M(P) = \begin{cases} 1, & P \in M, \\ 0, & P \notin M \end{cases}$$

is measurable, then M is said to be a measurable subset of U. In particular, if χ_M always takes value 0 on U_1 except on a set of measure zero, M is said to be a set of measure 0.

With those definitions, one can show that the sum and the product of measurable functions are measurable and that continuous functions are measurable. The sum and the product of measurable sets are measurable, and open sets and closed sets are measurable. We do not distinguish two measurable functions that take the same value except on a set of measure 0. Therefore, a measurable function $f(P)$ on U need not be defined on a set M of measure zero. Note also that, if continuous functions f_1 and f_2 on U are the same function in the above sense, then f_1 and f_2 take the same value at every point without exception.

§3. Differentials and integrals on Riemann surfaces

One can define the notion of differentials on an R-surface $\mathfrak{R}$ as we defined H-differentials in §5 of the previous chapter. Let $z = f(P)$ and $w = g(P)$ be analytic functions on $\mathfrak{R}$. For an arbitrary point P and a locally uniformizing variable t at P, $w(\partial z/\partial t)$ has a power series expansion in t in a neighborhood of P since f and g have power series expansions in t. We say the pairs (w, z) and (w_1, z_1) are equivalent if the t-expansion of $w(\partial z/\partial t)$ coincides with the t-expansion of $w_1(\partial z_1/\partial t)$ for an arbitrary point P and an arbitrary locally uniformizing variable t. The class of equivalent pairs of analytic functions is said to be a *differential* on $\mathfrak{R}$. The differential determined by (w, z) is denoted by $w\,dz$.

As in Chapter 2, $w\,dz$ may be regarded as the product of the function w and the differential $dz = 1 \cdot dz$.

We can similarly define differentials on a domain U in R by considering analytic functions $z = f(P)$ and $w = g(P)$ in U.

Let z be a nonconstant analytic function on $\mathfrak{R}$, whose existence will be proved in the following section. From Theorem 3.3, dz/dt is a nonzero power series in t. Hence, for an arbitrary differential $w'\,dz'$,

$$w'\frac{dz'}{dt} \Big/ \frac{dz}{dt} \tag{3.1}$$

also has a power series expansion in t. For another locally uniformizing variable t' at P, we have

$$\frac{dz'}{dt'} = \frac{dz'}{dt} \cdot \frac{dt}{dt'} \quad \text{and} \quad \frac{dz}{dt'} = \frac{dz}{dt} \cdot \frac{dt}{dt'}.$$

Therefore, the value of (3.1) is determined only by P, i.e., independent of the choice of a locally uniformizing variable. That is, (3.1) determines an analytic function $w = g(P)$ on $\mathfrak{R}$. Furthermore,

$$w' \frac{dz'}{dt} = w \frac{dz}{dt}$$

implies $w'\,dz' = w\,dz$. Hence, an arbitrary differential on $\mathfrak{R}$ may be written as $w\,dz$ using the above z. For a fixed dz, such an expression as $w\,dz$ for a differential is unique. Then define the sum of differentials as

$$w_1\,dz + w_2\,dz = (w_1 + w_2)\,dz.$$

With this definition, the totality of all the differentials on $\mathfrak{R}$ forms a $K(\mathfrak{R})$-module of dimension one. The differential $0 \cdot dz = 0$ is called the zero differential. For another dz', which is not the zero differential, an arbitrary differential can be written uniquely as $w'\,dz'$ for some $w' \in K(\mathfrak{R})$. Then we have $w_1'\,dz + w_2'\,dz' = (w_1' + w_2')\,dz'$ as in (3.2). See the paragraph preceding Definition 2.15.

Let

$$w \frac{dz}{dt} = \sum b_n t^n \tag{3.3}$$

be the t-expansion of $w\,dz$ at a point P. Here the power series on the right-hand side contains at most finitely many negative exponents. The least exponent n of t with nonzero coefficient is denoted by

$$v_P(w\,dz).$$

This integer is called the order of the differential $w\,dz$ at P. The notions of a pole and a zero can be defined similarly as for functions. For another locally uniformizing variable t' at P, we have from Theorem 3.1

$$t' = a_1 t + a_2 t^2 + \cdots, \qquad \frac{dt'}{dt} = a_1 + 2a_2 t + \cdots, \qquad a_1 \neq 0.$$

Therefore, it follows from

$$w \frac{dz}{dt} = w \frac{dz}{dt'} \cdot \frac{dt'}{dt}$$

that the smallest exponent of t' with nonzero coefficient in $w\,dz/dt'$ coincides with the above n. That is, $v_P(w\,dz)$ is defined independently of the choice of a locally uniformizing variable at P. Moreover, the coefficient b_{-1} of t^{-1} in (3.3) does not depend upon the choice of t as was shown in

Theorem 2.16. We call b_{-1} the residue of the differential $w\,dz$ at P. The residue is denoted by

$$\mathrm{Res}_P(w\,dz).$$

See Definition 2.12.

In order to introduce the notion of an integral of a differential $w\,dz$, we first speak of a path on an R-space $\mathfrak{R}$. That is, a path is a continuous curve: the image γ of a continuous map P from the closed interval $[a, b]$ to $\mathfrak{R}$, i.e.,

$$\gamma = \{P(s)\,;\, a \leq s \leq b\}. \tag{3.4}$$

The points $P(a)$ and $P(b)$ are called the initial point and the end point of γ, respectively. In particular, γ is called a closed path if $P(a) = P(b)$. Furthermore, if $P(s) = P_0$ for all s, then γ is called a constant path. When the interval $[a, b]$ is mapped onto the closed interval $[a', b']$ by an order preserving homeomorphism ψ, then we regard the path (3.4) to be the same as

$$\{Q(s')\,;\, a' \leq s' \leq b',\ Q(\psi(s)) = P(s)\}.$$

That is to say, they represent the same path with different parameters. Among infinitely many parameters for γ, we may choose $[0, 1]$ for the closed interval.

For the path γ defined by (3.4), the path

$$\{P'(s) = P(-s)\,;\, -b \leq s \leq -a\}$$

is called the inverse path of γ, denoted by γ^{-1}. Let

$$\gamma_1 = \{P_1(s)\,;\, a_1 \leq s \leq b_1\} \quad \text{and} \quad \gamma_2 = \{P_2(s)\,;\, a_2 \leq s \leq b_2\}$$

be such that

$$P_1(b_1) = P_2(a_2).$$

Then define $P_3(s)$ by

$$P_3(s) = P_1(s), \qquad a_1 \leq s \leq b_1,$$

and

$$P_3(s) = P_2(s - b_1 + a_2), \qquad b_1 \leq s \leq b_1 + b_2 - a_2.$$

The path $\gamma_3 = \{P_3(s)\,;\, a_1 \leq s \leq b_1 + b_2 - a_2\}$ is called the product of γ_1 and γ_2, and is denoted by

$$\gamma_3 = \gamma_2\gamma_1.$$

Note that one can show the independency of the choice of a parameter for the definitions of the inverse and product. When $\gamma_2\gamma_1$ and $\gamma_3\gamma_2$ are defined, we have

$$\gamma_3(\gamma_2\gamma_1) = (\gamma_3\gamma_2)\gamma_1.$$

Hence, we are allowed to write the product without parentheses as

$$\gamma_r \cdots \gamma_2\gamma_1,$$

when defined.

Consider a continuous deformation from one path to the other keeping fixed the initial and terminal points. More precisely, let γ and γ' be paths on $\mathfrak{R}$. Then γ and γ' are called *homotopic*, denoted $\gamma \approx \gamma'$, if there exists a continuous map ψ from a square $\{(x, y); 0 \le x \le 1, 0 \le y \le 1\}$ to $\mathfrak{R}$ such that

$$\begin{cases} \psi(x, 0) = P(0) = P'(0), & \psi(x, 1) = P(1) = P'(1), \\ \gamma = \{\psi(0, y); 0 \le y \le 1\}, & \gamma' = \{\psi(1, y);\ 0 \le y \le 1\}. \end{cases} \tag{3.5}$$

The homotopic relation is an equivalence relation. We classify all the paths on $\mathfrak{R}$ into homotopic classes of paths. Moreover, $\gamma \approx \gamma'$ implies $\gamma^{-1} \approx \gamma'^{-1}$, and we also have $\gamma\gamma^{-1} \approx \gamma_0$, the constant path. When $\gamma_2\gamma_1$ is defined, $\gamma_1 \approx \gamma_1'$ and $\gamma_2 \approx \gamma_2'$ imply $\gamma_2\gamma_1 \approx \gamma_2'\gamma_1'$. Therefore, the product of classes $\{\gamma_1\}$ and $\{\gamma_2\}$,

$$\{\gamma_2\}\{\gamma_1\} = \{\gamma_2\gamma_1\},$$

is well defined. In particular, the set of closed paths with initial point P_0 forms a group under the product defined above. This group, denoted $G(P_0)$, is called the *fundamental group.*

The preceding definitions and properties apply to any topological space (in particular, a manifold). But in the case of an R-surface, there exists a particular path as follows. Suppose a path γ, defined by (3.4), is contained in an analytical domain U of $\mathfrak{R}$. Let $t = x + iy$ be an analytical variable in U, and let

$$P(s) = t(s) = x(s) + iy(s).$$

If the continuous functions $x(s)$ and $y(s)$ are real analytic functions of s such that their derivatives do not vanish simultaneously, then γ is called an *analytic path in the strict sense.* If γ is also contained in another analytical domain U', then γ' is still an analytic curve in the above sense for U' and an analytical variable t' in U'. This is because the one-to-one map between t and t' is conformal. Therefore, the above definition of γ is independent of the choice of U.

In general, a product of finitely many analytic curves in the strict sense

$$\gamma = \gamma_r \cdots \gamma_2\gamma_1$$

is called an *analytic path.* For an analytic path we will consider only the parameters $x(s)$ and $y(s)$ that are regular on each interval.

Let γ' be an arbitrary path on $\mathfrak{R}$. Then, one can choose finitely many points $P_1, \dots, P_m$ on γ' and locally uniformizing variables t_i at P_i, $i = 1, \dots, m$, so that the neighborhoods $U_i = \{P; |t_i(P)| < 1\}$ of P_i cover γ'. Then decompose γ' as

$$\gamma' = \gamma_m' \cdots \gamma_2'\gamma_1'$$

such that $\gamma_i' \subset U_i$, $i = 1, \dots, m$, holds. For the initial point t_i^0 and the end point t_i^1 of γ_i',

$$P_i(s) = t_i(s) = t_i^0 + s(t_i^1 - t_i^0), \qquad 0 \le s \le 1,$$

determines an analytic path γ_i in the strict sense. Notice that γ_i' may be deformed continuously to γ_i in U_i. Hence,

$$\gamma_i' \approx \gamma_i, \qquad i = 1, \dots, m.$$

Consequently,

$$\gamma = \gamma_m \cdots \gamma_2\gamma_1 \approx \gamma' = \gamma_m' \cdots \gamma_2'\gamma_1'.$$

That is, we have the following lemma.

LEMMA 3.1. *For an arbitrary path γ' on an R-surface $\mathfrak{R}$, there exists an analytic path γ which is homotopic to γ'.*

In the above proof, we approximated each γ_i' by a straight line for t_i. We could have approximated it better with segments. That is, for a given path γ', there exists an analytic path which approximates γ' as closely as one wishes.

Let γ be an arbitrary analytic path on $\mathfrak{R}$, let $w\,dz$ be a differential that is regular at each point on γ. Then, we define the integral along γ, denoted by

$$\int_\gamma w\,dz,$$

as follows. In the case when γ is an analytic path in the strict sense, given as in (3.4), define

$$\int_\gamma w\,dz = \int_a^b w\frac{dz}{ds}\,ds, \quad \text{where } z = z(P(s)) \text{ and } w = w(P(s)).$$

If we regard γ as a curve in the t-plane, then the above integral is nothing but the line integral

$$\int_\gamma w\frac{dz}{dt}\,dt.$$

This last expression shows that $\int_\gamma w\,d\bar{z}$ is independent of the choice of parameter, and furthermore, we have

$$\int_\gamma w\,dz = \int_{\gamma_1} w\,dz + \int_{\gamma_2} w\,dz \tag{3.6}$$

for the product $\gamma = \gamma_2\gamma_1$. In general, if γ is the product $\gamma = \gamma_m \cdots \gamma_1$ of paths that are analytic in the strict sense, then we define

$$\int_\gamma w\,dz = \int_{\gamma_1} w\,dz + \cdots + \int_{\gamma_m} w\,dz. \tag{3.7}$$

Let $\gamma = \gamma_n' \cdots \gamma_1'$ be another decomposition of γ, and then let $\gamma = \gamma_r'' \cdots \gamma_1''$ be the decomposition obtained from initial points and end points of both γ_i

and γ'_j, $1 \le i \le m$, $1 \le j \le n$. Since γ_i is a product of γ''_k, we have from (3.6) that

$$\sum_{i=1}^{m} \int_{\gamma_i} w\,dz = \sum_{k=1}^{r} \int_{\gamma''_k} w\,dz.$$

Similarly, we have

$$\sum_{j=1}^{n} \int_{\gamma'_j} w\,dz = \sum_{k=1}^{r} \int_{\gamma''_k} w\,dz.$$

Therefore,

$$\sum_{i=1}^{m} \int_{\gamma_i} w\,dz = \sum_{j=1}^{n} \int_{\gamma'_j} w\,dz$$

holds, which indicates that definition (3.7) is independent of the decomposition of γ.

The integral of $w\,dz$, as is defined above, satisfies the usual properties of the line integral on the complex plane. For example, one can verify

$$\int_{\gamma_2\gamma_1} w\,dz = \int_{\gamma_1} w\,dz + \int_{\gamma_2} w\,dz, \tag{3.8}$$

and

$$\int_{\gamma^{-1}} w\,dz = -\int_{\gamma} w\,dz. \tag{3.9}$$

Note also that for an arbitrary point P_0 on $\mathfrak{R}$ the residue of $w\,dz$ at P_0 is given by

$$\operatorname{Res}_{P_0}(w\,dz) = \frac{1}{2\pi i}\int_{\gamma} w\,dz,$$

where γ is a sufficiently small simple curve going around P counterclockwise. This formula suggests that the residue does not depend upon the choice of a locally uniformizing variable.

THEOREM 3.6. *If the residue of $w\,dz$ is always 0 at every point in the domain U in $\mathfrak{R}$, then for homotopic analytic paths γ and γ' in U on which $w\,dz$ is regular, we have*

$$\int_{\gamma} w\,dz = \int_{\gamma'} w\,dz. \tag{3.10}$$

PROOF. Given γ and γ' there exists a continuous map $\psi(x, y)$ from the square $T = \{(x, y);\ 0 \le x \le 1,\ 0 \le y \le 1\}$ to U satisfying (3.5). For an arbitrary natural number N, subdivide T into N^2 squares

$$T_{mn} = \left\{(x, y);\ \frac{m-1}{N} \le x \le \frac{m}{N},\ \frac{n-1}{N} \le y \le \frac{n}{N}\right\},$$

where $n, m = 1, 2, \ldots, N$. One can take N large enough so that $\psi(T_{mn})$ is always contained in a neighborhood $U_P = \{Q;\ |t(Q)| < \varepsilon\}$ of a point P

in U. Note that $t(Q)$ is a locally uniformizing variable at P. Let γ'_{mn} be the image of the perimeter of T_{mn} under ψ, i.e., $\gamma'_{mn} = \gamma'_4\gamma'_3\gamma'_2\gamma'_1$, where

$$\gamma'_1 = \left\{\psi\left(\left(x, \frac{n-1}{N}\right)\right); \frac{m-1}{N} \le x \le \frac{m}{N}\right\},$$

$$\gamma'_2 = \left\{\psi\left(\left(\frac{m}{N}, y\right)\right); \frac{n-1}{N} \le y \le \frac{n}{N}\right\},$$

$$\gamma'_3 = \left\{\psi\left(\left(-x, \frac{n}{N}\right)\right); -\frac{m}{N} \le x \le -\frac{m-1}{N}\right\},$$

$$\gamma'_4 = \left\{\psi\left(\left(\frac{m-1}{N}, -y\right)\right); -\frac{n}{N} \le y \le -\frac{n-1}{N}\right\}.$$

The integral of $w\,dz$ along γ'_{mn} may be computed as the line integral

$$\int_{\gamma'_{mn}} w\,dz = \int_{\gamma'_{mn}} w\frac{dz}{dt}\,dt \tag{3.11}$$

in the t-plane. From our assumption on the residue, the residue theorem for analytic functions implies that the integrals in (3.11) have value 0. When one considers the sum of the integrals in (3.11) over all the T_{mn}, one can observe that the integrals over the images of the segments inside T cancel out in pairs. Consequently, we obtain

$$\int_{\psi(0,0)}^{\psi(1,0)} + \int_{\psi(1,0)}^{\psi(1,1)} + \int_{\psi(1,1)}^{\psi(0,1)} + \int_{\psi(0,1)}^{\psi(0,0)} w\,dz = 0.$$

The first and the third terms on the left are 0, and the second and the fourth terms equal $\int_{\gamma'} w\,dz$ and $-\int_{\gamma} w\,dz$, respectively. This basically finishes the proof of (3.10). However, we still need to observe the following remark. On the compact set $\psi(T)$, $w\,dz$ has at most finitely many poles, and furthermore, any path can be approximated as accurately as needed by analytic paths. Therefore, there exists a net of analytic paths γ''_{mn}, $m, n = 1, 2, \ldots, N$, like γ_{mn} such that $w\,dz$ has no poles on γ''_{mn}. Repeating the same argument for γ''_{mn}, we complete the proof of (3.10).

In the above theorem, in the case where U is simply connected, i.e., any two paths with the same initial and end points are homotopic in U, let U' be the domain obtained by removing all the poles of $w\,dz$ from U. Let P_0 be a fixed point in U', and let P be an arbitrary point in U'. For a path γ from P_0 to P in U', define

$$f(P) = \int_\gamma w\,dz.$$

Then $f(P)$ is determined by P only, independently of the choice of γ by the preceding theorem. That is, $f(P)$ is a single-valued function on U'. Let t be a locally uniformizing variable at P, and let

$$w\frac{dz}{dt} = a_0 + a_1 t + a_2 t^2 + \cdots$$

be the t-expansion in a neighborhood of P. Then f has expansion

$$f = f(Q) = c + a_0 t + \frac{a_1}{2} t^2 + \frac{a_3}{3} t^3 + \cdots$$

in a neighborhood of P. Therefore, f is a regular analytic function on U'. Next let P_1 be a pole of $w\,dz$ in U. For a locally uniformizing variable t_1 at P_1, let

$$w\frac{dz}{dt_1} = b_{-n}t_1^{-n} + \cdots + b_{-2}t_1^{-2} + b_0 + b_1 t_1 + b_2 t_1^2 + \cdots$$

be the expansion of $w\,dz$. As before, we have

$$f = f(Q_1) = -\frac{b_{-n}}{n-1}t_1^{-n+1} - \cdots - b_{-2}t_1^{-1} + c + b_0 t_1 + \frac{b_1}{2}t_1^2 + \frac{b_2}{3}t_1^3 + \cdots$$

in a neighborhood of P_1. If we define $f(P_1) = \infty$, then $f(P)$ becomes an analytic function in U. For a locally uniformizing variable t at each point P in U, we have

$$\frac{dw'}{dt} = w\frac{dz}{dt}$$

for $w' = f(P)$. Hence, we obtain the following theorem.

THEOREM 3.7. *For any differential $w\,dz$ whose residues are 0 in a simply connected domain U in an R-surface $\mathfrak{R}$, there exists an analytic function w' on U such that at each point P of U*

$$\frac{dw'}{dt} = w\frac{dz}{dt} \tag{3.12}$$

for a locally uniformizing variable t at P.

For any domain U in $\mathfrak{R}$, not necessarily simply connected, if an analytic function w' satisfies (3.12), then w' is said to be an *integrating function* of the differential $w\,dz$ in U. If w' is an integrating function of $w\,dz$, all other integrating functions are obtained by adding constant number c to w'. Hence, by Theorem 3.7, a differential having residues 0 in a simply connected domain possesses an integrating function. We will need this result later.

§4. Existence of analytic functions

We have described properties of analytic functions and differentials on a Riemann surface. In this section we prove that nontrivial analytic functions and differentials indeed exist on a Riemann surface. As we noted in our introduction, we will give a proof that does not use the Dirichlet principle. We will follow H. Weyl, *Method of orthogonal projections in potential theory*, Duke Math. J. 7 (1940), and K. Kodaira, *Über die harmonischen Tensorfelder in Riemannschen Mannigfaltigkeiten*, II, Proc. Imp. Acad. Tokyo **20** (1944).

Let $w\,dz$ be a differential on an R-surface $\mathfrak{R}$, and let t be an analytical variable in a domain containing a point P on $\mathfrak{R}$. By separating the real part and the imaginary part of $w\,dz/dt$ as follows:

$$w\frac{dz}{dt} = v_1(P, t) - iv_2(P, t), \tag{4.1}$$

we have the real-valued functions

$$v_1(P, t) \quad \text{and} \quad v_2(P, t) \tag{4.2}$$

for P and t. If $t_1 = x_1 + iy_1$ and $t_2 = x_2 + iy_2$ are two analytical variables at the same point P, then

$$v_1(P, t_1) - iv_2(P, t_1) = w\frac{dz}{dt_1} = w\frac{dz}{dt_2}\frac{dt_2}{dt_1} = (v_1(P, t_2) - iv_2(P, t_2))\frac{dt_2}{dt_1},$$

and

$$\frac{dt_2}{dt_1} = \frac{\partial x_2}{\partial x_1} + i\frac{\partial y_2}{\partial x_1} = \frac{1}{i}\frac{\partial x_2}{\partial y_1} + \frac{\partial y_2}{\partial y_1}$$

hold. Therefore, we obtain

$$\begin{cases} v_1(P, t_1) = v_1(P, t_2)\dfrac{\partial x_2}{\partial x_1} + v_2(P, t_2)\dfrac{\partial y_2}{\partial x_1}, \\ v_2(P, t_1) = v_1(P, t_2)\dfrac{\partial x_2}{\partial y_1} + v_2(P, t_2)\dfrac{\partial y_2}{\partial y_1}. \end{cases} \tag{4.3}$$

These are the transformation formulas between $(v_1(P, t_1), v_2(P, t_1))$ and $(v_1(P, t_2), v_2(P, t_2))$. Next, we will construct a differential and an analytic function from a pair (4.2) of real-valued functions of P and t satisfying (4.3).

DEFINITION 3.11. Consider a pair of real-valued functions

$$v = (v_1(P, t), v_2(P, t)) \tag{4.4}$$

defined for a point P in a domain U of $\mathfrak{R}$ and analytical coordinates $t = x + iy$ at P. Then v is said to be a *vector field* if, for analytical coordinates $t_1 = x_1 + iy_1$ and $t_2 = x_2 + iy_2$ at P, the transformation formulas (4.3) hold. For a vector field $v = (v_1(P, t), v_2(P, t))$, we call $v_1(P, t)$ and $v_2(P, t)$ the x-coordinate and the y-coordinate of v at P, respectively.

As one may observe from the above definition, the pair

$$v^* = (v_2(P, t), -v_1(P, t))$$

is also a vector field when $v = (v_1(P, t), v_2(P, t))$ is a vector field. Notice also that if the x-coordinate and the y-coordinate of v at P are both 0, then for another analytical variable $t' = x' + iy'$ the x'-coordinate and the y'-coordinate of v at P are 0. In this case we say that the vector field v is 0 at P. A vector field v is called the zero vector field in U, denoted by 0, when v is 0 at all the points in U.

Let v be a vector field in U, and let U' be an analytical domain contained in U. Since analytical coordinates $t = x + iy$ in U' are analytical coordinates at every point P in U', we obtain real-valued functions

$$v_1(t) = v_1(P, t) \quad \text{and} \quad v_2(t) = v_2(P, t) \tag{4.5}$$

of t in U', therefore, $v_1(t) = v_1(x, y)$ and $v_2(t) = v_2(x, y)$ are well-defined functions of (x, y) in U'. A vector field v is said to be measurable when the above functions $v_1(x, y)$ and $v_2(x, y)$ are measurable with respect to (x, y) for an arbitrary analytical domain U' contained in U. For an arbitrary analytical domain U', if two measurable vector fields $v = (v_1, v_2)$ and $v' = (v_1', v_2')$ satisfy

$$v_1(x, y) = v_1'(x, y) \quad \text{and} \quad v_2(x, y) = v_2'(x, y)$$

for all (x, y) except on a set of measure 0, the vector fields v and v' are said to coincide everywhere except on a set of measure 0. As is the case for functions, two vector fields are regarded as the same when they differ only on a set of measure 0. We can also define a continuous vector field and an n-times continuously differentiable vector field as follows. When $v_1(t)$ and $v_2(t)$ in (4.5) are n-times continuously differentiable as functions of (x, y), the vector field v is said to be n-times continuously differentiable. For example, if $f(P)$ is an n-times continuously differentiable real-valued function in U, then

$$v_1(P, t) = \frac{\partial f}{\partial x} \quad \text{and} \quad v_2(P, t) = \frac{\partial f}{\partial y},$$

for $t = x + iy$ and $f(P) = f(t)$, satisfy (4.3), and we obtain an $(n-1)$-times continuously differentiable vector field. This vector field is called the gradient of the function f, denoted by

$$\operatorname{grad} f = \left(\frac{\partial f}{\partial x}, \frac{\partial f}{\partial y}\right).$$

Similarly, we obtain another vector field from f,

$$\operatorname{grad}^* f = \left(\frac{\partial f}{\partial y}, -\frac{\partial f}{\partial x}\right).$$

Let v be a continuously differentiable vector field in U. For analytical coordinates $t = x + iy$ at P in U, define

$$\operatorname{div}_t v = \frac{\partial v_1(P, t)}{\partial x} + \frac{\partial v_2(P, t)}{\partial y}.$$

Then $\operatorname{div}_t v$ depends upon P and t. However, for analytical variables t_1 and t_2 at P, we have

$$\operatorname{div}_{t_1} v = \operatorname{div}_{t_2} v \cdot \left|\frac{dt_2}{dt_1}\right|^2. \tag{4.6}$$

Therefore, the condition as to whether $\operatorname{div}_t v = 0$ or not, does not depend upon the choice of t, and is determined by v. We may write

$$(\operatorname{div} v)_P = 0.$$

When $\operatorname{div} v = 0$ at every point in U, v is called a solenoidal vector field. Note also that similar results hold for

$$\operatorname{div}_t^* v = \operatorname{div}_t v^* = \frac{\partial v_2(P, t)}{\partial x} - \frac{\partial v_1(P, t)}{\partial y}.$$

If U is an analytical domain in $\mathfrak{R}$ and $t = x + iy$ is an analytical variable for U, then for a twice continuously differentiable function f and vector field v, one can verify the following formulae:

$$\operatorname{div} \operatorname{grad}^* f = \operatorname{div}^* \operatorname{grad} f = 0, \tag{4.7}$$

$$\operatorname{div} \operatorname{grad} f = -\operatorname{div}^* \operatorname{grad}^* f = \Delta f = \frac{\partial^2 f}{\partial x^2} + \frac{\partial^2 f}{\partial y^2}, \tag{4.8}$$

$$\operatorname{grad} \operatorname{div} v - \operatorname{grad}^* \operatorname{div}^* v = \Delta v = \left(\frac{\partial^2 v_1}{\partial x^2} + \frac{\partial^2 v_1}{\partial y^2}, \frac{\partial^2 v_2}{\partial x^2} + \frac{\partial^2 v_2}{\partial y^2} \right). \tag{4.9}$$

We also have the following well-known lemma.

LEMMA 3.2 (Gauss and Green Theorem). *Let L be a simple closed analytic path in U, and let G be the domain in U which L encloses. For twice continuously differentiable functions f and g, and a vector field v, we have*

$$\begin{aligned} \iint_G \operatorname{div} v \, dx \, dy &= \int_L (v_1 \, dy - v_2 \, dx), \\ \iint_G \operatorname{div}^* v \, dx \, dy &= \int_L (v_1 \, dx + v_2 \, dy). \end{aligned} \tag{4.10}$$

$$\iint_G (f \cdot \Delta g - g \cdot \Delta f) \, dx \, dy = \int_L \left(f \frac{\partial g}{\partial r} - g \frac{\partial f}{\partial r} \right) ds, \tag{4.11}$$

where $\int_L$ indicates the line integral for the positive direction, ds denotes the infinitesimal length of L, and $\partial / \partial r$ is the outer normal derivative at a point on L.

Formulae (4.10) and (4.11) imply

LEMMA 3.3. *Let $v = (v_1, v_2)$ be a solenoidal vector field in U, and let f be an arbitrary continuously differentiable function, and let G and L be as in Lemma 3.2. Then we have*

$$\iint_G \left(v_1 \frac{\partial f}{\partial x} + v_2 \frac{\partial f}{\partial y} \right) dx \, dy = \int_L f(v_1 \, dy - v_2 \, dx). \tag{4.12}$$

Similarly, for $\operatorname{div}^* v = 0$ we have

$$\iint_G \left(v_2 \frac{\partial f}{\partial x} - v_1 \frac{\partial f}{\partial y} \right) dx \, dy = \int_L f(v_1 \, dx + v_2 \, dy). \tag{4.13}$$

Now let $v = (v_1, v_2)$ be a measurable vector field on $\mathfrak{R}$, and let E be a measurable set in an analytical domain U. Define

$$\|v\|_E^2 = \iint_E (v_1(P, t)^2 + v_2(P, t)^2)\, dx\, dy \tag{4.14}$$

for an analytical variable $t = x + iy$ in U. Let U' be an arbitrary analytical domain containing E. Then for an analytical variable $t' = x' + iy'$ in U', we have the following from transformation formulae (4.3):

$$\begin{aligned}(v_1(P, t)^2 + v_2(P, t)^2)\, dx\, dy &= (v_1(P, t')^2 + v_2(P, t')^2)\left|\frac{dt'}{dt}\right|^2 dx\, dy \\ &= (v_1(P, t')^2 + v_2(P, t')^2)\, dx'\, dy',\end{aligned}$$

$$\|v\|_E^2 = \iint_E (v_1(P, t')^2 + v_2(P, t')^2)\, dx'\, dy'.$$

Therefore, definition (4.14) for $\|v\|_E$ does not depend upon the choice of U containing E. Furthermore, if E is the sum of mutually disjoint measurable subsets E_1 and E_2, then we have

$$\|v\|_E^2 = \|v\|_{E_1}^2 + \|v\|_{E_2}^2. \tag{4.15}$$

Such an E, which is a measurable subset in an analytical domain U, is called a fundamental set in $\mathfrak{R}$. We denote the collection of all subsets that can be expressed as a finite sum of fundamental sets by

$$\mathfrak{S} = \{M\}.$$

Note that the sum, the difference, and the intersection of two sets in $\mathfrak{S}$ also belong to $\mathfrak{S}$. Moreover, any M in $\mathfrak{S}$ can be decomposed as a finite sum

$$M = E_1 + \cdots + E_n,$$

where the E_i are mutually disjoint fundamental sets. Since the above decomposition is not unique, for another decomposition

$$M = E_1' + \cdots + E_m',$$

let $E_{ij} = E_i \cap E_j'$. Then $M = \sum_{i=1}^n \sum_{j=1}^m E_{ij}$. Hence, for an arbitrary measurable vector field on $\mathfrak{R}$, we have from (4.15)

$$\|v\|_{E_i}^2 = \sum_{j=1}^m \|v\|_{E_{ij}}^2 \quad \text{and} \quad \|v\|_{E_j'}^2 = \sum_{i=1}^n \|v\|_{E_{ij}}^2.$$

Therefore,

$$\sum_{i=1}^n \|v\|_{E_i}^2 = \sum_{j=1}^m \|v\|_{E_j'}^2.$$

We define $\|v\|_M^2$ to be this common value. That is, $\|v\|_M^2$ is determined independently of the decomposition of M, i.e., $\|v\|_M^2$ is determined by M only.

DEFINITION 3.12. We define the *norm* of a measurable vector field v by

$$\|v\| = \sup_M \|v\|_M,$$

where M runs through the family $\mathfrak{S}$.

Notice that $\|v\|$ is a nonnegative real number, or possibly $+\infty$. The norm $\|v\|$ is 0 if and only if v is 0 everywhere in $\mathfrak{R}$ except on a set of measure 0. That is, v is the zero vector field on $\mathfrak{R}$.

In general, for arbitrary vector fields $v = (v_1, v_2)$ and $v' = (v_1', v_2')$ on $\mathfrak{R}$, we have a vector field on $\mathfrak{R}$

$$(av_1(P, t) + bv_1'(P, t), av_2(P, t) + bv_2'(P, t)),$$

where a and b are real numbers. We write this vector field as

$$av + bv'.$$

Since $|\alpha+\beta|^2 \leq 2(|\alpha|^2+|\beta|^2)$ for any two real numbers α and β, we obtain from definition (4.14) that

$$\|av + bv'\|_E^2 \leq 2|a|^2\|v\|_E^2 + 2|b|^2\|v'\|_E^2$$

for any fundamental set E. Therefore, for an arbitrary $M \in \mathfrak{S}$

$$\|av + bv'\|_M^2 \leq 2|a|^2\|v\|_M^2 + 2|b|^2\|v'\|_M^2.$$

Consequently,

$$\|av + bv'\|^2 \leq 2|a|^2\|v\|^2 + 2|b|^2\|v'\|^2. \tag{4.16}$$

LEMMA 3.4. *For $\|v\| < \infty$, there exists a countable family of fundamental sets $E_1, E_2, \ldots$ such that*

$$\|v\| = \lim_{n\to\infty} \|v\|_{M_n}, \qquad M_n = \sum_{i=1}^{n} E_i, \tag{4.17}$$

and such that v is 0 everywhere outside $M_\infty = \sum_{i=1}^\infty E_i$, except on a set of measure 0.

PROOF. Since $\|v\| < \infty$, there exists M_n' in $\mathfrak{S}$ such that

$$\|v\| = \lim_{n\to\infty} \|v\|_{M_n'}.$$

Since each M_n' is a finite sum of fundamental sets, one can take all of the fundamental sets $E_1, E_2, \ldots$ such that (4.17) holds. Let E be a fundamental set outside $M_\infty = \sum_{i=1}^\infty E_i$. Then $E + M_n'$ also belongs to $\mathfrak{S}$. Therefore,

$$\|v\|^2 \geq \lim_{n\to\infty} \|v\|_{E+M_n'}^2 = \lim_{n\to\infty} (\|v\|_E^2 + \|v\|_{M_n'}^2) = \|v\|_E^2 + \|v\|^2$$

implies $\|v\|_E = 0$. Hence, v is 0 everywhere, except on a set of measure 0, outside M_∞.

LEMMA 3.5. *Let* $\|v\| < \infty$ *and* $\|v'\| < \infty$. *Then*

$$(v, v') = \frac{1}{4}(\|v+v'\|^2 - \|v-v'\|^2) \tag{4.18}$$

is finite, and we have

$$(v, v) = \|v\|^2, \tag{4.19}$$

$$(v, v') = (v', v), \tag{4.20}$$

$$(av + bv', v'') = a(v, v'') + b(v', v''), \tag{4.21}$$

where $\|v''\| < \infty$, *and* a *and* b *are arbitrary real numbers.*

PROOF. From (4.16), we have $\|v+v'\| < \infty$ and $\|v-v'\| < \infty$. Therefore, (v, v') is a finite real number. Equations (4.19) and (4.20) follow from $\|2v\|^2 = 4\|v\|^2$ and $\|-v\|^2 = \|v\|^2$. Next we will prove (4.21). Notice $\|av + bv'\| < \infty$ follows from (4.16). Then from Lemma 3.4, choose fundamental sets $E_1, E_2, \dots$ as follows. The vector fields v, v' and v'' are zero outside $M_\infty = \sum_{i=1}^{\infty} E_i$. For $M_n = \sum_{i=1}^{n} E_i$,

$$\begin{cases} \|av + bv' \pm v''\| = \lim_{n\to\infty} \|av + bv' \pm v''\|_{M_n}, \\ \|v \pm v''\| = \lim_{n\to\infty} \|v \pm v''\|_{M_n}, \\ \|v' \pm v''\| = \lim_{n\to\infty} \|v' \pm v''\|_{M_n}. \end{cases} \tag{4.22}$$

For a fundamental set E and $M \in \mathfrak{S}$, let

$$(v, v')_E = \frac{1}{4}(\|v+v'\|_E^2 - \|v-v'\|_E^2) = \iint_E (v_1 v_1' + v_2 v_2')\, dx\, dy,$$

$$(v, v')_M = \frac{1}{4}(\|v+v'\|_M^2 - \|v-v'\|_M^2).$$

Then over E, we have

$$(av + bv', v'')_E = a(v, v'')_E + b(v', v'')_E.$$

Therefore, we obtain over M

$$(av + bv', v'')_M = a(v, v'')_M + b(v, v'')_M.$$

Then (4.21) follows from (4.22) by letting $M = M_n$.

LEMMA 3.6. *Let* $v_1, v_2, \dots$ *be vector fields on* $\mathfrak{R}$ *such that* $\|v_i\| < \infty$. *If we have*

$$\|v_m - v_n\| \to 0 \quad \text{as } m, n \to \infty,$$

then there exists a vector field v *such that* $\|v\| < \infty$ *and*

$$\lim_{n\to\infty} \|v - v_n\| = 0.$$

PROOF. From Lemma 3.4, one can choose fundamental sets $E_1, E_2, \dots$ so that each v_n is 0 outside $M_\infty = \sum_{i=1}^{\infty} E_i$. We may assume that the E_i are mutually disjoint. Let v be the vector field that we need to construct. Let

v be 0 outside M_∞. Next, let U_l be an analytical domain containing E_l, and let $t_l = x_l + iy_l$ be an analytical variable in U_l. From our assumption,

$$\|v_m - v_n\|_{E_l}^2 = \iint_{E_l} \{(v_1^{(m)}(P, t_l) - v_1^{(n)}(P, t_l))^2 + (v_2^{(m)}(P, t_l) - v_2^{(n)}(P, t_l))^2\}\, dx_l\, dy_l$$

converges to 0 as m and n approach ∞. Therefore, one can find subsequences of $\{v_1^{(n)}(P, t_l)\}$ and $\{v_2^{(n)}(P, t_l)\}$ so that these subsequences converge to measurable functions $v_1(P, t_l)$ and $v_2(P, t_l)$, respectively, on E_l except on a set of measure 0. Therefore, we have

$$\lim_{n\to\infty} \iint_{E_l} \{(v_1^{(n)}(P, t_l) - v_1(P, t_l))^2 + (v_2^{(n)}(P, t_l) - v_2(P, t_l))^2\}\, dx_l\, dy_l = 0. \tag{4.23}$$

(That is, the function space L^2 over E_l is complete.) Now let $v_1(P, t_l)$ and $v_2(P, t_l)$ be the x_l-coordinate and the y_l-coordinate of v at P, respectively. Given any analytical coordinates $t' = x' + iy'$, we can choose the coordinates of v, $v_1(P, t')$ and $v_2(P, t')$, so that the transformation formulae (4.3) hold. Thus, we have obtained a measurable vector field v on $\mathfrak{R}$. We have from (4.23) that

$$\lim_{m\to\infty} \|v^{(m)} - v^{(n)}\|_{E_l}^2 = \|v - v^{(n)}\|_{E_l}^2.$$

For an arbitrary $\varepsilon > 0$, take n_0 sufficiently large so that

$$\|v^{(m)} - v^{(n)}\| < \varepsilon$$

for m and $n \geq n_0$. Then for such an n and an arbitrary natural number s, we have

$$\sum_{l=1}^{s} \|v - v^{(n)}\|_{E_l}^2 = \lim_{m\to\infty} \sum_{l=1}^{s} \|v^{(m)} - v^{(n)}\|_{E_l}^2 \leq \lim_{m\to\infty} \|v^{(m)} - v^{(n)}\|^2 \leq \varepsilon^2.$$

Therefore, as $s \to \infty$,

$$\|v - v^{(n)}\|^2 = \sum_{l=1}^{\infty} \|v - v^{(n)}\|_{E_l}^2 \leq \varepsilon^2.$$

Hence, we have $\|v\| < \infty$ from (4.16) and

$$\lim_{n\to\infty} \|v - v^{(n)}\| = 0.$$

Let $\mathfrak{H}$ be the set of all vector fields of finite norm on $\mathfrak{R}$. By (4.16), $\mathfrak{H}$ is a linear space over the field of real numbers. An inner product (v, v') is defined in $\mathfrak{H}$, by Lemma 3.5, with properties (4.19)–(4.21). Furthermore, Lemma 3.6 implies that $\mathfrak{H}$ is complete for the metric induced by the norm. That is, we have the following theorem.

THEOREM 3.8. *The set $\mathfrak{H}$ of all measurable vector fields of finite norm on a Riemann surface $\mathfrak{R}$ forms a real Hilbert space.*

Let Γ be the totality of twice continuously differentiable functions f on $\mathfrak{R}$ with the following property. For a neighborhood $U = \{Q;\, t(Q) < \varepsilon\}$ of a suitably chosen point P_0 on $\mathfrak{R}$ ($t(Q)$ is a locally uniformizing variable at P_0) and for some ε_1, $0 < \varepsilon_1 < \varepsilon$, f is 0 outside the set $\{Q;\, t(Q) < \varepsilon_1\}$. Notice that for f in Γ, $\operatorname{grad} f$ and $\operatorname{grad}^* f$ belong to $\mathfrak{H}$. Let $\mathfrak{G}$ be the closed linear subspace of $\mathfrak{H}$ spanned by these vector fields $\operatorname{grad} f$, i.e.,

$$\mathfrak{G} = [\operatorname{grad} f;\, f \in \Gamma],$$

similarly,

$$\mathfrak{G}^* = [\operatorname{grad}^* f;\, f \in \Gamma].$$

LEMMA 3.7. *The closed subspaces $\mathfrak{G}$ and $\mathfrak{G}^*$ are orthogonal, i.e., for arbitrary vector fields v and v' in $\mathfrak{G}$ and $\mathfrak{G}'$, respectively, we have*

$$(v\,,\, v') = 0.$$

PROOF. It is sufficient to prove

$$(\operatorname{grad} f\,,\, \operatorname{grad}^* g) = 0$$

for $f, g \in \Gamma$. Our assumption implies that for some point Q_0 and a locally uniformizing variable $t_1 = t_1(Q) = x + iy$ at Q_0, g, and therefore $\operatorname{grad}^* g$, are 0 outside $\{Q;\, |t_1(Q)| < \varepsilon_1'\}$. Therefore, we have

$$(\operatorname{grad} f\,,\, \operatorname{grad}^* g) = \iint_{|t_1|<\varepsilon_1'} \left(\frac{\partial f}{\partial x}\frac{\partial g}{\partial y} - \frac{\partial f}{\partial y}\frac{\partial g}{\partial x}\right) dx\, dy.$$

For $v = \operatorname{grad} f$, $\operatorname{div}^* v = \operatorname{div}^* \operatorname{grad} f = 0$ holds for $|t_1| < \varepsilon'$ from (4.7). Hence, Lemma 3.3 implies

$$(\operatorname{grad} f\,,\, \operatorname{grad}^* g) = \int_{|t_1|=\varepsilon_1'} g(v_1\, dy + v_2\, dx).$$

Since $g = 0$ holds on $|t_1| = \varepsilon_1'$, the right-hand side of the above equation becomes 0, which is what was to be proved.

Let the totality of vector fields in $\mathfrak{H}$ that are orthogonal to $\mathfrak{G}$ be

$$\mathfrak{M} = \mathfrak{H} \ominus \mathfrak{G}, \quad \text{i.e., } \mathfrak{H} = \mathfrak{G} \oplus \mathfrak{M}.$$

Similarly, let $\mathfrak{M}^*$ be the set of all vector fields that are orthogonal to $\mathfrak{G}^*$, i.e.,

$$\mathfrak{M}^* = \mathfrak{H} \ominus \mathfrak{G}^*, \qquad \mathfrak{H} = \mathfrak{G}^* \oplus \mathfrak{M}^*.$$

Here $\oplus$ denotes the orthogonal direct sum. From Lemma 3.7, we have $\mathfrak{M} \supseteq \mathfrak{G}^*$ and $\mathfrak{M}^* \supseteq \mathfrak{G}$. If we let

$$\mathfrak{E} = \mathfrak{M} \cap \mathfrak{M}^*,$$

then

$$\mathfrak{E} = \mathfrak{M} \ominus \mathfrak{G}^* = \mathfrak{M}^* \ominus \mathfrak{G},$$

$$\mathfrak{H} = \mathfrak{G} \oplus \mathfrak{M} = \mathfrak{G}^* \oplus \mathfrak{M}^* = \mathfrak{G} \oplus \mathfrak{G}^* \oplus \mathfrak{E}.$$

We now have the following important theorem about $\mathfrak{E}$.

THEOREM 3.9. *Any vector field in $\mathfrak{E}$ is continuously differentiable up to any order.*

PROOF. Let U be an arbitrary analytical domain in a Riemann surface $\mathfrak{R}$, and let t be an analytical variable for U. For t_0 in U, choose a sufficiently small $\varepsilon > 0$ such that $|t - t_0| < \varepsilon$ is contained in U. Then we will prove that given any vector field v in $\mathfrak{E}$ there exists a pair of functions

$$(v_1'(P, t), v_2'(P, t)) \tag{4.24}$$

which is continuously differentiable up to any order and coincides with

$$v = (v_1(P, t), v_2(P, t))$$

in $|t-t_0| < \varepsilon$ everywhere except on a set of measure zero. Suppose that this is proved. Then for each choice of t_0 and $\varepsilon > 0$, we obtain $(v_1'(P, t), v_2'(P, t))$ as above. For each choice of t_0 and $\varepsilon > 0$, $(v_1'(P, t), v_2'(P, t))$ coincides with $(v_1(P, t), v_2(P, t))$ everywhere except on a set of measure zero. Since v_1' and v_2' are continuous, they are the same function where their domains overlap. Thus, one obtains a pair of functions $(v_1'(P, t), v_2'(P, t))$ that coincides with $(v_1(P, t), v_2(P, t))$ everywhere in U except on a set of measure zero. Let U_1 and t_1 be another analytical domain and an analytical variable in U_1, respectively. Then similarly, one would obtain a pair of functions $(v_1'(P, t_1), v_2'(P, t_1))$ on U_1, continuously differentiable up to any order. Where U and U' intersect, the transformation formulae (4.3) for $(v_1(P, t), v_2(P, t))$ and $(v_1'(P, t_1), v_2'(P, t_1))$ hold everywhere except on a set of measure zero. Since all those functions are continuous, the equality must hold without exceptional points. Therefore,

$$v' = (v_1(P, t), v_2(P, t))$$

is a vector field on $\mathfrak{R}$ which coincides with v everywhere except on a set of measure zero. Thus, we may identify v with v' and v is continuously differentiable up to any order. Therefore, to prove Theorem 3.9, it is sufficient for us to prove the existence of (4.24) in $|t - t_0| < \varepsilon$.

Let $w = (w_1, w_2)$ be a vector field which is three times continuously differentiable and which is 0 outside $|t - t_0| < \varepsilon_1$, $0 < \varepsilon_1 < \varepsilon$. Then $\operatorname{div}_t w$ and $\operatorname{div}_t^* w$ belong to Γ. Therefore, we have

$$\operatorname{grad} \operatorname{div}_t w \in \mathfrak{G} \quad \text{and} \quad \operatorname{grad}^* \operatorname{div}_t^* w \in \mathfrak{G}^*.$$

Our assumption implies that v is orthogonal to both $\mathfrak{G}$ and $\mathfrak{G}^*$. From (4.9), we obtain

$$(v, \Delta w) = 0.$$

That is,

$$\iint_{|t-t_0|<\varepsilon} (v_1 \Delta w_1 + v_2 \Delta w_2)\, dx\, dy = 0, \quad \text{where } \Delta w_i = \frac{\partial^2 w_i}{\partial x^2} + \frac{\partial^2 w_i}{\partial y^2}.$$

Since we can let $w_1 = 0$ or $w_2 = 0$ in the above, the following lemma will complete the proof of Theorem 3.9.

Lemma 3.8. *Let U be the interior of the unit circle C, $|t| < 1$, in the t-plane where $t = x + iy$. Let $w = w(x, y)$ be an arbitrary function which is three times continuously differentiable in U and is 0 in a neighborhood of C. If a measurable real-valued function $v = v(x, y)$, with the property*

$$\iint_U |v^2|\,dx\,dy < \infty,$$

satisfies

$$\iint_U v\Delta w\,dx\,dy = 0, \tag{4.25}$$

then v coincides with a harmonic function $v' = v'(x, y)$ everywhere in U except on a set of measure 0.

Proof. We will give a proof in several steps, which is based on K. Kawada, Kōdai Math. Sem. Report, vol. 1, no. 3, Tokyo Institute of Technology.

(i) Let C_r be a circle of radius r with the center at the origin and let U_r be its interior. In general, for a measurable and integrable function $p(x, y)$ in $U = U_1$, we let

$$M_r p(x, y) = \frac{1}{\pi r^2}\iint_{U_r} p(x + x_1, y + y_1)\,dx_1\,dy_1$$

for $(x, y) \in U_{1-r}$. As is well known, $M_r p(x, y)$ is continuous on U_{1-r}. Furthermore, if $p(x, y)$ is continuous or continuously differentiable, then $M_r p(x, y)$ is continuously differentiable or twice continuously differentiable.

Take r and s sufficiently small, and let (x, y) be a point in U_{1-r-s}. Then we have

$$M_r M_s p(x, y)$$
$$= \frac{1}{\pi y^2}\frac{1}{\pi s^2}\int_{U_r}\left\{\int_{U_s} p(x + x_1 + x_2, y + y_1 + y_2)\,dx_2\,dy_2\right\}dx_1\,dy_1.$$

From Fubini's theorem, we can interchange the order of integrals on the right-hand side, i.e.,

$$M_r M_s p(x, y) = M_s M_r p(x, y), \qquad (x, y) \in U_{1-r-s}. \tag{4.26}$$

Next we will prove the following. If $M_r p(x, y)$ is a harmonic function in U_{1-r} for all $0 < r < 1$, then $p(x, y)$ itself coincides with a harmonic function everywhere in U except on a set of measure 0. Since $M_r p(x, y)$ is assumed to be harmonic, the Gauss theorem implies that the mean value over the disc equals the value at the center. That is, for a sufficiently small s we have

$$M_s M_r p(x, y) = M_r p(x, y).$$

Therefore, from (4.26) we have

$$M_r M_s p(x, y) = M_r p(x, y), \qquad M_r(M_s p(x, y) - p(x, y)) = 0. \tag{4.27}$$

Hence, for a fixed s, when r converges to 0 inside U_{1-s}, the left-hand side of (4.27) converges everywhere to $M_s p(x, y) - p(x, y)$ except on a set of measure 0 by a theorem on differentiation of indefinite integrals. Therefore,

$$M_s p(x, y) - p(x, y) = 0,$$

i.e., $p(x, y)$ coincides with the harmonic function $M_s p(x, y)$ everywhere in U_{1-s} except on a set of measure 0. Therefore, for $0 < t < s$, $M_s p(x, y)$ and $M_t p(x, y)$ coincide with $p(x, y)$ everywhere in U_{1-s} except on a set of measure zero. Since $M_s p(x, y)$ and $M_t p(x, y)$ are continuous, they must coincide everywhere in U_{1-s}. We can conclude that, for an arbitrary $0 < s < 1$, there exists a harmonic function $p'(x, y)$ in U such that $p'(x, y)$ coincides with $M_s(p(x, y))$ in U_{1-s}. This $p'(x, y)$ clearly differs from $p(x, y)$ only on a set of measure zero in U.

(ii) Let $v(x, y)$ be an arbitrary measurable function satisfying condition (4.25) as in Lemma 3.8. Let r be any positive real number less than 1, and let $w(x, y)$ be a three times continuously differentiable function in U such that $w(x, y)$ vanishes everywhere outside $U_{r'}$ for some $r' < 1 - r$. Then Fubini's theorem implies

$$\begin{aligned}
&\int_{U_{1-r}} M_r v(x, y) \Delta w(x, y)\, dx\, dy \\
&= \frac{1}{\pi r^2} \int_{U_{1-r}} \left\{ \int_{U_r} v(x + x_1, y + y_1) \Delta w(x, y)\, dx_1\, dy_1 \right\} dx\, dy \\
&= \frac{1}{\pi r^2} \int_{U_r} \left\{ \int_{U_{1-r}} v(x + x_1, y + y_1) \Delta w(x, y)\, dx\, dy \right\} dx_1\, dy_1 \\
&= \frac{1}{\pi r^2} \int_{U_r} \left\{ \int_{U'_{1-r}} v(x, y) \Delta w(x - x_1, y - y_1)\, dx\, dy \right\} dx_1\, dy_1,
\end{aligned}$$

where U'_{1-r} is the interior of the circle of radius $1 - r$ with the center (x_1, y_1). Note that $w(x - x_1, y - y_1)$ is identically zero outside the circle of radius r' with center at (x_1, y_1). Hence,

$$\int_{U'_{1-r}} v(x, y) \Delta w(x - x_1, y - y_1)\, dx\, dy = \int_U v(x, y) \Delta w(x - x_1, y - y_1)\, dx\, dy.$$

The right-hand side integral is 0 by (4.25). As a consequence we obtain

$$\int_{U_{1-r}} M_r v(x, y) \Delta w(x, y)\, dx\, dy = 0.$$

That is, if $v(x, y)$ satisfies a condition similar to (4.25), then $M_r v(x, y)$ also satisfies the condition on U_{1-r}.

(iii) Finally, we are ready to complete the proof of our lemma. First consider the case when $v(x, y)$ is twice continuously differentiable in U. Let $w(x, y)$ be an arbitrary function such as described in the lemma. When

r', $r' < 1$, is sufficiently close to 1, $w(x, y)$ is identically 0 outside $U_{r'}$. Choose r so that $r' < r < 1$. Then use (4.11) for v and w and for U_r and C_r. Since the line integral over C_r is 0, we have

$$\iint_{U_r} (v\Delta w - w\Delta v)\, dx\, dy = 0.$$

Moreover, w and Δw are identically 0 outside U_r. Hence,

$$\iint_{U} (v\Delta w - w\Delta v)\, dx\, dy = 0.$$

Then (4.25) implies

$$\iint_{U} w\Delta v\, dx\, dy = 0.$$

Recall that w is taken to be an arbitrary three times continuously differentiable function such that w is 0 in a neighborhood of C. Therefore, the above equation implies

$$\Delta v = 0 \quad \text{in } U.$$

That is, v is a harmonic function.

(iv) Let $v(x, y)$ be any function satisfying the conditions of the lemma. For sufficiently small positive numbers r, s, and t, let

$$v_3(x, y) = M_t M_s M_r v(x, y).$$

As we noted in part (i), $v_3(x, y)$ is twice continuously differentiable in $U_{1-r-s-t}$. Furthermore, from part (ii), $v_3(x, y)$ satisfies an equation similar to (4.25) in $U_{1-r-s-t}$. Then we know $v_3(x, y)$ is harmonic in $U_{1-r-s-t}$ from part (iii). Then replacing $U = U_1$ with U_{1-r-s}, apply the result of (i) to $p(x, y) = M_s M_r v(x, y)$. Then $M_s M_r v(x, y)$ coincides with a function harmonic everywhere in U_{1-r-s} except on a set of measure 0. Since $M_s M_r v(x, y)$ is continuous, $M_s M_r v(x, y)$ itself is a harmonic function. Next take $p(x, y) = M_r v(x, y)$ in U_{1-r}. Just as in the above, $M_r v(x, y)$ is harmonic. Therefore, by part (i), for $v(x, y)$ in U, we finally conclude that $v(x, y)$ coincides with a harmonic function $v'(x, y)$ everywhere in U except on a set of measure zero. We have proved Lemma 3.8 and hence Theorem 3.9.

NOTE. The above proof is considerably simpler than that of Weyl mentioned at the beginning of this section.

Let $v = (v_1, v_2)$ be an arbitrary vector field in $\mathfrak{E}$. We may consider that v is continuously differentiable up to any order by Theorem 3.9. Let P_0 be an arbitrary point on $\mathfrak{R}$, and let t be a locally uniformizing variable at P_0. For an arbitrary twice continuously differentiable function $f = f(P)$ that is 0 outside $|t| < \varepsilon_1$, $0 < \varepsilon_1 < \varepsilon$, we let

$$fv = (fv_1, fv_2).$$

Then we have

$$\operatorname{div}_t(fv) = f \operatorname{div}_t v + \left(v_1 \frac{\partial f}{\partial x} + v_2 \frac{\partial f}{\partial y} \right).$$

Therefore,

$$\iint_{|t|<\varepsilon} \operatorname{div}_t(fv)\, dx\, dy = \iint_{|t|<\varepsilon} f \operatorname{div}_t v\, dx\, dy + \iint_{|t|<\varepsilon} \left(v_1 \frac{\partial f}{\partial x} + v_2 \frac{\partial f}{\partial y} \right) dx\, dy.$$

The left-hand side may be computed by the line integral over $|t| = \varepsilon$ by (4.10). But $f = 0$ on $|t| = \varepsilon$. Therefore, the integral is also 0. The second term on the right-hand side equals $(v, \operatorname{grad} f)$. Because of our assumptions $f \in \Gamma$ and $v \in \mathfrak{E}$; $(v, \operatorname{grad} f)$ is 0. Therefore, we have

$$\iint_{|t|<\varepsilon} f \operatorname{div}_t v\, dx\, dy = 0,$$

where f is any twice continuously differentiable function which is 0 outside $|t| < \varepsilon_1$. Hence, $\operatorname{div}_t v = 0$.

Since P_0 was taken to be arbitrary, we have

$$\operatorname{div} v = 0.$$

Similarly, we have also

$$\operatorname{div}^* v = 0.$$

Conversely, if v is continuously differentiable and $\operatorname{div} v = \operatorname{div}^* v = 0$, then the above discussion implies $(v, \operatorname{grad} f) = (v, \operatorname{grad}^* f) = 0$ for any $f \in \Gamma$. We have $v \in \mathfrak{E}$.

THEOREM 3.9 (corollary). *If $v \in \mathfrak{E}$, then we have*

$$\operatorname{div} v = 0, \quad \operatorname{div}^* v = 0, \quad \textit{and} \quad \Delta v = 0. \tag{4.28}$$

Conversely, (4.28) *implies $v \in \mathfrak{E}$.*

DEFINITION 3.13. When a vector field defined on a domain U in a Riemann surface $\mathfrak{R}$ is continuously differentiable up to any order and $\operatorname{div} v = \operatorname{div}^* v = 0$ (hence $\Delta v = 0$) holds, then v is called a *harmonic vector field.*

REMARK. Let v be a vector field on $\mathfrak{R}$. From the proof of Theorem 3.9 and its corollary, we have the following. If for any arbitrary function f in Γ, which is 0 outside a domain U in $\mathfrak{R}$,

$$(v, \operatorname{grad} f) = (v, \operatorname{grad}^* f) = 0,$$

then v is continuously differentiable up to any order in U and satisfies (4.28). That is, v is a harmonic vector field in U.

Let P_0 be an arbitrary point on $\mathfrak{R}$, and let $t = x + iy$ be an arbitrary locally uniformizing variable at P_0. Choose a and b sufficiently small such that $0 < a < b$ holds. Let U_1 be the domain satisfying $|t| < b$, and let

U_2 be the complement in $\mathfrak{R}$ of the domain $|t| < a$. Notice that U_1 and U_2 both contain the annulus domain $a < |t| < b$. Let $v^{(0)}$ be a vector field which is 0 outside U_1, and in U_1 equals

$$\operatorname{grad}^* \frac{x}{x^2+y^2} = \operatorname{grad} \frac{y}{x^2+y^2}, \qquad |t| < b.$$

Let f_1 and f_2 be any functions in Γ which coincide with $x/(x^2+y^2)$ and $y/(x^2+y^2)$ in $a < |t| < b$. Then define

$$v = \begin{cases} \operatorname{grad}^* f_1, & P \in U_1, \\ \operatorname{grad} f_2, & P \in U_2. \end{cases} \tag{4.29}$$

We have $\operatorname{grad}^* f_1 = \operatorname{grad} f_2 = v^{(0)}$ in $U_1 \cap U_2$. Therefore, from (4.29), we obtain a vector field in $\mathfrak{R}$ which is continuously differentiable up to any order. Notice that this v belongs to $\mathfrak{H}$. Since $\mathfrak{H} = \mathfrak{G} \oplus \mathfrak{M}$,

$$v = v^{(1)} + v^{(2)}, \qquad v^{(1)} \in \mathfrak{G} \text{ and } v^{(2)} \in \mathfrak{M}.$$

Let

$$w^{(1)} = \begin{cases} v^{(1)}, & P \in U_1, \\ 0, & P \notin U_1, \end{cases} \qquad w^{(2)} = \begin{cases} v^{(2)}, & P \in U_2, \\ 0, & P \notin U_2. \end{cases}$$

Then both $w^{(1)}$ and $w^{(2)}$ belong to $\mathfrak{H}$. For any function f in Γ which is 0 outside U_1, we have

$$\begin{aligned}(w^{(1)}, \operatorname{grad} f) = (v^{(1)}, \operatorname{grad} f) &= (v - v^{(2)}, \operatorname{grad} f) \\ &= (v, \operatorname{grad} f) - (v^{(2)}, \operatorname{grad} f).\end{aligned}$$

Since $\operatorname{grad} f$ vanishes outside U_1, from Lemma 3.7, the first term of the far right-hand side becomes $(v, \operatorname{grad} f) = (\operatorname{grad}^* f_1, \operatorname{grad} f) = 0$. Furthermore, since $v^{(2)}$ belongs to $\mathfrak{M}$, we have $(v^{(2)}, \operatorname{grad} f) = 0$. Therefore,

$$(w^{(1)}, \operatorname{grad} f) = 0.$$

On the other hand, since $v^{(1)} \in \mathfrak{G}$ and $\mathfrak{G} \subseteq \mathfrak{M}^*$,

$$(w^{(1)}, \operatorname{grad}^* f) = (v^{(1)}, \operatorname{grad}^* f) = 0.$$

By the remark following Definition 3.13, $w^{(1)}$ is a harmonic vector field in U_1.

Next let g be any function in Γ which is 0 outside U_2. Since $v^{(2)} \in \mathfrak{M}$, we have

$$(w^{(2)}, \operatorname{grad} g) = (v^{(2)}, \operatorname{grad} g) = 0.$$

On the other hand, we also have

$$\begin{aligned}(w^{(2)}, \operatorname{grad}^* g) = (v^{(2)}, \operatorname{grad}^* g) &= (v - v^{(1)}, \operatorname{grad}^* g) \\ &= (v, \operatorname{grad}^* g) - (v^{(1)}, \operatorname{grad}^* g).\end{aligned}$$

Since $\text{grad}^* g$ is 0 outside U_2, Lemma 3.7 implies $(v, \text{grad}^* g) = (\text{grad} f_2, \text{grad}^* g) = 0$. Since $v^{(1)} \in \mathfrak{G}$ and $\mathfrak{G} \subseteq \mathfrak{M}^*$, we have $(v^{(1)}, \text{grad}^* g) = 0$. Hence, $(w^{(2)}, \text{grad}^* g) = 0$. As above, $w^{(2)}$ is a harmonic vector field in U_2. Then define

$$w = \begin{cases} v^{(0)} - w^{(1)}, & P \in U_1, \\ w^{(2)}, & P \in U_2. \end{cases}$$

Over $U_1 \cap U_2$ we have $v^{(0)} - w^{(1)} = v - v^{(1)} = v^{(2)} = w^{(2)}$. Hence, the above definition is consistent. We know that $w^{(1)}$ and $w^{(2)}$ are harmonic vector fields in U_1 and U_2, respectively. Therefore, we have proved the following theorem.

THEOREM 3.10. *Let P_0 be an arbitrary point on $\mathfrak{R}$, and let $t = x + iy$ be any locally uniformizing variable at P_0. Then there exists a vector field w on $\mathfrak{R}$ such that w is harmonic except at P_0, and such that*

$$w - \text{grad}^* \frac{x}{x^2 + y^2}$$

is harmonic in a neighborhood of P_0.

Let Γ_1 be the set of all continuously differentiable functions $f_1 = f_1(P)$ on $\mathfrak{R}$ such that $\|\text{grad} f_1\| < \infty$ holds. Denote the closed subspace of $\mathfrak{H}$ spanned by such $\text{grad} f_1$ by

$$\mathfrak{G}_1 = [\text{grad} f_1 ; f_1 \in \Gamma_1].$$

Since $\Gamma \subseteq \Gamma_1$, we have $\mathfrak{G} \subseteq \mathfrak{G}_1$. In fact, one can prove $\mathfrak{G} = \mathfrak{G}_1$. One only needs to observe the following. That is, keep $\mathfrak{G}^*$ as it was and replace $\mathfrak{G}$ by $\mathfrak{G}_1$. Then all the results remain valid if $\mathfrak{M}$ is replaced by $\mathfrak{M}_1 = \mathfrak{H} \ominus \mathfrak{G}_1$. The reader is encouraged to read through the proof of Lemma 3.7 again. For any function g in Γ_1 which is 0 in a neighborhood of P_0, write g as a sum of functions g_1 and g_2:

$$g = g_1 + g_2, \qquad g_1 \in \Gamma \text{ and } g_2 \in \Gamma_1,$$

where g_1 is 0 outside U_1, and g_2 is 0 outside U_2. As in the proof of Theorem 3.10, let

$$v = v^{(1)} + v^{(2)}, \qquad v^{(1)} \in \mathfrak{G}_1 \text{ and } v^{(2)} \in \mathfrak{M}_1,$$

and also define $w^{(1)}$, $w^{(2)}$ and w be as before. We obtain

$$\begin{aligned} (w, \text{grad} g) &= (w, \text{grad} g_1) + (w, \text{grad} g_2) \\ &= (v^{(0)} - w^{(1)}, \text{grad} g_1) + (w^{(2)}, \text{grad} g_2) \\ &= (v^{(0)}, \text{grad} g_1) - (w^{(1)}, \text{grad} g_1) + (w^{(2)}, \text{grad} g_2). \end{aligned} \tag{4.30}$$

Notice that w is not in $\mathfrak{H}$. That is, $(w, \text{grad} g)$ is not defined. Since $v^{(0)}$ is not a vector field on $\mathfrak{R}$, $(v^{(0)}, \text{grad} g_1)$ is not defined either. However, it should be clear what those inner products mean. For example, for $v^{(0)} = (v_1^{(0)}, v_2^{(0)})$,

$$(v^{(0)}, \operatorname{grad} g_1) = \iint_{U_1} \left(v_1^{(0)} \frac{\partial g_1}{\partial x} + v_2^{(0)} \frac{\partial g_2}{\partial y} \right) dx\, dy.$$

Even though $v^{(0)}$ is a discontinuous vector field at P_0, i.e., $v^{(0)} = (\infty, \infty)$ at P_0, the above integral becomes finite since g_1 is 0 in a neighborhood of P_0. Furthermore, since we have

$$v^{(0)} = \operatorname{grad}^* \frac{x}{x^2 + y^2},$$

one can prove, as in the proof of Lemma 3.7,

$$(v^{(0)}, \operatorname{grad} g_1) = 0.$$

We obtain $(w^{(1)}, \operatorname{grad} g_1) = 0$ and $(w^{(2)}, \operatorname{grad} g_2) = 0$ from the proof of Theorem 3.10. Now from (4.30), $(w, \operatorname{grad} g) = 0$.

We have proved the following useful corollary.

THEOREM 3.10 (corollary). *One can choose the vector field w in Theorem 3.10 so as to satisfy*

$$(w, \operatorname{grad} g) = 0$$

for any function g in Γ_1 that is 0 in a neighborhood of P_0.

At the beginning of this section we derived the notion of a vector field from the notion of a differential on $\mathfrak{R}$. We proved the existence of a particular vector field in Theorem 3.10. Conversely, we will prove the existence of nontrivial differentials and analytic functions on $\mathfrak{R}$ using that vector field. Namely, for the vector field w in Theorem 3.10, we will examine the following complex-valued function of P and t:

$$f(P, t_1) = w_1(P, t) - i w_2(P, t).$$

The exact converse computation to derive the transformation formulae (4.3) gives

$$f(P, t_1) = f(P, t_2) \frac{dt_2}{dt_1} \tag{4.31}$$

for analytical variables t_1 and t_2 at P. For $P \neq P_0$, we obtain the Cauchy-Riemann equations

$$\frac{\partial w_1}{\partial x} + \frac{\partial w_2}{\partial y} = \operatorname{div}_t w = 0 \quad \text{and} \quad \frac{\partial w_2}{\partial x} - \frac{\partial w_1}{\partial y} = \operatorname{div}_t^* w = 0,$$

where $t = x + iy$ is an analytical variable at P. That is, $f(P, t)$ is a regular analytic function for t. On the other hand, we have, in a neighborhood of P_0,

$$\begin{aligned}\operatorname{grad}^* \frac{x}{x^2+y^2} &= \operatorname{grad}^* \operatorname{Re}\left(\frac{1}{t}\right) = \left(\frac{\partial}{\partial y} \operatorname{Re}\left(\frac{1}{t}\right), -\frac{\partial}{\partial x} \operatorname{Re}\left(\frac{1}{t}\right)\right) \\ &= \left(-\frac{\partial}{\partial x} \operatorname{Im}\left(\frac{1}{t}\right), -\frac{\partial}{\partial x} \operatorname{Re}\left(\frac{1}{t}\right)\right)\end{aligned}$$

and

$$-\frac{\partial}{\partial x}\operatorname{Im}\left(\frac{1}{t}\right)+i\frac{\partial}{\partial x}\operatorname{Re}\left(\frac{1}{t}\right)=i\frac{\partial}{\partial x}\left(\operatorname{Re}\left(\frac{1}{t}\right)+i\operatorname{Im}\left(\frac{1}{t}\right)\right)$$
$$=i\frac{\partial}{\partial x}\left(\frac{1}{t}\right)=\frac{-i}{t^2}.$$

Hence, $f(P, t)+i/t^2$ is a regular function of t. That is, $f(P, t)$ has a pole of order 2 at P_0, i.e., at $t=0$.

Earlier in the proof of Theorem 3.10 we started with

$$v^{(0)}=\operatorname{grad}^*\operatorname{Re}\left(\frac{1}{t}\right)=-\operatorname{grad}\operatorname{Im}\left(\frac{1}{t}\right).$$

Quite similarly, if we use

$$\operatorname{grad}^*\operatorname{Re}\left(\frac{1}{t^n}\right)=-\operatorname{grad}\operatorname{Im}\left(\frac{1}{t^n}\right), \qquad n\geq 2,$$

we can show that there exists a vector field w' which is harmonic at $P\neq P_0$ and for which

$$w'-\operatorname{grad}^*\operatorname{Re}\left(\frac{1}{t^n}\right)$$

is harmonic in a neighborhood of P_0. Therefore, there exists $f'(P, t)$ that is regular in t at $P\neq P_0$ and has a pole of order $n+1$ at P_0.

Let P_1 and P_2 be distinct points on $\mathfrak{R}$. Applying the above remark to the cases $P_0=P_1$ and $P_0=P_2$, we obtain functions

$$f_{ij}(P, t), \qquad i=1, 2,\ j=2, 3$$

that are regular at $P\neq P_i$, $i=1, 2$ and have poles of order j at P_i. Then define

$$f_1(P, t)=f_{12}(P, t)+f_{23}(P, t), \qquad f_2(P, t)=f_{13}(P, t)+f_{22}(P, t).$$

Then f_1 has a pole of order 2 at P_1 and a pole of order 3 at P_2, and f_1 is regular at all other points. The function f_2 has a pole of order 3 at P_1 and a pole of order 2 at P_2, and f_2 is regular at all other points. Furthermore, $f_2(P, t)$, (and $f_1(P, t)$), can never be identically 0 in a neighborhood of any point P. Assuming otherwise, we let M_1 be the set of such points where f_2 is identically 0 in a neighborhood of that point. Let $\mathfrak{R}-M_1=M_2$. By the analyticity of $f_2(P, t)$, M_1 and M_2 are open sets. Since $\mathfrak{R}$ is connected, either $M_1=\varnothing$ or $M_2=\varnothing$ must hold. However, P_1 and P_2 belong to M_2. We conclude $M_1=\varnothing$. See the proof of Theorem 3.3. Let

$$z(P, t)=\frac{f_1(P, t)}{f_2(P, t)}.$$

Note that $z(P, t)$ has a power series expansion in t in a neighborhood of each point P. For two analytical variables t_1 and t_2 at the same point P, $f_{ij}(P, t)$, hence $f_1(P, t)$ and $f_2(P, t)$, satisfy (4.31). Therefore,

$$z(P, t_1)=z(P, t_2)$$

holds, i.e., z is a function of P only and is an analytic function of P. From the properties of $f_1(P, t)$ and $f_2(P, t)$, we conclude that $z = z(P)$ has a zero of order 1 at P_1 and has a pole of order 1 at P_2. We have obtained one of our goals in this section.

THEOREM 3.11. *Let P_1 and P_2 be arbitrary distinct points on a Riemann surface. Then there exists an analytic function $z = z(P)$ on $\mathfrak{R}$ which has a zero of order* 1 *at P_1 and has a pole of order* 1 *at P_2.*

The differential dz of the above $z(P)$ is a nonzero differential. That is, the t-expansion of dz/dt is not identically 0 in a neighborhood of every point P. Let $f(P, t)$ be a function with a pole of order 2 at P_0 and regular otherwise. As mentioned earlier, such a $f(P, t)$ exists. Define

$$g(P, t) = f(P, t)\left(\frac{dz}{dt}\right)^{-1}.$$

Then $f(P, t)$ and dz/dt satisfy the transformation formulae (4.31) for different analytical variables (in particular, locally uniformizing variables). Hence, $g(P, t)$ is an analytic function on $\mathfrak{R}$ depending on only P, as was seen in the proof of Theorem 3.11. Letting $w = g(P)$, we have

$$f(P, t) = w\frac{dz}{dt}.$$

That is, we have proved that there exists a differential $w\,dz$ which has a pole of order 2 at P_0 and is regular otherwise. Similarly, replacing $f(P, t)$ by $f'(P, t)$ having a pole of order $n+1$, $n \geq 2$, at P_0, we obtain the following theorem.

THEOREM 3.12. *Let P_0 be an arbitrary point on $\mathfrak{R}$. Then there exists a differential on $\mathfrak{R}$ which has a pole of order n, $n \geq 2$, at P_0 and is regular otherwise.*

As an application of Theorem 3.11, we have

THEOREM 3.13. *Let f' be a map from the underlying space S' of a Riemann surface $\mathfrak{R}'$ to the underlying space S of $\mathfrak{R}$. If for an arbitrary analytic function $g(P)$ on $\mathfrak{R}$, $g(f'(P'))$ is always analytic on $\mathfrak{R}'$, then f' is an analytic map from $\mathfrak{R}'$ to $\mathfrak{R}$.*

PROOF. Let P'_0 be an arbitrary point on $\mathfrak{R}'$, and let $P_0 = f'(P'_0)$. From Theorem 3.11, there is an analytic function $z(P)$ which has a zero of order 1 at P_0. From the assumption, $z(f'(P'))$ is an analytic function on $\mathfrak{R}'$. Hence, for a locally uniformizing variable t' at P', we have the expansion

$$z(f'(P')) = b_1 t' + b_2 t'^2 + \cdots \tag{4.32}$$

in a neighborhood of P'_0. Since $t = z(P)$ is a locally uniformizing variable at P_0, (4.32) indicates that f' is analytic at P'_0. Hence, f is an analytic map.

As an immediate consequence of the above theorem, we have

THEOREM 3.14. *Let f' be a one-to-one map from the underlying space S' of $\mathfrak{R}'$ onto the underlying space S of $\mathfrak{R}$. If*

$$g'(P') = g(f'(P')), \qquad g' \in K(\mathfrak{R}'), \ g \in K(\mathfrak{R}),$$

defines a one-to-one map between the totality of analytic functions on $\mathfrak{R}'$ and the totality of analytic functions on $\mathfrak{R}$, then f' is an isomorphism from $\mathfrak{R}'$ to $\mathfrak{R}$.

This theorem tells us that a Riemann surface is completely determined by the underlying set of points and the totality of analytic functions over the set. Therefore, in order to define a Riemann surface on an underlying space, one only needs to specify the family of complex-valued functions which are to be analytic functions of the Riemann surface. This is Weyl's idea in his book cited at the beginning of this section.

In a sense, the above theorem is the converse of Theorem 3.5. We will prove in the next chapter that for compact Riemann surfaces $\mathfrak{R}$ and $\mathfrak{R}'$, $K(\mathfrak{R}) \cong K(\mathfrak{R}')$ implies $\mathfrak{R} \cong \mathfrak{R}'$.

§5. Covering Riemann surfaces

Before our discussion of a covering Riemann surface, we will introduce the notion of a covering manifold. However, we restrict ourselves to describing only those results that are needed for what follows. For details see C. Chevalley, *The theory of Lie groups*, Princeton University Press, 1946. In this section, we primarily follow this book's method.

DEFINITION 3.14. Let f' be a map from a connected n-dimensional manifold S' to a connected n-dimensional manifold S. Then S' is called a *covering manifold* over the base manifold S, and f' is called its *covering map* if the following is satisfied: for each point P in S there exists a neighborhood U of P such that each connected component of $f'^{-1}(U)$ is mapped homeomorphically onto U by f'.

Note that the concept of a covering manifold includes the covering map. That is, for two different covering maps f_1' and f_2' from the same manifold S' to S, we consider the covering manifold S' with the covering map f_1' to be different from the covering manifold S' with the covering map f_2'. It is appropriate to write $\{S', f'\}$ or $\{S', f'; S\}$ to describe a covering manifold.

Let $\{S_1', f_1'; S_1\}$ and $\{S_2', f_2'; S_2\}$ be a pair of covering manifolds. If there exist a homeomorphism φ from S_1 onto S_2 and a homeomorphism φ' from S_1' onto S_2' such that

$$f_2'\varphi' = \varphi f_1', \tag{5.1}$$

then $\{S_1', f_1'; S_1\}$ and $\{S_2', f_2'; S_2\}$ are called isomorphic. We call such a pair (φ', φ) an isomorphism and write

$$\{S_1', f_1'; S_1\} \cong \{S_2', f_2'; S_2\}.$$

In particular, when $S_1 = S_2 = S$, unless specified, φ is always considered to be the identity map, and φ' is called an isomorphism from $\{S_1', f_1'; S\}$ to $\{S_2', f_2'; S\}$. Furthermore, if $S_1' = S_2' = S'$ and $f_1' = f_2' = f'$ hold, then the φ' above is an automorphism of $\{S', f'; S\}$. This is nothing but a homeomorphic transformation from S' onto S' such that

$$f'\varphi' = f'.$$

The set of these φ' forms a topological transformation group of S'. We call this set the automorphism group of $\{S', f'; S\}$, denoted by $A(S', f'; S)$.

Any connected n-dimensional manifold S has at least one covering manifold, namely, $\{S, f_e; S\}$, where f_e is the identity map. When $\{S, f_e; S\}$ is the only covering manifold up to isomorphism, S is called a *simply connected* manifold. In general, S is not necessarily simply connected, but there always exists a simply connected covering manifold. (We will come back to this topic.) Note that any two simply connected covering manifolds $\{S^*, f^*; S\}$ and $\{S_1^*, f_1^*; S\}$ of S are isomorphic

$$\{S^*, f^*; S\} \cong \{S_1^*, f_1^*; S\}. \tag{5.2}$$

That is, there is essentially only one simply connected covering manifold of S. The automorphism group $A(S^*, f^*; S)$ of S is called the *fundamental group*, or the *Poincaré group* of S. In particular, $A(S^*, f^*; S)$ and $A(S_1^*, f_1^*; S)$ are isomorphic by (5.2). Hence, the structure of the fundamental group is independent of the choice of a simply connected covering manifold over S.

The fundamental group $G = A(S^*, f^*; S)$, as we saw, is a topological transformation group of S^*. Moreover, G has the following property: let P^* be an arbitrary point on S^*. Then there exists a neighborhood U^* of P^* such that

$$\varphi_1(U^*) \cap \varphi_2(U^*) = \varnothing \tag{5.3}$$

holds for any distinct maps in G. Therefore, unless $\varphi \in G$ is the identity map, φ has no fixed points.

In general, for a topological space S^*, if the topological transformation group G of S^* has the above property, i.e., each point P^* has a neighborhood with the property (5.3), then G is called a properly discontinuous transformation group without fixed points. We call this group *the discontinuous topological transformation group.* For a general discussion of discontinuous transformation groups, see B. L. van der Waerden, *Gruppen von linearen Transformation*, Springer, Berlin, 1935. As noted above, the fundamental group $A(S^*, f^*; S)$ is a discontinuous transformation group of S^*. Conversely, any discontinuous transformation group G of a simply connected manifold S^* can be regarded as the fundamental group of a suitable manifold S. Here is how to construct such a manifold: let P^* and Q^* be arbitrary points on S^*. When $Q^* = \varphi(P^*)$ holds for a transformation φ in

G, define P^* and Q^* to be equivalent through G. This equivalence relation classifies the points on S^*. Let $S^*(G)$ be the set of classes. Let f_G^* denote the map that assigns a point P^* on S^* to its class $\overline{P^*}$:

$$\overline{P^*} = f_G^*(P^*).$$

Also define a neighborhood of $\overline{P^*}$ to be the image of a neighborhood of P^* under f_G^*. Then $S^*(G)$ is a connected manifold whose dimension equals that of S^*. One can see that $\{S^*, f_G^*; S^*(G)\}$ is a covering manifold having G as its fundamental group. Then $S^*(G)$ is said to be the manifold belonging to the discontinuous group G of S^*.

The above $S^*(G)$ gives, in a sense, a canonical form of a connected manifold. We have obtained the following lemma.

LEMMA 3.9. *Let S^* be the simply connected covering manifold of a connected manifold S, and let G be the fundamental group of $\{S^*, f^*; S\}$. Two points P^* and Q^* on S^* are equivalent through G if and only if*

$$f^*(P^*) = f^*(Q^*).$$

In this case, there exists a unique $\psi \in G$ such that $Q^* = \psi(P^*)$ holds. Therefore, there is a one-to-one map ρ from S onto $S^*(G)$ such that $\{S^*, f^*; S\}$ and $\{S^*, f_G^*; S^*(G)\}$ are isomorphic with respect to the map (f_e^*, ρ). Here, f_e^* is the identity map of S^*. Therefore, in particular, S and $S^*(G)$ are homeomorphic manifolds. Furthermore, the manifolds $S^*(G_1)$ and $S^*(G_2)$, belonging to discontinuous transformation groups G_1 and G_2, respectively, are homeomorphic if and only if G_1 and G_2 are conjugate in the total topological transformation group $T(S^*)$. Then, for a homeomorphism φ from $S^*(G_1)$ to $S^*(G_2)$, there exists a topological transformation φ^* of S^* such that $\{S^*, f_{G_1}^*; S^*(G_1)\}$ and $\{S^*, f_{G_2}^*; S^*(G_2)\}$ are isomorphic via (φ^*, φ), and

$$\varphi^* G_1 {\varphi^*}^{-1} = G_2.$$

Consider the family of manifolds S which have S^* as a simply connected covering manifold. By this lemma, if one divides S into classes of homeomorphic manifolds, then there is a one-to-one correspondence between these classes and the classes of discontinuous subgroups of $T(S^*)$ that are conjugate.

A general covering manifold $\{S', f'; S\}$, i.e., not necessarily simply connected is closely related to S^* and G. In particular, we have the following lemma.

LEMMA 3.10. *Let S, S^*, and G be as in the previous lemma, and let $\{S', f'; S\}$ be an arbitrary covering manifold of S. Then there exists a subgroup H of G such that*

$$\{S', f'; S\} \cong \{S^*(H), f'_{G,H}; S^*(G)\},$$

where $S^*(H)$ *is the manifold belonging to* H, *and* $f'_{G,H}$ *is a uniquely determined map from* $S^*(H)$ *onto* $S^*(G)$ *satisfying*

$$f^*_G = f'_{G,H} f^*_H. \tag{5.4}$$

Furthermore, there is a one-to-one correspondence between the points in $f'^{-1}(P)$, *for each* P *on* S, *and the cosets of* G *reduced* mod H.

Conversely, let H be any subgroup of G. Then $\{S^*(H), f'_{G,H}; S^*(G)\}$ is a covering manifold of $S^*(G)$. For subgroups H_1 and H_2 of G, the covering manifolds $\{S^*(H_1), f'_{G,H_1}; S^*(G)\}$ and $\{S^*(H_2), f'_{G,H_2}; S^*(G)\}$ are isomorphic if and only if H_1 and H_2 are conjugate subgroups of G. Thus, classes of isomorphic covering manifolds correspond to classes of conjugate subgroups in a one-to-one fashion.

Lemma 3.9 and Lemma 3.10 give us a group-theoretic method of constructing covering manifolds from simply connected manifolds and their discontinuous topological transformation groups. This is similar to Galois theory, where intermediate fields of a Galois extension are captured by the Galois group and its subgroups.

Our main result is that there exists at least one simply connected covering manifold S^* for a given connected n-dimensional manifold S. Such an S^* can be constructed as follows. Fix a point P_0 on S, and let S^*_0 be the set of all the homotopic classes of paths with initial point P_0. Let $\{\gamma\}$ be an arbitrary point on S^*_0, that is, a class of homotopic paths in S with the initial point P_0. From the definition of homotopy, all the paths in $\{\gamma\}$ have the same terminal point P. We call P the terminal point of $\{\gamma\}$. Next, take a neighborhood U of P which is homeomorphic to an n-dimensional open ball, and for any point Q in U, let γ' be any path in U connecting P with Q. Then let U^* be the set of those $\{\gamma'\}\{\gamma\} = \{\gamma\gamma'\}$. Define a topology on S^*_0 by taking all those U^* as neighborhoods of $\{\gamma\}$. Then S^*_0 is a simply connected manifold of dimension n. If one defines a map f^*_0 from $\{\gamma\}$ to its terminal point P by

$$P = f^*_0(\{\gamma\}),$$

then one can show that $\{S^*_0, f^*_0; S\}$ is a simply connected covering manifold of S. We call S^*_0 the *homotopy* space of S with initial point P_0. Since any simply connected covering manifold $\{S^*, f^*; S\}$ is isomorphic to the above $\{S^*_0, f^*_0; S\}$, we can prove the following lemma from the properties of S^*_0.

LEMMA 3.11. *Let* $\{S^*, f^*; S\}$ *be an arbitrary simply connected covering manifold of* S, *let* G *be the fundamental group, let* P^*_0 *be an arbitrary initial point of* S^*, *and let* $P_0 = f^*(P^*_0)$. *Notice that the image* $\gamma = f^*(\gamma^*)$ *of any path* γ^* *in* S^* *with the initial point* P^*_0 *is a path in* S *with initial point* P_0. *Conversely, for an arbitrary path in* S *with initial point* P_0, *there is a unique path* γ^* *in* S^* *with initial point* P^*_0 *such that* $\gamma = f^*(\gamma^*)$. *For* γ_1 *to* *be homotopic to* γ_2 *in* S *it is necessary and sufficient that the corresponding paths* γ_1^* *and* γ_2^* *in* S^* *have the same terminal point. In particular, for a closed path* γ, *let* Q^* *be the terminal point of the corresponding* γ^*. *Since* $f^*(Q^*) = f^*(P_0^*)$ *holds, from Lemma* 3.10 *there exists a unique* $\varphi \in G$ *such that we have* $Q^* = \varphi(P_0^*)$. *The correspondence between the class* $\{\gamma\}$ *of* γ *and this* φ:

$$\{\gamma\} \leftrightarrow \varphi \tag{5.5}$$

gives rise to an isomorphism between the homotopy group of S *with initial point* P_0 *and the fundamental group* G.

From the above lemma, or by a direct verification, the homotopy groups with different initial points are isomorphic to each other. From Lemma 3.10, S is simply connected if and only if the fundamental group G of S is the identity group. Therefore, from the above lemma, this is also equivalent to saying that all the closed paths in S are homotopic to the constant path. In fact we used the notion of simply connectedness in §3 in this sense. According to this characterization, it is clear that an n-dimensional Euclidean space and an n-dimensional open ball are simply connected manifolds.

We are now ready to apply the above general theory of covering manifolds to the theory of Riemann surfaces. The next theorem plays a fundamental role in what follows.

THEOREM 3.15. *For a covering manifold* $\{S', f'; S\}$, *when* S *is an underlying space of a Riemann surface* $\mathfrak{R}$, *there exists a unique Riemann surface* $\mathfrak{R}'$ *having* S' *as the underlying space so that* f' *is an analytic map from* $\mathfrak{R}'$ *to* $\mathfrak{R}$.

PROOF. Suppose there exist Riemann surfaces $\mathfrak{R}$ and $\mathfrak{R}'$ with underlying spaces S and S', respectively, such that f' is an analytic map between them. For an arbitrary point P' on $\mathfrak{R}'$, let $P = f'(P')$. From Theorem 3.2, one can choose locally uniformizing variables t and t' at P and P', respectively, so that

$$t = f'(t') = (t')^n, \qquad n \geq 1. \tag{5.6}$$

By the definition of a covering map, f' maps a neighborhood of P' homeomorphically onto a neighborhood of P. Hence, we have $n = 1$ in (5.6), i.e.,

$$t = f'(t') = t'.$$

We now begin the proof. Let $\mathfrak{F} = \{\varphi_P, U_P\}$ be an R-function system defining $\mathfrak{R}$, where U_P is chosen small enough so that f' maps each connected component of $f'^{-1}(U_P)$ homeomorphically onto U_P. Let P' be an arbitrary point in $f'^{-1}(P)$. Let $U'_{P'}$ be the connected component of $f'^{-1}(U_P)$ containing P'. On $U'_{P'}$ define

$$\varphi'_{P'} = \varphi_P f',$$

then $\mathfrak{F}' = \{\varphi'_{P'}, U'_{P'}\}$ is an R-function system on S'. It is clear that $\mathfrak{R}' = \{S', \{\mathfrak{F}'\}\}$ satisfies the properties in the theorem and that uniqueness follows from our remark at the beginning of the proof. Q.E.D.

DEFINITION 3.15. The Riemann surface $\mathfrak{R}'$ in Theorem 3.15 is called the covering Riemann surface of $\mathfrak{R}$ for the covering map f', denoted by $\{\mathfrak{R}', f'; \mathfrak{R}\}$, or $\{\mathfrak{R}', f'\}$.

THEOREM 3.16. *Let* $\{\mathfrak{R}'_1, f'_1 : \mathfrak{R}_1\}$ *and* $\{\mathfrak{R}'_2, f'_2; \mathfrak{R}_2\}$ *be covering Riemann surfaces, let* S_1, S_2, S'_1, S'_2 *be the underlying spaces of* $\mathfrak{R}_1$, $\mathfrak{R}_2$, $\mathfrak{R}'_1$, $\mathfrak{R}'_2$, *respectively, and let*

$$\{S'_1, f'_1; S_1\} \cong \{S'_2, f'_2; S_2\}. \tag{5.7}$$

Denote the above isomorphism by (φ', φ). *Then if one of* φ' *and* φ *is an analytic map, then so is the other. In this case, we say that* $\{\mathfrak{R}'_1, f'_1; \mathfrak{R}_1\}$ *and* $\{\mathfrak{R}'_2, f'_2; \mathfrak{R}_2\}$ *are isomorphic and write*

$$\{\mathfrak{R}'_1, f'_1; \mathfrak{R}_1\} \cong \{\mathfrak{R}'_2, f'_2; \mathfrak{R}_2\}.$$

PROOF. By the definition of a covering Riemann surface, f'_i $(i = 1, 2)$ are analytic maps. Each f'_i conformally maps a small enough neighborhood of P'_i on $\mathfrak{R}'_i$ onto a neighborhood of P_i on $\mathfrak{R}_i$ in a one-to-one fashion. If φ' is an analytic map such that φ' conformally maps a neighborhood of P'_1 on $\mathfrak{R}'_1$ onto a neighborhood of $P'_2 = \varphi'(P'_1)$ in a one-to-one fashion, then from (5.1) φ also maps conformally a neighborhood of P_1 on $\mathfrak{R}_1$ onto a neighborhood of $P_2 = \varphi(P_1)$ in a one-to-one fashion. Here $P_1 = f'_1(P'_1)$ and $P_2 = f'_2(P'_2)$. Therefore, φ is an analytic map. The converse can be proved in the same way.

In particular, if $\mathfrak{R}_1 = \mathfrak{R}_2 = \mathfrak{R}$ and $S_2 = S_1 = S$ hold, then by our earlier agreement φ is the identity map and φ is clearly an analytic map. Therefore, φ' is also an analytic map. Furthermore, if $\mathfrak{R}'_1 = \mathfrak{R}'_2 = \mathfrak{R}'$ and $S'_1 = S'_2 = S'$ hold, then any φ' in the automorphism group $A(S', f'; S)$ of $\{S', f'; S\}$ is analytic and is an automorphism of the Riemann surface $\mathfrak{R}'$. We denote the automorphism group $A(S', f'; S)$ by $A(\mathfrak{R}', f'; \mathfrak{R})$ and call it the *automorphism group* of the covering Riemann surface $\{\mathfrak{R}', f'; \mathfrak{R}\}$. As we indicated above, $A(\mathfrak{R}', f', \mathfrak{R})$ is a subgroup of the automorphism group $A(\mathfrak{R}')$.

Let $\mathfrak{R}^*$ be an arbitrary simply connected Riemann surface and let S^* be the underlying space. Let G be any discontinuous subgroup of $A(\mathfrak{R}^*)$, and let $S^*(G)$ be the manifold belonging to G. Then we will prove the following: there exists a unique Riemann surface having $S^*(G)$ as its underlying space such that the covering map f^*_G from S^* to $S^*(G)$ is an analytic map. Uniqueness follows from the proof of Theorem 3.15. We will prove its existence. Let P^* be an arbitrary representative from a class of equivalent points in S^* with respect to G. Then let $\varphi^*_{P^*}$ be a function which maps a sufficiently small neighborhood U_{P^*} of P^* conformally into the interior of a unit disc in a one-to-one fashion. For $Q^* = \varphi(P^*)$, $\varphi \in G$, let

$$\overset{*}{\varphi}_{Q^*} = \overset{*}{\varphi}_{P^*}\varphi^{-1}, \qquad U_{Q^*} = \varphi(U_{P^*}).$$

Since φ is an automorphism of $\mathfrak{R}^*$, it is clear that

$$\mathfrak{F}^* = \{\overset{*}{\varphi}_{P^*}, U_{P^*}\}$$

is an R-function system on $\mathfrak{R}^*$. For the point $P = f_G^*(P^*)$ let

$$\varphi_P = \overset{*}{\varphi}_P \overset{*}{f_G}{}^{-1}, \qquad U_P = f_G^*(U_{P^*}).$$

Then φ_P and U_P are defined independently of the choice of P^* satisfying $P = f_G^*(P^*)$. Then $\{\varphi_P, U_P\}$ is an R-function system on $S^*(G)$, and $\{\varphi_P, U_P\}$ satisfies the required properties. This Riemann surface determined by G is called the Riemann surface belonging to G and is denoted by $\mathfrak{R}^*(G)$.

For a general Riemann surface $\mathfrak{R}$ with the underlying space S, let $(S^*, f^*; S)$ be a simply connected covering manifold of S. Then, by Theorem 3.15, there is a unique simply connected covering Riemann surface $(\mathfrak{R}^*, f^*; \mathfrak{R})$ of $\mathfrak{R}$ having S^* as its underlying space. The automorphism group $G = A(\mathfrak{R}^*, f^*; \mathfrak{R})$ of $\{\mathfrak{R}^*, f^*; \mathfrak{R}\}$ is precisely the fundamental group of S for S^*, and as we noted earlier, it is a discontinuous subgroup of the automorphism group $A(\mathfrak{R}^*)$ of $\mathfrak{R}^*$. Then let $\mathfrak{R}^*(G)$ be the Riemann surface belonging to G. We have

$$\{S^*, f^*; S\} \cong \{S^*, f_G^*; S^*(G)\}$$

from Lemma 3.10. Furthermore, since the map from S^* to S^* is the identity map, Theorem 3.16 implies

$$\{\mathfrak{R}^*, f^*; \mathfrak{R}\} \cong \{\mathfrak{R}^*, f_G^*; \mathfrak{R}^*(G)\}. \tag{5.8}$$

In particular, $\mathfrak{R}^*$ and $\mathfrak{R}^*(G)$ are isomorphic Riemann surfaces. That is, any Riemann surface $\mathfrak{R}$ is isomorphic to the Riemann surface $\mathfrak{R}^*(G)$ belonging to a discontinuous subgroup G of the automorphism group $A(\mathfrak{R}^*)$ of a simply connected Riemann surface $\mathfrak{R}^*$.

Let $\mathfrak{R}_1^*$ and $\mathfrak{R}_2^*$ be simply connected Riemann surfaces, and let $\mathfrak{R}_1^*(G_1)$ and $\mathfrak{R}_2^*(G_2)$ be the Riemann surfaces belonging to the discontinuous subgroups G_1 and G_2 of $A(\mathfrak{R}_1^*)$ and $A(\mathfrak{R}_1^*)$, respectively. Suppose that $\mathfrak{R}_1^*(G_1) \cong \mathfrak{R}_2^*(G_2)$, and let φ be an isomorphism from $\mathfrak{R}_1^*(G_1)$ onto $\mathfrak{R}_2^*(G_2)$. Needless to say, φ is a homeomorphism from the underlying space $S_1^*(G_1)$ onto $S_2^*(G_2)$. Therefore, $\{S_1^*, \varphi f_{G_1}^*; S_2^*(G_2)\}$ is a covering manifold of $S_2^*(G_2)$, and consequently $\{\mathfrak{R}_1^*, \varphi f_{G_1}^*; \mathfrak{R}_2^*(G_2)\}$ is a simply connected covering Riemann surface of $\mathfrak{R}_2^*(G_2)$. The notation S_1^* and $f_{G_1}^*$ is self-explanatory. Since simply connected covering manifolds of $S_2^*(G_2)$ are isomorphic to each other, we have

$$\{S_1^*, \varphi f_{G_1}^*; S_2^*(G_2)\} \cong \{S_2^*, f_{G_2}^*; S_2^*(G_2)\}.$$

Therefore, by Theorem 3.16, we have

$$\{\mathfrak{R}_1^*, \varphi f_{G_1}^*; \mathfrak{R}_2^*(G_2)\} \cong \{\mathfrak{R}_2^*, f_{G_2}^*; \mathfrak{R}_2^*(G_2)\}.$$

In particular, $\mathfrak{R}_1^* \cong \mathfrak{R}_2^*$ must hold.

Next for the case $\mathfrak{R}_1^* = \mathfrak{R}_2^* = \mathfrak{R}^*$ and $S_1^* = S_2^* = S^*$, we ask when $\mathfrak{R}^*(G_1)$ and $\mathfrak{R}^*(G_2)$ are isomorphic. As above, let φ be an isomorphism from $\mathfrak{R}^*(G_1)$ onto $\mathfrak{R}^*(G_2)$. Since $S^*(G_1)$ is homeomorphic to $S^*(G_2)$ via φ, Lemma 3.10 implies that there is an isomorphism (φ^*, φ) between $\{S^*, f_{G_1}^*; S^*(G_1)\}$ and $\{S^*, f_{G_2}^*; S^*(G_2)\}$ such that

$$\varphi^* G_1 \varphi^{*^{-1}} = G_2, \qquad \varphi^* \in T(S^*). \tag{5.9}$$

The analyticity of φ implies that of φ^* by Theorem 3.16. Therefore, G_1 and G_2 are conjugate subgroups of $A(\mathfrak{R}^*)$. Conversely, if G_1 and G_2 are conjugate in $A(\mathfrak{R}^*)$, then from the definitions, $\mathfrak{R}^*(G_1)$ and $\mathfrak{R}^*(G_2)$ are isomorphic.

The next theorem summarizes what we have obtained thus far.

THEOREM 3.17. *When all Riemann surfaces are divided into classes of isomorphic Riemann surfaces, a complete set of representatives for those classes can be obtained in the following steps. Choose a representative from each class of simply connected Riemann surfaces. Let $\mathfrak{R}^*$ be such a representative. Then choose a representative G from each class of conjugate discontinuous subgroups of the automorphism group $A(\mathfrak{R}^*)$ of $\mathfrak{R}^*$. When $\mathfrak{R}^*$ and G run through the sets of representatives as above; the totality of Riemann surfaces $\mathfrak{R}^*(G)$ forms a complete set of representatives for the classes of all Riemann surfaces.*

Using this theorem, we will determine all the types of simply connected Riemann surfaces and then find the canonical forms of Riemann surfaces.

§6. Simply connected Riemann surfaces and canonical forms of Riemann surfaces

We first consider a simply connected Riemann surface $\mathfrak{R} = \mathfrak{R}^*$. In the following we will mainly follow §19 of the book by Weyl cited earlier. Fix a point P_0 on $\mathfrak{R}$, and let $w = (w_1(P, t), w_2(P, t))$ be the vector field in Theorem 3.10. As in §4, let

$$f(P, t) = w_1(P, t) - iw_2(P, t) = z_1 \frac{dz_2}{dt},$$

where $z_1\, dz_2$ is a differential on $\mathfrak{R}$ which has a pole of order 2 at P_0 and is regular otherwise. Since $\mathfrak{R}$ is simply connected, from Theorem 3.7,

$$\frac{dz}{dt} = z_1 \frac{dz_2}{dt}$$

holds for some analytic function $z = z(P)$ on $\mathfrak{R}$. Notice that $z(P)$ is regular for $P \neq P_0$, and in a neighborhood of P_0, we have

$$z = \frac{i}{t} + a_0 + a_1 t + a_2 t^2 + \cdots . \tag{6.1}$$

Separate $z(P)$ into its real part and complex part

$$z(P) = u(P) + iv(P).$$

Then we have

$$\frac{\partial u}{\partial x} = \frac{\partial v}{\partial y} = w_1(P, t) \quad \text{and} \quad \frac{\partial u}{\partial y} = -\frac{\partial v}{\partial x} = w_2(P, t).$$

Furthermore, for any twice continuously differentiable real-valued function $g(P)$ which is 0 in a neighborhood of P_0 satisfying

$$\|\operatorname{grad} g\|^2 = \iint \left(\left(\frac{\partial g}{\partial x}\right)^2 + \left(\frac{\partial g}{\partial y}\right)^2 \right) dx\, dy < \infty, \tag{6.2}$$

we have, by Theorem 3.10 (corollary),

$$(w, \operatorname{grad} g) = \iint \left(\frac{\partial u}{\partial x} \cdot \frac{\partial g}{\partial x} + \frac{\partial u}{\partial y} \cdot \frac{\partial g}{\partial y} \right) dx\, dy = 0. \tag{6.3}$$

Note that the right-hand side integrals in (6.2) and (6.3) are taken over the entire $\mathfrak{R}$. See §4 for the precise definition.

First we prove

(1) let v_0 be an arbitrary real number, let $D(v_0)$ be the set of points P on $\mathfrak{R}$ satisfying $v(P) > v_0$, and let $D'(v_0)$ be the set of points on $\mathfrak{R}$ satisfying $v(P) < v_0$. Then $D(v_0)$ and $D'(v_0)$ are domains, i.e., connected open sets.

From (6.1), a neighborhood of P_0 is conformally mapped to a neighborhood of $z = \infty$ on the z-sphere $\mathfrak{R}_0^*$. Therefore, for a sufficiently small $\varepsilon > 0$, the intersection of $D(v_0)$ and the neighborhood $\{t; |t| < \varepsilon\}$ of P_0 is a domain. Suppose that $D(v_0)$ has more than two connected components. Then one of them, call it U, does not intersect $\{t; |t| < \varepsilon\}$. Choose twice continuously differentiable functions $\xi(\lambda)$ and $\eta(\lambda)$ defined on the real line with the property that

$$\eta(v_0) = \eta'(v_0) = \eta''(v_0) = 0, \quad \text{where } \eta' = \frac{d\eta}{d\lambda}, \ \eta'' = \frac{d^2\eta}{d\lambda^2}. \tag{6.4}$$

Let

$$g(P) = \begin{cases} \xi(u(P))\eta(v(P)), & P \in U, \\ 0, & P \notin U. \end{cases}$$

Then $g(P)$ is twice continuously differentiable by (6.4), and for a locally uniformizing variable $t = x + iy$ at P in U

$$\frac{\partial g}{\partial x} = \xi'(u)\eta(v)\frac{\partial u}{\partial x} + \xi(u)\eta'(v)\frac{\partial v}{\partial x} = \xi'(u)\eta(v)\frac{\partial u}{\partial x} - \xi(u)\eta'(v)\frac{\partial u}{\partial y},$$
$$\frac{\partial g}{\partial y} = \xi'(u)\eta(v)\frac{\partial u}{\partial y} + \xi(u)\eta'(v)\frac{\partial u}{\partial x},$$
$$\left(\frac{\partial g}{\partial x}\right)^2 + \left(\frac{\partial g}{\partial y}\right)^2 = (\xi'^2\eta^2 + \xi^2\eta'^2)\left(\left(\frac{\partial u}{\partial x}\right)^2 + \left(\frac{\partial u}{\partial y}\right)^2\right).$$

Hence, if ξ, ξ', η and η' are bounded functions of λ, then $g(P)$ satisfies (6.2). Notice from the choice of the vector field w that the integral of

$$\left(\frac{\partial u}{\partial x}\right)^2 + \left(\frac{\partial u}{\partial y}\right)^2 = (w_1(P, t)^2 + w_2(P, t)^2)$$

over any domain outside a neighborhood of P_0 is finite. Therefore, the equality in (6.3) holds for the above g. Since we have

$$\frac{\partial u}{\partial x}\cdot\frac{\partial g}{\partial x} + \frac{\partial u}{\partial y}\cdot\frac{\partial g}{\partial y} = \xi'(u)\eta(v)\left(\left(\frac{\partial u}{\partial x}\right)^2 + \left(\frac{\partial u}{\partial y}\right)^2\right),$$

if one chooses $\xi'(\lambda)$ and $\eta(\lambda)$ to be positive (except at $\lambda = v_0$ for η), then equality (6.3) is obviously false. This contradiction shows that $D(v_0)$ is a domain. A proof for $D'(v_0)$ is exactly the same.

Next, we prove

(2) the differential dz has no zeros on $\mathfrak{R}$.

Suppose dz has a zero of order $n-1$, $n \geq 2$, at P_1. Let $z(P_1) = z_1$. Then $z = z(P)$ maps a neighborhood of P_1 to an n-folded neighborhood of z_1 on $\mathfrak{R}_0^*$. Since the same argument holds for the case $n > 2$, for the sake of simplicity, we will prove the case when $n = 2$. Then $t = \sqrt{z - z_1}$ is a locally uniformizing variable at P_1. The neighborhood $|t| < \varepsilon$ of P_1 is divided into parts according to the values of $v(P)$ as shown in Figure 3.1. That is, the shaded parts indicate $v(P) > v_0 = \operatorname{Im}(z_1)$, and the unshaded parts indicate $v(P) < v_0$. Since t_1 and t_3 in Figure 3.1 belong to the domain $D(v_0)$ mentioned in part (1), there is an analytic path connecting them. Then one gets a simple closed curve γ consisting of this curve and

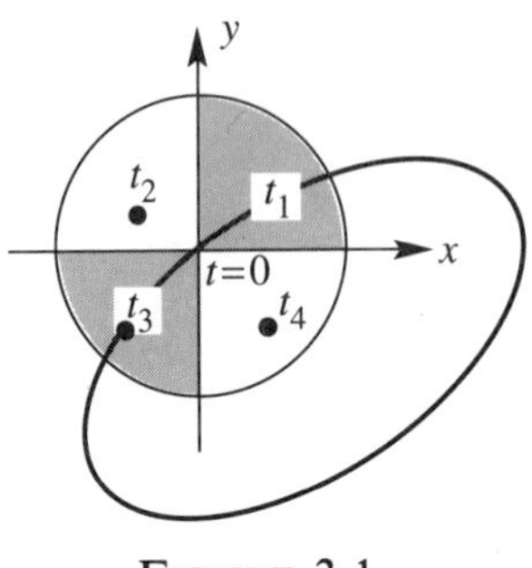

FIGURE 3.1

the segment connecting t_1 and t_2 in Figure 3.1. Let S_1 be the complement of γ in the underlying space S. Then S_1 is a connected set. This can be shown as follows. For an arbitrary point P in S_1, connect P and P_1 by a path γ_1 in S. Starting with P along γ_1, follow γ leaving γ_1 just before γ_1 intersects with γ. Then proceed along γ in a neighborhood of γ in S_1. Then one can connect P in S_1 with t_2 or t_4 in the figure. Since t_2 and t_4 belong to $D'(v_0)$, they can be connected by a path γ_2 in $D'(v_0)$. Then γ_2 does not intersect with γ. Therefore S_1 is connected.

Next, let $S^{(1)}$ and $S^{(2)}$ be manifolds of dimension 2 to which S is homeomorphic, and let φ_i be homeomorphisms from S to $S^{(i)}$, $i = 1, 2$,

$$S^{(1)} = \varphi_1(S) \quad \text{and} \quad S^{(2)} = \varphi_2(S).$$

Also let

$$\gamma^{(1)} = \varphi_1(\gamma), \quad \gamma^{(2)} = \varphi_2(\gamma), \quad S_1^{(1)} = \varphi_1(S_1), \quad S^{(2)} = \varphi_2(S_1).$$

Then we will define a neighborhood system for each point in the union S^* of $\gamma^{(1)}$, $\gamma^{(2)}$, $S_1^{(1)}$, and $S_1^{(2)}$ as follows. For a point $P^{(1)}$ in $S_1^{(1)}$, a neighborhood system of $P^{(1)}$ in S^* is defined by the images $\{\varphi_1(U)\}$ of a neighborhood system of $\varphi_1^{-1}(P^{(1)})$ in S_1. A similar neighborhood system can be defined for a point in $S_1^{(2)}$. Let $P^{(1)}$ be a point on $\gamma^{(1)}$. Since γ is an analytic path on $\mathfrak{R}$, there exists a small enough neighborhood U of $\varphi_1^{-1}(P^{(1)})$ that can be mapped conformally and in a one-to-one fashion into the interior of the unit disc. Then U is divided by γ into right side and the left side parts U_1 and U_2. (See Figure 3.2.) Let γ' be the intersection of γ and U. Take the set of

$$U' = \varphi_1(\gamma') + \varphi_1(U_1) + \varphi_2(U_2)$$

as a neighborhood system of $P^{(1)}$ in S^*.

Then exactly the same construction for $P^{(2)}$ on $\gamma^{(2)}$ provides a neighborhood system

$$U'' = \varphi_2(\gamma') + \varphi_2(U_1) + \varphi_1(U_2)$$

for $P^{(2)}$ in S^*. We have obtained a connected manifold S^* of dimension 2. With f^* the map which assigns $\varphi_1^{-1}(P^{(1)})$ and $\varphi_2^{-1}(P^{(2)})$ for $P^{(1)}$ and $P^{(2)}$ in S^*, $\{S^*, f^*; S\}$ is a covering manifold of S. A rigorous topological

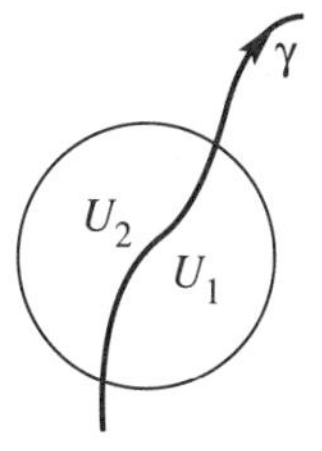

FIGURE 3.2

argument is left for the reader. However, S is simply connected by our assumption. Hence any covering manifold of S is isomorphic to $\{S, f_e; S\}$. Therefore, our assumption that $dz = 0$ leads to a contradiction. Assertion (2) has been proved.

We can give an explicit presentation of the map $z = z(P)$ from $\mathfrak{R}$ to the z-sphere $\mathfrak{R}_0^*$ using (1) and (2). For an arbitrary real number v_0, consider the set $W(v_0)$ of points P in $\mathfrak{R}$ satisfying $v(P) = v_0$. Take N sufficiently large so that a small neighborhood U of P_0 may be mapped conformally and injectively by $z(P)$ onto the neighborhood $|z| > N$ of ∞ on $\mathfrak{R}_0^*$. Since $W(v_0)$ is the preimage of the circle $v = v_0$ on $\mathfrak{R}_0^*$ by the map $z(P)$, $\gamma' = W(v_0) \cap U$ is an analytic path in $\mathfrak{R}$ in the strict sense, going through P_0, which is conformally and injectively mapped onto the open arc

$$|u| > \sqrt{N^2 - v_0^2}, \qquad v = v_0, \qquad z = u + iv.$$

Next starting from a point Q_1 on γ' satisfying $u(Q_1) < -\sqrt{N^2 - v_0^2}$, move P on $W(v_0)$ in the direction where $u(P)$ increases. Since a sufficiently small neighborhood of each point P into $\mathfrak{R}$ is mapped, conformally and in a one-to-one fashion, into $\mathfrak{R}_0^*$, $W(v_0)$ is an open analytic path without end points in such a neighborhood. Therefore, P traces an analytic path on $\mathfrak{R}$ as the value $u(P)$ increases. If P returns inside U for a sufficiently large $u(P)$, then P would have to be on γ', and the preimage of the circle $v = v_0$ in $\mathfrak{R}_0^*$ under the map $z(P)$ is the closed analytic path $\gamma(v_0)$, e.g., this happens for $|v_0| > N$. In the case when P does not return inside U, since P_0 is the only point satisfying $u(P) = \infty$, we get an analytic path γ_1' in $\mathfrak{R}$, without an endpoint, which is conformally mapped in a one-to-one fashion onto

$$u(Q_1) \le u < u_1, \qquad v = v_0 \tag{6.5}$$

for some real number u_1, possibly $u_1 = +\infty$. Note that $W(v_0)$ cannot be extended further along γ_1'. Then starting from Q_2 on γ' satisfying $u(Q_2) > \sqrt{N^2 - v_0^2}$, we also obtain γ_2' by moving P in the direction for which $u(P)$ decreases so that γ_2' may be mapped conformally and injectively on

$$u_2 < u \le u(Q_2), \qquad v = v_0 \tag{6.6}$$

for some u_2. Let $\gamma(v_0)$ be the analytic path connecting γ_2' with the above γ_1' through γ'.

In either case, the points on $W(v_0)$ are all on $\gamma(v_0)$. Here is the proof. Let S_1' be the complement of $\gamma(v_0)$ in S. Let us suppose that S_1' contains a point P such that $v(P) = v_0$. Then choose P_1 and P_2 in a neighborhood of P with $v(P_1) > v_0$ and $v(P_2) < v_0$, and also choose P_3 and P_4 in a neighborhood of P_0 with $v(P_3) > v_0$ and $v(P_4) < v_0$. Then by (1), one can connect P_1 and P_3 by a path in $D(v_0)$, i.e., without intersecting $\gamma(v_0)$, and similarly connect P_2 and P_4 by a path without intersecting $\gamma(v_0)$. One

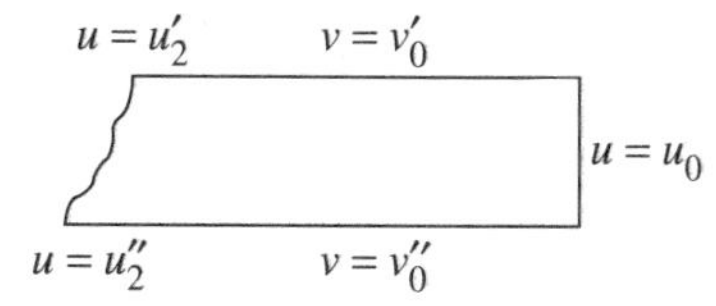

FIGURE 3.3

can clearly connect P_1 and P_2 by a path in S_1'. Therefore, P_3 and P_4 are connected in S_1'. Just as in (2), cut S open along $\gamma(v_0)$ to obtain another covering manifold. This contradicts the simply connectedness of S. We have $W(v_0) = \gamma(v_0)$.

We will prove that there exists at most one real number v_0 for which $\gamma(v_0)$ is not a closed path. Let both $\gamma(v_0')$ and $\gamma(v_0'')$ not be closed for $v_0' \neq v_0''$, and then let u_1', u_2', u_1'', u_2'' be the u_1 and u_2 in (6.5) and (6.6), respectively. Choose u_0 so that $u_0 > N$, then let γ^* be the preimage of the bold segments in Figure 3.3. From the choice of u_0, the preimage of $u = u_0$ is a closed analytic path. Hence, γ^* is an analytic curve without endpoints in $\mathfrak{R}$. Let S_2' be the complement of γ^* in S.

If one assumes the connectedness of S_2', then exactly the same method as above would lead to a contradiction of the simply connectedness of S. Therefore S_2 is not connected, and there exists a connected component of S_2 that does not contain P_0. Let U^* be such a connected component. Then define, as in (1),

$$g(P) = \begin{cases} \xi(u(P))\eta(v(P)), & P \in U^*, \\ 0, & P \notin U^*, \end{cases}$$

where $\xi(\lambda)$ and $\eta(\lambda)$ are twice continuously differentiable real-valued functions satisfying

$$\xi(u_0) = \xi'(u_0) = \xi''(u_0) = 0, \qquad \eta(v_0') = \eta'(v_0') = \eta''(v_0') = 0,$$
$$\eta(v_0'') = \eta'(v_0'') = \eta''(v_0'') = 0.$$

We also require that ξ, ξ', η, η' are bounded and that ξ' and η are positive except at $\lambda = u_0$ and $\lambda = v_0', v_0''$, respectively. We can clearly find such functions $\xi(\lambda)$ and $\eta(\lambda)$. Then we reach a contradiction just as in (1) to (6.3). Hence, there is at most one v_0 such that $\gamma(v_0)$ is not closed. When such a v_0 exists, u_1 and u_2 in (6.5) and (6.6) must satisfy $u_1 \leq u_2$ because of the following. If $u_2 < u_1$, let $u_2 < u_3 < u_1$. There are two points P such that

$$z(P) = u_3 + iv_0.$$

Since $z(P)$ is an analytic map, for v_1 sufficiently close to v_0, there exist also two points P' satisfying

$$z(P') = u_3 + iv_1. \tag{6.7}$$

However, since $\gamma(v_1)$ is closed if $v_i \neq v_0$, there should exist a unique P' satisfying (6.7). We are led to a contradiction.

From what we have found, $\mathfrak{R}$ is mapped conformally and injectively by $z = z(P)$ onto either

(1′) the entire z-sphere $\mathfrak{R}_0^*$,
(2′) the domain $\mathfrak{R}_0^* - z_0$, or
(3′) the complementary domain of the segment $u_1 \leq u \leq u_2$, $v = v_0$ $(u_1 < u_2)$ in $\mathfrak{R}_0^*$.

Furthermore, the domain (2′) is conformally mapped by the linear transformation $z' = 1/(z - z_0)$ to the finite z-plane $\mathfrak{R}_\infty^*$. The domain (3′) is mapped by a suitable linear transformation to the complementary domain of the segment $-1 \leq u \leq 1$, $v = 0$ in $\mathfrak{R}_0^*$, which is conformally mapped by $z' = \psi(z)$ to the interior of the unit circle $\mathfrak{R}_1^*$, where $z' = \psi(z)$ is given by

$$z = \frac{1}{2}\left(z' + \frac{1}{z'}\right).$$

Therefore, we have proved that any simply connected Riemann surface $\mathfrak{R}$ is isomorphic to either

(1) the z-sphere $\mathfrak{R}_0^*$, i.e., the Riemann sphere,
(2) the finite z-plane $\mathfrak{R}_\infty^* = \{z\,;\,|z| < \infty\}$, or
(3) the interior of unit circle $\mathfrak{R}_1^* = \{z\,;\,|z| < 1\}$.

Conversely, all of $\mathfrak{R}_0^*$, $\mathfrak{R}_\infty^*$, and $\mathfrak{R}_1^*$ are simply connected Riemann surfaces, but no one of them is isomorphic to another one. The underlying space of $\mathfrak{R}_0^*$ is compact, and the underlying spaces of $\mathfrak{R}_\infty^*$ and $\mathfrak{R}_1^*$ are not compact. Hence, $\mathfrak{R}_0^*$ cannot be isomorphic to either $\mathfrak{R}_\infty^*$ or $\mathfrak{R}_1^*$. Suppose $\mathfrak{R}_\infty^*$ and $\mathfrak{R}_1^*$ were isomorphic. Let

$$w = f(z)$$

be a regular function that maps $\mathfrak{R}_\infty^*$ conformally onto $\mathfrak{R}_1^*$. Then $|f(z)| < 1$ for $|z| < \infty$. By Liouville's theorem, $f(z)$ would be a constant function, which contradicts $f(\mathfrak{R}_\infty^*) = \mathfrak{R}_1^*$.

We have obtained the following

THEOREM 3.18. *There are three distinct types of simply connected Riemann surfaces, and their representatives are given by* $\mathfrak{R}_0^*$, $\mathfrak{R}_\infty^*$, *and* $\mathfrak{R}_1^*$.

We have completely answered the first step of Theorem 3.17. Next, we will find the automorphism groups of these simply connected Riemann surfaces and also their discontinuous subgroups. We will also examine the Riemann surfaces belonging to such discontinuous subgroups.

We will first consider $\mathfrak{R}_0^*$. Let φ be an arbitrary map belonging to the automorphism group $A(\mathfrak{R}_0^*)$ of $\mathfrak{R}_0^*$. Then $\varphi = \varphi(z)$ is an analytic function which is a one-to-one map from the z-sphere onto itself. Therefore, it must be a linear function in z. Conversely, any linear function $\varphi(z)$ clearly belongs to $A(\mathfrak{R}_0^*)$. That is, $A(\mathfrak{R}_0^*)$ is the linear transformation group

$$A(\mathfrak{R}_0^*) = \left\{ \varphi(z) = \frac{az+b}{cz+d};\ ad-bc \neq 0,\ a, b, c, d \in k \right\}.$$

Let G be an arbitrary discontinuous subgroup of $A(\mathfrak{R}_0^*)$. Then any nonidentity map in G has no fixed points. However, since any linear transformation has a fixed point on $\mathfrak{R}_0^*$, G can contain only the identity element. That is, the only discontinuous subgroup of $A(\mathfrak{R}_0^*)$ is the identity group. Therefore, $\mathfrak{R}_0^*$ is the only Riemann surface, up to isomorphism, that has $\mathfrak{R}_0^*$ as a simply connected covering Riemann surface.

Let us next consider $\mathfrak{R}_\infty^*$. Any $\varphi \in A(\mathfrak{R}_\infty^*)$ is also a linear function of z. Since $\varphi(z)$ is finite for a finite z, we must have

$$A(\mathfrak{R}_\infty^*) = \{\varphi(z) = az+b;\ a \neq 0,\ a, b \in k\}.$$

The only elements in $A(\mathfrak{R}_\infty^*)$ that do not have fixed points are parallel transformations $\varphi(z) = z+b$. The discontinuous subgroups of $A(\mathfrak{R}_\infty^*)$ are the following ones:

(1) $G =$ identity group,
(2) $G = G(\alpha) = \{\varphi(z) = z + n\alpha,\ \alpha \neq 0,\ \alpha \in k,\ n = 0, \pm 1, \pm 2, \dots\}$,
(3) $G = G(\omega_1, \omega_2) = \{\varphi(z) = z + m_1\omega_1 + m_2\omega_2;\ \operatorname{Im}(\omega_1/\omega_2) > 0,$ $\omega_1, \omega_2 \in k,\ m_1, m_2 = 0, \pm 1, \pm 2, \dots\}$.

First, $\mathfrak{R}_0^*$ itself is the only Riemann surface belonging to (1).

Next all the $G(\alpha)$ in (2) are conjugate. For example, let $\psi(z) = (\alpha/2\pi i)z$, then we have

$$\psi^{-1}G(\alpha)\psi = G(2\pi i).$$

Therefore, all the Riemann surfaces belonging to $G(\alpha)$ are isomorphic, e.g., to $\mathfrak{R}_\infty^*(G(2\pi i))$. By the map $w = e^z$, this Riemann surface is isomorphic to the Riemann surface obtained from the w-sphere, with the 0 and ∞ removed.

Next consider the conjugate relationship among groups of type (3). Let

$$G(\omega_1', \omega_2') = \varphi^{-1}G(\omega_1, \omega_2)\varphi, \qquad \varphi(z) = \alpha z + \beta.$$

Then $\varphi^{-1}G(\omega_1, \omega_2)\varphi = G(\omega_1/\alpha, \omega_2/\alpha)$ holds. Therefore,

$$\omega_1' = a\left(\frac{\omega_1}{\alpha}\right) + b\left(\frac{\omega_2}{\alpha}\right),$$

$$\omega_2' = c\left(\frac{\omega_1}{\alpha}\right) + d\left(\frac{\omega_2}{\alpha}\right), \qquad ad - bc = \pm 1,\ a, b, c, d = \text{rational integers}.$$

If we let

$$\tau = \frac{\omega_2}{\omega_1} \quad \text{and} \quad \tau' = \frac{\omega_2'}{\omega_1'},$$

then we have

$$\tau' = \frac{c + d\tau}{a + b\tau}. \tag{6.8}$$

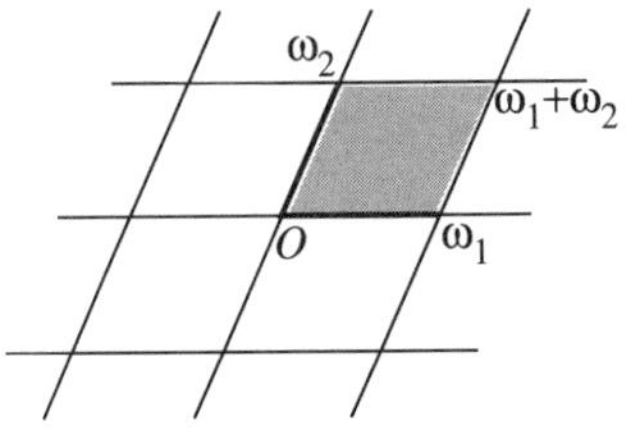

FIGURE 3.4

By our assumption, $\mathrm{Im}(\tau) > 0$, $\mathrm{Im}(\tau') > 0$, and we also have

$$\tau' = \frac{(a+b\bar{\tau})(c+d\tau)}{|a+b\tau|^2} = \frac{ac+bd|\tau|^2+ad\tau+bc\bar{\tau}}{|a+b\tau|^2}.$$

Hence, $ad-bc>0$, i.e.,

$$ad-bc=1 \qquad (a\,,b\,,c\,,d = \text{rational integers}). \tag{6.9}$$

Conversely, for $G(\omega_1, \omega_2)$ and $G(\omega_1', \omega_2')$, if $\tau = \omega_2/\omega_1$ and $\tau' = \omega_2'/\omega_1'$ are related as in (6.8) with a, b, c, d satisfying (6.9), then the above remark implies that these groups are conjugate in $A(\mathfrak{R}_\infty^*)$. Therefore, all the complex numbers τ, $\mathrm{Im}\,\tau > 0$, that are not equivalent by means of the transformation (6.8) satisfying (6.9), generate distinct types of Riemann surfaces $\mathfrak{R}_\infty^*(\omega_1, \omega_2) = \mathfrak{R}_\infty^*(G(\omega_1, \omega_2))$. A fundamental domain in $\mathfrak{R}_\infty^*$ of the discontinuous group $G(\omega_1, \omega_2)$ is the shaded area of Figure 3.4.

In general, a fundamental domain D in a topological space S^* of a discontinuous topological transformation group G is a domain in S^* such that for $\varphi_1, \varphi_2 \in G$, $\varphi_1 \neq \varphi_2$, $\varphi_1(D) \cap \varphi_2(D) = \varnothing$, and such that any point in S^* is contained in the closure of $\varphi(D)$ for some $\varphi \in G$. Hence, there is a one-to-one correspondence between $\varphi \in G$ and the image $\varphi(D)$ of D. The union of all the $\varphi(\overline{D})$, $\varphi \in G$, covers the whole space S^*, overlapping only the images of the boundary of D. Therefore, in order to get the space $S^*(G)$, consisting of classes of equivalent points through G, we need only to identify suitable boundary points of $\overline{D}$. The partition of S^* obtained from the fundamental domain D and its images $\varphi(D)$ is important for seeing G and $S^*(G)$ in an intuitive manner. This importance can be observed from the above $G = G(\omega_1, \omega_2)$. In particular, $S^*(G(\omega_1, \omega_2))$ is obtained by identifying $\overline{O\omega_1}$ and $\overline{\omega_2\omega_3}$ and identifying $\overline{O\omega_2}$ and $\overline{\omega_1\omega_3}$, where $\omega_3 = \omega_1 + \omega_2$. Hence, $\mathfrak{R}_\infty^*(\omega_1, \omega_2)$ is a compact Riemann surface homeomorphic to a torus. We will come back to this topic in Chapter 4.

It is clear that $G(\alpha)$ and $G(\omega_1, \omega_2)$ can never be conjugate since they are not isomorphic. Hence, the conjugate relations mentioned above are the only such relations among discontinuous subgroups of $A(\mathfrak{R}_\infty^*)$. Therefore, we have found all the representatives of Riemann surfaces having $\mathfrak{R}_\infty^*$ as their covering Riemann surface.

Lastly we consider $\mathfrak{R}_1^*$. In this case, $\varphi \in A(\mathfrak{R}_1^*)$ is an analytic function that maps the interior of a unit circle onto itself in a one-to-one fashion.

Therefore, we have

$$A(\mathfrak{R}_1^*) = \left\{ \varphi(z) = e^{i\alpha} \frac{z - z_0}{1 - \bar{z}_0 z}, \ |z_0| < 1, \ \alpha = \text{real number} \right\}. \tag{6.10}$$

There are various kinds of discontinuous subgroups of $A(\mathfrak{R}_1^*)$. In this case it is not as simple to classify those subgroups as in the case $\mathfrak{R}_\infty^*$. The following view may be useful to understand the properties of the subgroups. Let γ be an analytic path in $\mathfrak{R}_1^*$. Then the integral along γ

$$\rho(\gamma) = \int_\gamma \frac{|dz|}{1 - |z|^2}$$

is called the non-Euclidean length of γ. For an arbitrary map $z' = \varphi(z)$ in $A(\mathfrak{R}_1^*)$, simple computation gives

$$\frac{|dz'|}{1 - |z'|^2} = \frac{|dz|}{1 - |z|^2}. \tag{6.11}$$

For the image $\gamma' = \varphi(\gamma)$ of γ, we have

$$\rho(\gamma') = \rho(\gamma).$$

In $A(\mathfrak{R}_1^*)$ there is a φ that maps the initial point z_1 of γ to the origin O and the terminal point z_2 of γ to α, $0 \le \alpha < 1$, on the real axis. Then γ' is an analytic path connecting O and α. From the general inequalities

$$|z - z'| \ge ||z| - |z'||, \qquad |dz| \ge d|z|,$$

one obtains

$$\rho(\gamma) = \rho(\gamma') = \int_{\gamma'} \frac{|dz|}{1 - |z|^2} \ge \int_0^\alpha \frac{d|z|}{1 - |z|^2} = \frac{1}{2} \log \frac{1 + \alpha}{1 - \alpha}.$$

The value of the far right side equals $\rho(\gamma_1')$, where γ_1' is the straight line from O to α. Let $\gamma_1 = \varphi^{-1}(\gamma_1')$. Then γ_1 is an arc through z_1 and z_2 normally intersecting with the unit circle such that $\rho(\gamma) \ge \rho(\gamma_1)$ holds. That is, γ_1 has the shortest length among analytic curves connecting z_1 and z_2. In fact, the above proof implies $\rho(\gamma) > \rho(\gamma_1)$ for $\gamma \ne \gamma_1$. Since the anharmonic ratio of -1, 0, α, 1 is precisely $(1 + \alpha)/(1 - \alpha)$ and since the anharmonic ratio is invariant under a linear transformation, for $z_\infty = \varphi^{-1}(-1)$ and $z'_\infty = \varphi^{-1}(1)$ we have

$$\rho(\gamma_1) = \frac{1}{2} \log r(z_\infty, z_1, z_2, z'_\infty),$$

where $r(z_\infty, z_1, z_2, z'_\infty)$ denotes the anharmonic ratio of the four points.

From what we have observed, if one defines a Riemannian geometry with line element given by

$$ds = \frac{|dz|}{1 - |z|^2},$$

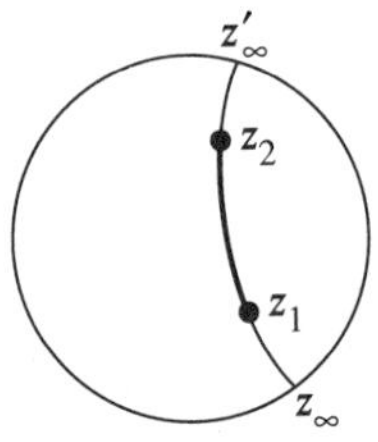

FIGURE 3.5

then the "line" i.e., geodesic curve, connecting z_1 and z_2 in this geometry is the arc in Figure 3.5 connecting z_1 and z_2. The "distance" between z_1 and z_2 is given by

$$\rho(z_1, z_2) = \frac{1}{2}\log r(z_\infty, z_1, z_2, z'_\infty). \tag{6.12}$$

Furthermore, the transformations belonging to $A(\mathfrak{R}_1^*)$ do not change the "distance". That is, $A(\mathfrak{R}_1^*)$ is the group of motions for the geometry. Note also that the axioms for non-Euclidean geometry (Bolyai-Lobatschefski hyperbolic geometry) are satisfied for these "lines" and "group of motions." In this sense, $\mathfrak{R}_1^*$ is a non-Euclidean space. For detail, see D. Hilbert, *Grundlagen der Geometrie*, Anhang III, Teubner, Leipzig. We need only to notice that the topologies on $\mathfrak{R}_1^*$ induced by the non-Euclidean distance $\rho(z_1, z_2)$ and the Euclidean distance $|z_1 - z_2|$ are equivalent. In what follows, "lines" and "distance" are used for the non-Euclidean geometry, whereas lines and distance are reserved for Euclidean geometry.

When $A(\mathfrak{R}_1^*)$ is interpreted as above, a discontinuous subgroup G is nothing but a discontinuous "group of motions" of $\mathfrak{R}_1^*$. This corresponds to discontinuous groups of motions $G(\alpha)$ and $G(\omega_1, \omega_2)$ when $\mathfrak{R}_\infty^*$ is considered as a Euclidean plane. By Theorem 3.17, for discontinuous "*groups of motions*" G_1 and G_2, $\mathfrak{R}_1^*(G_1)$ and $\mathfrak{R}_1^*(G_2)$ are isomorphic if and only if we have

$$\psi^{-1} G_1 \psi = G_2, \qquad \psi \in A(\mathfrak{R}_1^*). \tag{6.13}$$

Since ψ is a "motion" of the space $\mathfrak{R}_1^*$, ψ may be interpreted as a coordinate transformation on $\mathfrak{R}_1^*$ (the reader should recall the relation between a motion and a coordinate transformation in Euclidean space). Thus, (6.13) means that G_1 and G_2 are presentations of the same "group of motions" with respect to two different coordinates. Therefore, for each discontinuous "group of motions" of $\mathfrak{R}_1^*$, determined up to coordinate transformations, there corresponds a representative of a Riemann surface having $\mathfrak{R}_1^*$ as the covering Riemann surface.

We have obtained the main theorem of this section.

THEOREM 3.19. *An arbitrary Riemann surface is isomorphic to one of the following types of Riemann surfaces*:

(1) *the Riemann sphere,*

(2) *the following surfaces having the finite complex plane* $\mathfrak{R}^*_\infty$ *as the covering Riemann surface.*

(2.1) $\mathfrak{R}^*_\infty$,

(2.2) $\mathfrak{R}^*_{0,\infty} = \{z\,;\, 0 < |z| < \infty\}$,

(2.3) $\mathfrak{R}^*_\infty(\omega_1, \omega_2) = \mathfrak{R}^*_\infty(G(\omega_1, \omega_2))$, *where*

$$G(\omega_1, \omega_2) = \left\{\varphi(z) = z + m_1\omega_1 + m_2\omega_2\,; \right.$$
$$\left. \operatorname{Im}\left(\frac{\omega_2}{\omega_1}\right) > 0, \quad m_1, m_2 = 0, \pm 1, \pm 2, \cdots \right\}.$$

If, for (ω_1, ω_2) *and* (ω_1', ω_2'),

$$\frac{\omega_2'}{\omega_1'} = \frac{c\omega_1 + d\omega_2}{a\omega_1 + b\omega_2}, \qquad ad - bc = 1, \ a, b, c, d = \textit{rational integers},$$

then

$$\mathfrak{R}^*_\infty(\omega_1', \omega_2') \cong \mathfrak{R}^*_\infty(\omega_1, \omega_2).$$

(3) *Those having the interior of the unit disc* $\mathfrak{R}^*_1$ *as the covering Riemann surface*:

$$\mathfrak{R}^*_1(G),$$

where G *is an arbitrary discontinuous subgroup of*

$$A(\mathfrak{R}^*_1) = \left\{\varphi(z) = e^{i\alpha}\frac{z - z_0}{1 - \bar{z}_0 z}, \ |z_0 < 1, \ \alpha \textit{ a real number}\right\}.$$

Here, $A(\mathfrak{R}^*_1)$ *is the "group of motions" of the non-Euclidean space* $\mathfrak{R}^*_1$. *If* G_1 *can be transformed to* G_2 *by a coordinate transformation, then we have*

$$\mathfrak{R}^*_1(G_1) \cong \mathfrak{R}^*_1(G_2).$$

The above Riemann surfaces $\mathfrak{R}^*_0$, $\mathfrak{R}^*_\infty$, $\mathfrak{R}^*_{0,\infty}$, $\mathfrak{R}^*_\infty(\omega_1, \omega_2)$, $\mathfrak{R}^*_1(G)$ are called canonical types. There are no isomorphisms between those Riemann surfaces other than those described in Theorem 3.19. An immediate result of the above theorem is

THEOREM 3.20. *The underlying space* S *of any Riemann surface satisfies the second countability axiom.*

NOTE. One can prove the above theorem without using Theorem 3.19. That is, one can directly prove Theorem 3.20 from other results, especially Theorem 2.10, in §4. In that section, we proved that there exists a vector field of finite norm which is harmonic everywhere on a Riemann surface $\mathfrak{R}$ except in a neighborhood of a point P_0. Then the above theorem can be proved by using the definition of norm and the property of a harmonic vector field that such a vector field $v \neq 0$ cannot be identically zero in any neighborhood of a point.

We will prove a few assertions about a discontinuous "group of motions" of $\mathfrak{R}_1^*$ which will play an important role in the following chapter.

We will determine a fundamental domain of G. For an arbitrary point z_0 in $\mathfrak{R}_1^*$, let

$$\overline{D} = \{z\,;\, \rho(z\,,\, z_0) \leq \rho(z\,,\, \psi(z_0))\,,\ \ \psi \in G\}.$$

We can actually find $\overline{D}$ as follows. For each $\psi \in G$, which is not the identity map f_e, let γ_ψ be the "orthogonally bisecting line" of the "straight line" connecting z_0 and its image $\psi(z_0)$. Then let $\overline{D}_\psi$ be the closed domain containing z_0 when $\mathfrak{R}_1^*$ is divided into two closed domains by γ_ψ, i.e.,

$$\overline{D}_\psi = \{z\,;\, \rho(z\,,\, z_0) \leq \rho(z\,,\, \psi(z_0))\}.$$

Then clearly

$$\overline{D} = \bigcap_{\psi \neq f_e} \overline{D}_\psi. \tag{6.14}$$

On the other hand, for any positive real number ρ_0, the number of γ_ψ that intersects the closed "disc" $\overline{U}_{\rho_0} = \{z\,;\, \rho(z\,,\, z_0) \leq \rho_0\}$ of radius ρ_0 with center z_0 is finite. We can show this as follows. Let z_1 be a point of intersection between γ_ψ and $\overline{U}_{\rho_0}$. We have

$$\rho(z_1\,,\, \psi(z_0)) \leq \rho(z_1\,,\, z_0) \leq \rho_0\,,$$
$$\rho(\psi(z_0)\,,\, z_0) \leq \rho(\psi(z_0)\,,\, z_1) + \rho(z_1\,,\, z_0) \leq 2\rho_0.$$

Since the set $\{\psi(z_0)\,;\, \psi \in G\}$ cannot have an accumulation point in $\mathfrak{R}_1^*$ because of the discontinuity of G, the number of $\psi(z_0)$ that are in the compact set $\overline{U}_{2\rho_0}$ must be finite. Therefore, there are only finitely many γ_ψ that go through a neighborhood of a point in $\mathfrak{R}_1^*$. Therefore, the boundary of $\overline{D}$ consists of "segments", i.e., parts of γ_ψ, and $\overline{D}$ is a "convex polygon" enclosed by these segments (infinitely many in general). From this, the interior of $\overline{D}$ is given by

$$D = \{z\,;\, \rho(z\,,\, z_0) < \rho(z\,,\, \psi(z_0))\,,\ \ \psi \in G\,,\ \ \psi \neq f_e\}\,,$$

and $\overline{D}$ is the closure of D.

Next we will prove that D is a fundamental domain of G. First, D is clearly a domain since D is the interior of the "convex polygon" $\overline{D}$. Let φ be an arbitrary map in G. Then

$$\varphi(\overline{D}) = \{z\,;\, \rho(z\,,\, \varphi(z_0)) \leq \rho(z\,,\, \psi(z_0)si \in G\} \tag{6.15}$$

and

$$\varphi(D) = \{z\,;\, \rho(z\,,\, \varphi(z_0)) < \rho(z\,,\, \psi(z_0))\,,\, \psi \in G\,,\, \psi \neq \varphi\}.$$

Hence, $\varphi_1(D)$ and $\varphi_2(D)$ do not intersect for $\varphi_1 \neq \varphi_2$. Lastly, let z be any point in $\mathfrak{R}_1^*$. Since there are only finitely many points $\psi(z_0)\,,\ \psi \in G$ in any bounded region of $\mathfrak{R}_1^*$, there exists, among all points $\psi(z_0)\,,\ \psi \in G\,,$

at least one point $\varphi(z_0)$ closest to z. Then from (6.15) we have $z \in \varphi(\overline{D})$. Hence D is a fundamental domain of D.

Since $G(\omega_1, \omega_2)$ partitions $\mathfrak{R}_\infty^*$ into congruent parallelograms, $\mathfrak{R}_1^*$ is packed by congruent polygons

$$\varphi(\overline{D}), \qquad \varphi \in G$$

induced by G. Following Fricke, we call these "polygons" the *normal "polygons"*. We denote the net of the partition of $\mathfrak{R}_1^*$ by $\mathfrak{N}$. As mentioned above, each normal polygon corresponds, in a one-to-one fashion, to an element of G. One can obtain the Riemann surface $\mathfrak{R}_1^*(G)$ belonging to G by identifying sides of a normal "polygon", e.g., $\overline{D} = f_e(\overline{D})$.

Next we will consider the case where $\mathfrak{R}_1^*(G)$ is a compact Riemann surface we will need this case in a later chapter.

First note that the above $\overline{D}$ is a "bounded set" for ρ. To see this, let t be an arbitrary point in $\mathfrak{R}_1^*$, and let

$$U(t, \varepsilon) = \{z ; \rho(z, t) < \varepsilon\}$$

be a sufficiently small neighborhood of t which is conformally and injectively mapped onto a neighborhood of $f_G^*(t)$ by the covering map f_G^* from $\mathfrak{R}_1^*$ to $\mathfrak{R}_1^*(G)$. (Here ε is a sufficiently small number, possibly depending on t.) Then $\mathfrak{R}_1^*(G)$ is covered by such open sets $f_G^*(U(t, \varepsilon))$. Since $\mathfrak{R}_1^*(G)$ is compact, there are finitely many $t_1, \dots, t_n$ so that

$$f_G^*(U(t_i, \varepsilon_i)), \qquad i = 1, \dots, n \tag{6.16}$$

cover $\mathfrak{R}_1^*(G)$. Since each $U(t_i, \varepsilon_i)$ is a "bounded set", the union U of $U(t_i, \varepsilon_i)$ is also bounded with respect to ρ. Furthermore, since the sets in (6.16) cover $\mathfrak{R}_1^*(G)$, for any point z in $\overline{D}$, U contains a point $\psi(z)$ equivalent through G. Noting that the "distance" ρ is invariant under G, we have, by the definition $\overline{D}$, that

$$\rho(z, z_0) \le \rho(z, \psi^{-1}(z_0)) = \rho(\psi(z), z_0), \qquad \psi(z) \in U.$$

Since U is "bounded", the right-hand side does not exceed a certain positive real number. Hence, $\overline{D}$ is also "bounded".

As we noted before, there are only finitely many γ_ψ that intersect a given bounded set. Hence, the number of sides of $\overline{D}$ is also finite. Let γ be such a side, and let $\varphi(z_0)$, $\varphi \in G$ be the point located "mirror opposite" of z_0 with respect to γ. Then γ is the common side of $\overline{D}$ and $\varphi(\overline{D})$. But, since $\overline{D}$ is a "convex polygon", we have

$$\gamma = \overline{D} \cap \varphi(\overline{D}).$$

Hence,

$$\varphi^{-1}\gamma = \varphi^{-1}(\overline{D}) \cap \overline{D}.$$

That is, $\delta = \varphi^{-1}(\gamma)$ is also a side of $\overline{D}$. Note that γ and δ come in pairs, and $\gamma = \delta$ never happens. Suppose $\varphi(\gamma)$ coincides with γ, even with the opposite orientation. Since φ is a homeomorphism and also γ is a closed "segment", there must be a fixed point of φ on γ. This cannot happen since G is a discontinuous group. Hence, on γ, all the sides of $\overline{D}$ are divided into pairs. Let γ_i, δ_i; $\delta_i = \varphi(\gamma_i)$, $i = 1, 2, \dots, m$ be those pairs. Then we have the following lemma.

LEMMA 3.12. *The above* $\varphi_1, \dots, \varphi_m$ *are generators for the discontinuous "group of motions". That is, any element* φ *of* G *can be written as*

$$\varphi = \varphi_{i_1}^{\pm 1} \cdots \varphi_{i_r}^{\pm 1}.$$

Furthermore, by a suitable choice of φ_i, *the number* r *of* φ_i *needed to express* φ *as above can be made to satisfy*

$$r \leq \lambda\rho(z, \varphi(z)) + \mu$$

for any point z *in* $\mathfrak{R}_1^*$, *where* λ *and* μ *are real numbers independent of* φ *and* z.

NOTE. The reader is referred to H. Poincaré, *Sur les fonctions fuchsiennes*, Acta Math. **1** (1982).

PROOF. Notice first that the number of polygons that share a common vertex of the net consisting of the normal "polygons" is bounded. This is because the net $\mathfrak{N}$ is invariant under the maps of G, and an arbitrary "polygon" of $\mathfrak{N}$ is congruent to $\overline{D}$ with respect to G. If λ_1 denotes the maximum of the number of normal "polygons" sharing each vertex of $\overline{D}$, the number of "polygons" sharing any vertex of $\mathfrak{N}$ cannot exceed λ_1.

The "distance" between arbitrary sides in $\mathfrak{N}$ which have no point in common has the minimum value $\rho_1 > 0$. Since the maps in G do not change "distance", we may assume one of the two sides is a side of $\overline{D}$. Since the number of $\psi(z_0)$ and therefore the number of $\psi(\overline{D})$ which lie within a bounded region are finite, for a fixed side γ of $\overline{D}$ the shortest distance ρ_γ to every side of $\mathfrak{N}$ from γ is determined. We can let ρ_1 be the minimum of ρ_γ for $2m$ sides γ of $\overline{D}$.

Next let Γ be the "line" from an arbitrary point z to $\varphi(z)$. Then we can choose an analytic path Γ' from z', sufficiently near z, to $\varphi(z')$ with the following conditions. First, Γ' does not go through vertices of $\mathfrak{N}$, Γ' does not share a segment with the sides of $\mathfrak{N}$, and

$$\rho(\Gamma') \leq \rho(\Gamma) + 1 = \rho(z_0, \varphi(z_0)) + 1.$$

Let

$$\pi_1, \dots, \pi_{s+1}; \qquad \pi_i = \psi_i(\overline{D}), \qquad i = 1, \dots, s+1 \tag{6.19}$$

be the normal "polygons" that the point z_1 of $\mathfrak{R}_1^*$ passes through as z_1 travels from z' to $\varphi(z')$ along Γ'. Since each $\psi(\overline{D})$ is a convex polygon, the "line" Γ does not cross the same $\psi(\overline{D})$ twice. Hence, we may assume that all the "polygons" $\pi_1, \dots, \pi_{s+1}$ that Γ' goes through are distinct. Let Γ_i

be the common side of π_i and π_{i+1}. When Γ_{i-1} and Γ_{i+1} have a common point, the common point must be on $\pi_i \cap \pi_{i+1} = \Gamma_i$. Hence, Γ_{i-1}, Γ_i, and Γ_{i+1} have a common vertex. Therefore, if Γ_{j-1} and Γ_{j+1}, Γ_j and Γ_{j+2}, Γ_{j+1} and Γ_{j+3}, ..., Γ_{j+l-2}, and Γ_{j+l}, Γ_{j+l-1}, and Γ_{j+l+1} have a common point, then $\Gamma_{j-1}, \Gamma_j, \Gamma_{j+1}, \dots, \Gamma_{j+l}, \Gamma_{j+l+1}$ have a common vertex in such a way that $\pi_j, \pi_{j+1}, \dots, \pi_{j+l}$ meet around the vertex. Hence, $l \leq \lambda_1$ must hold. Therefore, among the pairs Γ_1 and Γ_3, Γ_2 and $\Gamma_4, \dots, \Gamma_{s-2}$ and Γ_s, there are at least

$$\left[\frac{s}{\lambda_1 + 1}\right]$$

many pairs which do not have a common point. Let Γ_{i-1} and Γ_{i+1} be a pair without a common point. Let z'_{i-1} be the point of intersection of Γ' and Γ_{i-1}, and let z'_{i+1} be the point of intersection of Γ' and Γ_{i+1}. By the definition of ρ_1, we have

$$\rho_1 \leq \rho(z'_{i-1}, z'_{i+1}).$$

Therefore,

$$\rho_1 \left[\frac{s}{\lambda_1 + 1}\right] \leq \rho(\Gamma') \quad \text{and} \quad \rho_1 \left(\frac{s}{\lambda_1 + 1} - 1\right) \leq \rho(z, \varphi(z)) + 1.$$

Hence, if we set

$$\lambda = \frac{\lambda_1 + 1}{\rho_1} \quad \text{and} \quad \mu = \frac{(\lambda_1 + 1)(\rho_1 + 1)}{\rho_1}, \tag{6.20}$$

then

$$s \leq \lambda \rho(z, \varphi(z)) + \mu. \tag{6.21}$$

Since $\Gamma_i = \pi_i \cap \pi_{i+1} = \psi_i(\overline{D}) \cap \psi_{i+1}(\overline{D})$ as we saw, $\psi_i^{-1}(\Gamma_i) = \overline{D} \cap \psi_i^{-1}\psi_{i+1}(\overline{D})$ is a side of $\overline{D}$. Hence, $\psi_i^{-1}(\Gamma_i)$ must coincide with either γ_j or δ_j $(j = 1, \dots, m)$, say $\psi_i^{-1}(\Gamma_i) = \gamma_j$. Then

$$\psi_i^{-1}\psi_{i+1}(\overline{D}) = \varphi_j(\overline{D}), \qquad \psi_i^{-1}\psi_{i+1} = \varphi_j.$$

If $\psi_i^{-1}(\Gamma_i) = \delta_j$, then we similarly obtain

$$\psi_i^{-1}\psi_{i+1} = \varphi_j^{-1}.$$

In either case

$$\psi_{i+1} = \varphi_{ji}^{\pm 1}\psi_i, \qquad i = 1, \dots, s.$$

Therefore,

$$\psi_{s+1} = \varphi_{ji}^{\pm 1} \cdots \varphi_{js}^{\pm 1}\psi_1. \tag{6.22}$$

Moreover, since $\varphi(z')$ and $\psi_{s+1}(\psi^{-1}(z'))$ are in the interior of $\psi_{s+1}(\overline{D})$,

$$\varphi(z') = \psi_{s+1}(\psi_1^{-1}(z')), \qquad \varphi = \psi_{s+1}\psi_1^{-1}.$$

Consequently, we obtain from (6.22)

$$\varphi = \varphi_{ji}^{\pm 1} \cdots \varphi_{js}^{\pm 1}. \tag{6.23}$$

Therefore, we have shown that for a given z, φ can be expressed by s of the $\varphi_i^{\pm 1}$, satisfying condition (6.21). Note that for a different z, a different product in (6.23) is required for φ, but s, the number of $\varphi_i^{\pm 1}$, always satisfies (6.21). Furthermore, as one can see from (6.20), λ and μ are independent of φ and z. If r denotes the minimum of the number of $\varphi_i^{\pm 1}$ required to express φ in the form of (6.17), then (6.18) is true for any z with the same r. The proof for the lemma is thus complete.

In the above, the fact that G is generated by the maps that map the normal "polygon" $\overline{D}$ to an adjacent normal "polygon" still holds for noncompact $\mathfrak{R}^*(G)$. This is clear from the proof above.

CHAPTER 4

Algebraic Function Fields and Closed Riemann Surfaces

§1. The Riemann surface for an algebraic function field

In this chapter and the next, the classical theory of algebraic functions will be explained using the results obtained in the preceding chapters. That is, we study algebraic functions over the field of complex numbers k. There are many books on this topic. In addition to Weyl's book mentioned in Chapter 4, there are, for example, Appell-Goursat, *Théorie des fonctions algébriques et de leurs intégrales.* I, Gauthier-Villars, Paris, 1929 and G. A. Bliss, *Algebraic functions* (Amer. Math. Soc. Colloq. Publ., vol. 16, Amer. Math. Soc., Providence, RI, 1933. The latter has an extensive bibliography.

In this section, we will first show the following: When an algebraic function field K over the coefficient field of complex numbers is given (in the algebraic sense), we will define a compact Riemann surface, i.e., *closed Riemann surface.* Then we will prove that K can be mapped isomorphically to the field of analytic functions on that closed Riemann surface. Let P be an arbitrary place on K, and let u be an arbitrary element of K. The residue class field of P coincides with k by Theorem 2.3. For $v_P(u) \geq 0$, i.e., u is in the valuation ring of P, there exists a unique a in k such that

$$u \equiv a \mod \mathfrak{p}$$

holds, where $\mathfrak{p}$ is the prime ideal belonging to P. Then let

$$\overline{u}(P) = \begin{cases} a & \text{if } v_P(u) \geq 0 \text{ and } v_P(u-a) > 0, \\ \infty & \text{if } v_P(u) < 0. \end{cases} \tag{1.1}$$

From this definition, one obtains a complex-valued function $\overline{u}(P)$ defined on the set S of all the places on K. Then for elements u and v in K

$$\overline{u}(P) + \overline{v}(P) = (\overline{u+v})(P) \quad \text{and} \quad \overline{u}(P) \cdot \overline{v}(P) = (\overline{uv})(P)$$

hold. For an element c in k, $\overline{c}$ is the function on S which always assigns the value c. We identify this function with c. Then the correspondence $u \mapsto \overline{u}(P)$ defines an isomorphism from K to the set

$$\overline{K} = \{\overline{u}(P) \,;\, u \in K\}.$$

That is, the abstractly (algebraically) defined K is expressed as the function field $\overline{K}$ on S.

The value of $\overline{u}(P)$ may also be obtained as follows. Let t be a prime element for P, and let

$$u = \sum a_i t^i, \qquad a_i \in k$$

be the expansion of u in the complete field K_P. When the right-hand side contains a term with a negative exponent, we have $v_P(u) < 0$, i.e., $\overline{u}(P) = \infty$. When the right-hand side is an integral power series, we have $\overline{u}(P) = a_0$.

Our major goal is to show that a Riemann surface exists over the set S of places having $\overline{K}$ as its analytic function field. Upon proving the above assertion, from Theorem 3.14 such a Riemann surface is unique. Then this Riemann surface is said to be the Riemann surface belonging to the algebraic function field K, denoted as $\mathfrak{R}(K)$. Therefore, we have

$$K(\mathfrak{R}(K)) \cong K. \tag{1.2}$$

In order to prove the existence of $\mathfrak{R}(K)$, first let $K = k(z, w)$, where the relation between z and w is given by

$$f(z, w) = a_0(z)w^n + a_1(z)w^{n-1} + \cdots + a_n(z) = 0.$$

Here $a_i(z)$ is a polynomial of z with coefficients in k. Since k is an algebraically closed field of characteristic zero, Theorem 2.6 tells us how to find places on K. We will make a detailed study of the construction of places on K. We call those points z on the Riemann sphere $\mathfrak{R}_0^*$ extraordinary points when z satisfies one of the following:

(i) $z = \infty$,

(ii) z is a root of $a_0(z) = 0$,

(iii) z is a root of the polynomial obtained by eliminating w from $f(z, w) = 0$ and $f_w(z, w) = 0$, i.e., z is a root of the discriminant $d(z) = 0$ of $f(z, w) = 0$.

Other points on the Riemann sphere are called ordinary points.

Since there are only a finite number of extraordinary points, we can denote them by $b_1, \ldots, b_r$. When a is an arbitrary ordinary point, for each z in a small neighborhood of a, there are always n distinct roots $w_1, \ldots, w_n$ of $f(z, w) = 0$. By the implicit function theorem of Weierstrass, for a sufficiently small $\varepsilon > 0$, one has an expansion of $w_i = w_i(z)$

$$w_i(z) = P_i(z, a) = \sum_{j=0}^{\infty} a_{ij}(z-a)^j$$

in the neighborhood $|z-a| < \varepsilon$. The right-hand side of the above equation is a power series converging for $|z-a| < \varepsilon$. Note also that $P_i(z, a) \neq P_j(z, a)$ for $i \neq j$. For $|z-a| < \varepsilon$ we have

$$f(z, w) = a_0(z) \prod_{i=1}^{n} (w - P_i(z, a)).$$

Therefore, if we let

$$z = a + t \quad \text{and} \quad w = P_i(t) = \sum_{j=0}^{\infty} a_{ij} t^j, \quad i = 1, \ldots, n, \tag{1.3}$$

then for $|t| < \varepsilon$

$$f(a + t, P_i(t)) = 0.$$

Hence, from Theorem 2.6, the place P_a on $k(z)$ has n distinct extensions $P_{a,1}, \ldots, P_{a,n}$ determined on K by (1.3). Then those $P_{a,i}$ on K do not ramify.

Next let a be an extraordinary point such that $a \neq \infty$. Choose $\varepsilon > 0$ small enough so as not to include in $|z - a| < \varepsilon$ any other extraordinary point. For each point b in $0 < |z - a| < \varepsilon$ there exist n distinct power series as shown above

$$w_i = P_i(z, b) = \sum_{j=0}^{\infty} a_{ij}(b)(z - b)^j, \qquad i = 1, \ldots, n$$

satisfying $f(z, w) = 0$. Since $P_i(z, b)$ are all the roots of $f(z, w) = 0$ in a neighborhood of b, the theorem on analytic relations implies that when $P_i(z, b)$ is continued analytically along any curve in $0 < |z-a| < \varepsilon$ to a point c in $0 < |z-a| < \varepsilon$, the set of the continuations of the n function elements $P_i(z, b)$, $i = 1, \ldots, n$ coincides with the set $P_i(z, c)$, $i = 1, \ldots, n$. In particular, let $P_i(z, b)$ be continued analytically to $P_{i'}(z, b)$ as starting from b around a in the counterclockwise direction to reach the starting point. Thus, we obtain a permutation of indices $1, 2, 3, \ldots, n$;

$$\sigma = \begin{pmatrix} 1 & 2 & 3 & \cdots & n \\ 1' & 2' & 3' & \cdots & n' \end{pmatrix}.$$

Decompose this permutation into the product of cycles:

$$\sigma = (1, 2, \ldots, e)(e + 1, \ldots, e + e')(e + e' + 1, \ldots) \cdots. \tag{1.4}$$

Then $P_1(z, b)$ and its function elements obtained from $P_1(z, b)$ by analytic continuation define a multivalued function $g(z)$ in $0 < |z-a| < \varepsilon$, and $g(z)$ has e branches $P_1(z, b), \ldots, P_e(z, b)$ at each point in the domain. Each branch takes the original value when z moves e times around the point a. Then let

$$(z - a) = t^e.$$

Then $0 < |t| < \varepsilon^{1/e}$ is mapped e-fold over $0 < |z - a| < \varepsilon$. Hence, $h(t) = g(a + t^e)$ becomes a single-valued regular function of t in $0 < |t| < \varepsilon^{1/e}$. If $a_0(a) \neq 0$ holds, $P_i(z, b)$ converges to a finite root of $f(z, w) = 0$ as $z \to 0$. Therefore, $h(t)$ is also a regular function at $t = 0$, by Riemann's theorem. If $a_0(a) = 0$ holds in $f(z, w)$, then let $a_0(z) = (z - a)^l a_0'(z)$, $a_0'(a) \neq 0$. Replacing $g(z)$ by $(z - a)^l g(z)$,

$$(z - a)^l g(z) = t^{el} h(t)$$

is finite at $t = 0$ and is therefore regular. That is, either case implies that $h(t)$ is an analytic function of t having at most a pole at $t = 0$.

Let

$$h(t) = c_1 t^{e_1} + c_2 t^{e_2} + \cdots, \qquad e_1 < e_2 < \cdots, \qquad c_i \neq 0$$

be the expansion of $h(t)$ in $|t| < \varepsilon^{1/e}$. Then we have

$$g(z) = \sum c_i (z-a)^{e_i/e}, \qquad |z-a| < \varepsilon.$$

Corresponding to e choices of the eth root $(z-a)^{1/e}$ of $(z-a)$ on the right, we obtain e Puiseux series

$$P_1(z, a), \ldots, P_e(z, a), \tag{1.5}$$

which coincide with e ramified elements $P_1(z, b), \ldots, P_e(z, b)$ of $g(z)$ in $0 < |z-a| < \varepsilon$.

For other cycles $(e+1, \ldots, e+e')$, $(e+e'+1, \ldots), \ldots$ of σ, we obtain similar groups of Puiseux series in $0 < |z-a| < \varepsilon$:

$$\begin{cases} P_{e+1}(z, a), \ldots, p_{e+e'}(z, a), \\ P_{e+e'+1}(z, a), \ldots, \\ \ldots, \end{cases} \tag{1.6}$$

while we have

$$f(z, w) = a_0(z) \prod_{i=1}^{n} (w - P_i(z, a)).$$

Therefore, the place P_a on $k(z)$ associated with the ideal $(z-a)$ has extensions on K with ramification indices $e, e', \ldots$, corresponding to the Puiseux series (1.5), (1.6).

If $a = \infty$, then consider a neighborhood of 0 of $z' = 1/z$. Take $z = t^{-e}$ instead of $z - a = t^e$ for this case. Then we can obtain similar results as above.

From the above we have obtained the following.

THEOREM 4.1. *Let $K = k(z, w)$ be an algebraic function field over the field complex numbers, let P be a place on K, and let e, $e \geq 1$, be the ramification index of P with respect to $k(z)$. Let*

$$z = \begin{cases} a + t^e, \\ t^{-e}, \end{cases} \qquad w = c_1 t^{e_1} + c_2 t^{e_2} + \cdots, \ e_1 < e_2 < \cdots, \ c_i \neq 0$$

be the expansions of z and w with respect to a prime element t for P as in Theorem 2.6. Then $c_1 t^{e_1} + c_2 t^{e_2} + \cdots$ is not only a formal power series in t, but also a converging power series in a neighborhood $|t| < \varepsilon^{1/e}$ of $t = 0$ with a possible exception at $t = 0$.

From what we have seen, when there are no extraordinary points of $\mathfrak{R}_0$ in $0 < |z-a| < \varepsilon$ or $0 < |1/z| < \varepsilon$ for a small ε, at an arbitrary point t_0

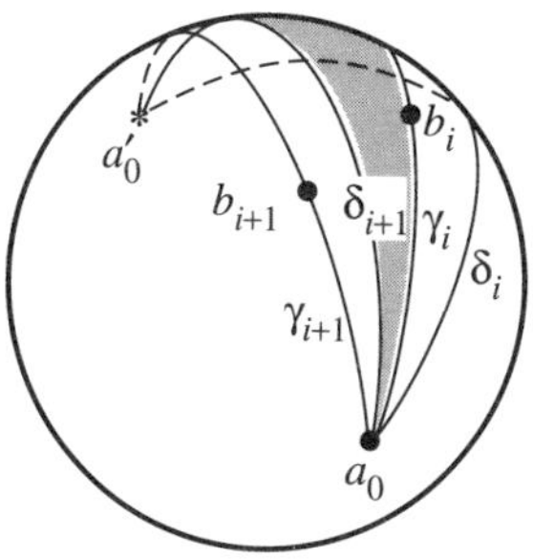

FIGURE 4.1

in $0 < |t| < \varepsilon^{1/e}$, where $(b - a) = t_0$, we have that $h(t) = c_1 t^{e_1} + c_2 t^{e_2} + \cdots$ determines a function element $P_i(z, b)$. Therefore, we obtain a place $P(t_0)$ on K which does not ramify for $K/k(z)$. Then if we let the place P in Theorem 4.1 be $P = P(0)$, then the set of places

$$U(P, \varepsilon) = \{P(t); |t| < \varepsilon^{1/e}\} \tag{1.8}$$

and the set of points in the open disc $|t| < \varepsilon^{1/e}$ in the t-plane correspond to each other in a one-to-one manner.

For a given place P on K, define the collection $\{U(P, \varepsilon)\}$ as a neighborhood system at P and introduce a topology on the set S of all places on K. Next we will prove S is a compact connected manifold of dimension 2 in this topology.

Let $\mathfrak{R}_0^*$ be the Riemann sphere as before, and let $b_1, \dots, b_r$ be extraordinary points on $\mathfrak{R}_0^*$. Let a_0 be an arbitrary point that differs from b_i and their antipodal points, and let a_0' be the antipodal point of a_0. (See Figure 4.1.) Then there exists a unique great circle going through a_0, b_i, and a_0', denoted as γ_i. One may assume that γ_{i-1} and γ_i are next to each other by renumbering the b_i. Let δ_i be the great circle through a_0 and a_0' bisecting the area bounded by γ_{i-1} and γ_i (we put $\delta_{-1} = \delta_r$ for $i = 1$). Let t_{2i-1} be the spherical triangle bounded by δ_i, $a_0 b_i$, and $b_i a_0'$, and let t_{2i} be the spherical triangle bounded by δ_{i+1}, $a_0 b_i$, and $b_i a_0'$. Then $\mathfrak{R}_0^*$ is decomposed into $2r$ triangles $t_1, \dots, t_{2r}$.

Next let $\overline{T}$ be a regular $2r$-gon in the plane with center a_1 and denote the vertices successively by $c_1, d_1, \dots, c_r, d_r$. Then there exists a continuous map (a simplicial map) from $\overline{T}$ to $\mathfrak{R}_0^*$ so that the $2r$ triangles $\Delta a_1 c_i d_i$, $\Delta a_1 d_i c_{i+1}$, $i = 1, \dots, r$, are mapped homeomorphically onto the $2r$ triangles t_{2i-1} and t_{2i} on $\mathfrak{R}_0^*$ so that

$$a_0' \leftrightarrow a_1, \qquad b_i \leftrightarrow d_i,$$

$$a_0' b_i \leftrightarrow a_1 d_i, \qquad a_0 b_i \leftrightarrow c_i d_i, \qquad \delta_i \leftrightarrow a_1 c_i, \qquad a_0 b_i \leftrightarrow d_i c_{i+1}.$$

Then $c_1, \dots, c_r$ in $\overline{T}$ correspond to a_0 on $\mathfrak{R}_0^*$, and $c_i d_i$ and $d_i c_{i+1}$ correspond to the arc $a_0 b_i$. Notice that the complement U_0 of $a_0 b_i$, $i =$

$1, \dots, r$, in $\mathfrak{R}_0^*$ is mapped homeomorphically onto the interior T of $\overline{T}$.

As before, let $P_i(z, a_0')$, $i = 1, \dots, n$, be the function elements determined by $f(z, w) = 0$ at $z = a_0'$. In the above U_0 each $P_i(z, a_0')$ can be analytically continued along any curve. Since U_0, like T, is simply connected, the analytic continuation determines n single-valued analtyic functions $g_i(z)$, $i = 1, \dots, n$. Each $g_i(z)$ may be continued analytically beyond the segment $a_0 b_j$, but then in general one gets another $g_{i'}(z)$ rather than $g_i(z)$. Let us suppose $g_i(z)$ becomes $g_{i'}(z)$ as going from t_{2j-1} to t_{2j} crossing γ_i. Then a permutation

$$\sigma_j = \begin{pmatrix} 1, 2, \dots, n \\ 1', 2', \dots, n' \end{pmatrix}$$

is defined. Let $\overline{T}, \overline{T}_2, \dots, \overline{T}_n$ be n copies of $\overline{T}$, and let $a_1^{(i)}$, $c_j^{(i)}$, $d_j^{(i)}$ be the points on $\overline{T}_i$ corresponding to a_1, c_j, d_j on $\overline{T}$. For the permutation σ_j, identify the side $(c_j^{(i)}, d_j^{(i)})$ of $\overline{T}_i$ with the side $(c_{j+1}^{(i')}, d_j^{(i')})$ of $\overline{T}_{i'}$ so that $c_j^{(i)}$ is identified with $c_{j+1}^{(i')}$ and $d_j^{(i)}$ with $d_j^{(i')}$. With this identification for each permutation σ_j, $j = 1, \dots, r$, one obtains a topological space S' from the above $\overline{T}_1, \dots, \overline{T}_n$. This topological space is a homeomorphic model of the space S of places on K. Let a be an ordinary point, $a \neq a_0$, on $\mathfrak{R}_0^*$. Then there are n function elements $P_i(z, a)$ of $f(z, w) = 0$ obtained from $g_i(z)$. For the place $P_{a,i}$ on K belonging to the function element $P_i(z, a)$, assign the image $a^{(i)}$ in $\overline{T}_i$ of a. Then the map defined by this assignment is a homeomorphism and in a neighborhood of $P_{a,i}$. The corresponding point on S' to $a = a_0$ may appear to be a point of a special role. However, if one considers that there are n distinct $P_i(z, a_0)$, $i = 1, \dots, n$, at a_0, one sees easily that there are exactly n points $a_0^{(i)}$, $i = 1, \dots, n$, on S' corresponding to $P_i(z, a_0)$, and that the correspondence $P_i(z, a) \leftrightarrow a^{(i)}$ also establishes a homeomorphism in a neighborhood of $a_0^{(i)}$. Finally, at $a = b_j$, decompose $\sigma = \sigma_j$ into cyclic parts as in (1.4). Then each of

$$d_j^{(1)} = d_j^{(2)} = \cdots = d_j^{(e)},$$
$$d_j^{(e+1)} = d_j^{(e+2)} = \cdots = d_j^{(e+e')},$$
$$\dots\dots$$

gives a point on S' to which one can assign a place on K determined from (1.6). Then this correspondence between S and S' is one-to-one and a homeomorphism at those points. (Details are left for the reader.)

From the above consequence, S and S' are homeomorphic to each other. Therefore, S is indeed a compact manifold of dimension 2. Next we will prove the connectedness of S. We need to show that S' is connected. That is, after the above identification, we will show that all the $\overline{T}_1, \dots, T_n$ are connected. Let us assume that with the identification $\overline{T}^{(1)}, \dots, \overline{T}^{(m)}$ and $\overline{T}^{(m+1)}, \dots, \overline{T}^{(n)}$, $1 \le m < n$, are not connected. Then the collection of function elements from $P_1(z, a_0'), \dots, P_m(z, a_0')$, obtained by analytic continuation along any curve avoiding extra ordinary points on $\mathfrak{R}_0^*$, must coincide with $P_1(z, b), \dots, P_m(z, b)$ belonging to $g_1(z), \dots, g_m(z)$. Therefore, the fundamental symmetric functions of $P_i(z, a)$, $i = 1, \dots, m$,

$$s_1(z) = \sum_{i=1}^{m} P_i(z, a), \quad s_2(z) = \sum_{i<j} P_i(z, a)P_j(z, a), \dots \tag{1.9}$$

are single-valued regular functions of z outside the extraordinary points. Moreover, each $P_i(z, a)$ can be expanded as a Puiseux series in $z - a_j$, in a neighborhood of an extraordinary point a_j. When $s_1(z)$, $s_2(z), \dots$ are multiplied by certain powers of $(z - a_j)$, they converge to 0 as $z \to a_j$. Hence, the only singularities of those functions at a_j are poles. That is, $s_i(z)$ is an analytic function on the Riemann sphere with possible singularities of poles. Therefore, $s_i(z)$ is a rational function of z. Hence,

$$f_1(z, w) = \prod_{i=1}^{m}(w - P_i(z, a)) = w^m - s_1(z)w^{m-1} + \cdots + (-1)^m s_m(z)$$

is a polynomial in w with coefficients in $k(z)$. One obtains a similar polynomial $f_2(z, w)$ from $P_{m+1}(z, a), \dots, P_n(z, a)$, and then one has

$$f(z, w) = a_0(z)\prod_{i=1}^{n}(w - P_i(z, a)) = a_0(z)f_1(z, a)f_2(z, a),$$

contradicting the irreducibility of $f(z, w)$. Therefore, S' is connected, which implies S is connected.

We have established that S is a compact connected manifold of dimension 2. Next we will define a Riemann surface, with underlying space S, by giving a system of R-functions on S. For an arbitrary point P in S choose a sufficiently small $\varepsilon_P > 0$. Then for the neighborhood defined in (1.8) set

$$U_P = U(P, \varepsilon_P).$$

Since the correspondence from a point Q in U_P to a point t in $|t| < \varepsilon^{1/e}$ is one-to-one, denoted as $t_P = t_P(Q)$, the map defined by

$$\varphi_P(Q) = \varepsilon_P^{-1/e} t_P(Q)$$

is a homeomorphism from U_P to the interior $|t| < 1$ of the unit disc. Next for $P \ne P'$ let Q be any point in $U_P \cap U_{P'}$, and let P_a, $P_{a'}$, and P_z be the projections on $k(z)$ of P, P', and Q, respectively. Then we have

$$t_P^e(Q) = z - a \quad \text{and} \quad t_{P'}^{e'}(Q) = z - a',$$

where e and e' are the ramification indices of P and P' with respect to $k(z)$, respectively. Eliminating z from the equations,

$$t_{P'}^{e'}(Q) = a - a' + t_P^e(Q). \tag{1.10}$$

This equation means that the coordinate $t_{P'}(Q)$ for $U_{P'}$ of Q in $U_P \cap U_{P'}$ is an analytic function of the coordinate $t_P(Q)$ for U_P. Since we know the correspondence between $t_{P'}(Q)$ and $t_P(Q)$ is one-to-one and continuous, equation (1.10) defines a single-valued regular function. Therefore, the function $\varphi_{P'}\varphi_P^{-1}$ is a one-to-one conformal map from $\varphi_P(U_P \cap U_{P'})$ onto $\varphi_{P'}(U_P \cap U_{P'})$. Thus,

$$\mathfrak{F} = \{\varphi_P, U_P\}$$

gives an R-function system on S. Hence, we have a Riemann surface

$$\mathfrak{R} = \{S, \{\mathfrak{F}\}\}.$$

We will prove that the analytic function field $K(\mathfrak{R})$ of the Riemann surface $\mathfrak{R}$ constructed above coincides with the function field defined earlier

$$\overline{K} = \{\overline{u}(P)\}.$$

Let u be an arbitrary element of K, and let P be an arbitrary point on $\mathfrak{R}$. Let $U_P = U(P, \varepsilon)$ and the locally uniformizing variable $t = t_P(Q)$ in U_P be as before. Since the element u in K can be written as a rational function of z and w (moreover, a rational integral function of w), let

$$u = r(z, w).$$

Substitute the t-expansions (1.7) of z and w into this $r(z, w)$. Then we have the t-expansion of u

$$u = \sum a_i t^i. \tag{1.11}$$

Note that (1.11) is the algebraic (formal) expansion of u with respect to the prime element t for P. Since (1.7) converges for $0 < |t| < \varepsilon$, there is ε_1, $0 < \varepsilon_1 < \varepsilon$, so that the right-hand side of (1.11) is a converging power series for $0 < |t| < \varepsilon_1$. Let $F(t)$ be the analytic function defined by the right-hand side for $|t| < \varepsilon_1$. Then by the definition of $\overline{u}(P)$ we have

$$\overline{u}(P) = F(0).$$

Then let P_1 be any point such that

$$0 < |t(P_1)| < \varepsilon_1.$$

Let $t_1 = t_{P_1}(Q)$ be a locally uniformizing variable for P_1 just as $t = t_P$ was defined for P in the above. Then

$$t_2(Q) = t(Q) - t(P_1) \tag{1.12}$$

is also a locally uniformizing variable in a neighborhood of P_1. Then, from Theorem 3.1, we have in a neighborhood of P_1 that

$$t_2(Q) = \sum_{i=1}^{\infty} b_i t_1(Q)^i, \qquad b_1 \neq 0.$$

That is, $t_2 = t_2(Q)$ is a prime element for P_1 in the algebraic sense as well as $t_1 = t_1(Q)$. Hence, the value of $\overline{u}(P_1)$ may be found from the t_2-expansion of u. Rewrite $u = F(t) = \sum a_i t^i$ using $t_2 = t - t(P_1)$ as a power series in t_2. Then since $F(t(P_1)) \neq \infty$,

$$u = \sum_{i=0}^{\infty} c_i t_2^i, \qquad c_0 = F(t(P_1)).$$

By the definition of $\overline{u}(P)$,

$$\overline{u}(P_1) = F(t(P_1)).$$

Replacing P_1 by Q, we have in the neighborhood $|t(Q)| < \varepsilon_1$ of P

$$\overline{u}(Q) = \sum a_i t(Q)^i. \tag{1.13}$$

That is, if one chooses a locally uniformizing variable $t = t_P(Q)$ for P on $\mathfrak{R}$ from the t in Theorem 4.1, then the algebraic expansion of the element u in K with respect to the prime element t of P and the analytic expansion of the function $\overline{u}(Q)$ from K for the locally uniformizing variable $t_P(Q)$ are given by the same power series. In particular, since the complex-valued function $\overline{u}(Q)$ in (1.13) is analytic in a neighborhood of P, $\overline{u}(P)$ must be an analytic function on the Riemann surface $\mathfrak{R}$. Therefore, we obtain

$$\overline{K} \subseteq K(\mathfrak{R}). \tag{1.14}$$

Conversely, let $g(P)$ be an arbitrary analytic function on $\mathfrak{R}$. For an ordinary point a on the Riemann sphere $\mathfrak{R}_0^*$, there are exactly n extensions on $K P_{a,1}, \dots, P_{a,n}$ of the place P_a on $k(z)$. Then each $P_{a,i}$ is unramified, and $t = z - a$ is a prime element (in the algebraic sense) and also a locally uniformizing variable (in the analytic sense). Therefore, $g(P)$ may be expanded in a neighborhood of $P_{a,i}$ as a power series in $(z - a)$:

$$g(P) = P_i(z, a) = \sum a_{ij}(z - a)^j,$$

where the right-hand side contains only finitely many terms with negative exponents. Consider the fundamental symmetric functions of $P_i(z, a)$

$$s_1'(z) = \sum_{i=1} P_i(z, a), \quad s_2'(z) = \sum_{i<j} P_i(z, a) P_j(z, a), \quad \dots .$$

These functions are single-valued analytic functions in the complementary domain of extraordinary points in $\mathfrak{R}_0^*$. Furthermore, as we showed in the paragraph containing equation (1.4), those symmetric functions can only have the singularities of poles in a neighborhood of an extraordinary point. That is, all the $s_i'(z)$ must be rational functions of z. In a neighborhood of an ordinary point a,

$$g(P)^n - s_1'(\overline{z}(P))g(P)^{n-1} + \cdots + (-1)^n s_n'(\overline{z}(P)) = 0.$$

Since the left-hand side is an analytic function on $\mathfrak{R}$, the above relation holds for every point P on $\mathfrak{R}$. Therefore, we have shown that an arbitrary function $g(P)$ in the analytic function field $K(\mathfrak{R})$ is always an algebraic element over the subfield $\overline{K}_0 = k(\overline{z}(P))$, with degree at most n.

Let $g_1 = g_1(P)$ be any function in $K(\mathfrak{R})$, and let $\overline{K}_1 = \overline{K}(g_1)$ be the field obtained by adjoining g_1 to $\overline{K}$. Since $[\overline{K} : \overline{K}_0] = [K : K_0] = n$, $\overline{K}_1$ is a finite extension of $\overline{K}_0$, and since the characteristic of K_0 is 0, there exists a function $g = g(P)$ satisfying

$$\overline{K}_1 = \overline{K}_0(g).$$

As proved above, the degree of g for $\overline{K}_0$ is at most n, i.e.,

$$[\overline{K}_1 : \overline{K}_0] \leq n.$$

On the other hand,

$$[\overline{K} : \overline{K}_0] = n \quad \text{and} \quad \overline{K} \subseteq \overline{K}_1.$$

Hence, $\overline{K} = \overline{K}_1$. In particular, $g_1 \in \overline{K}$. Since g_1 was an arbitrary element in $K(\mathfrak{R})$, we have $K(\mathfrak{R}) \subseteq \overline{K}$. Combining this result with (1.14), we have

$$K(\mathfrak{R}) = \overline{K}. \tag{1.15}$$

Therefore, we have proved the following theorem.

THEOREM 4.2. *Let K be an arbitrary algebraic function field over the field k of complex numbers. Then there is a unique closed Riemann surface $\mathfrak{R}$ such that the points of $\mathfrak{R}$ are the places on K and its analytic function field on $\mathfrak{R}$ coincides with $K = \{\overline{u}(P)\}$. This Riemann surface $\mathfrak{R}$ is said to belong to K and is denoted by $\mathfrak{R}(K)$. From* (1.2) *we have*

$$K(\mathfrak{R}(K)) \cong K. \tag{1.16}$$

§2. Analytic function fields on closed Riemann surfaces

In the preceding section we defined the closed Riemann surface belonging to a given algebraic function field K. We will consider the analytic function field $K(\mathfrak{R})$ over a given closed Riemann surface $\mathfrak{R}$, and show that $K(\mathfrak{R})$ is an algebraic function field over the field of complex numbers. Furthermore we will also examine various relations between algebraic function fields and closed Riemann surfaces.

From Theorem 3.11, $K(\mathfrak{R})$ contains a nonconstant analytic function. Let $z = z(P)$ be one of the analytic functions in $K(\mathfrak{R})$. Then $z(P)$ defines an analytic map from $\mathfrak{R}$ to the Riemann sphere $\mathfrak{R}_0^*$. Since z is not constant, Theorem 3.3 implies that the image $z(\mathfrak{R})$ is an open set in $\mathfrak{R}_0^*$. On the other hand, $\mathfrak{R}$ is compact by the assumption, and $z(P)$ is continuous. Therefore, $z(\mathfrak{R})$ is compact so that the subset $z(\mathfrak{R})$ in the connected space $\mathfrak{R}_0^*$ is simultaneously, open and closed. Hence, we have

$$z(\mathfrak{R}) = \mathfrak{R}_0^*.$$

Let $Q_1, \dots, Q_r$ be the branch points for the map $z(P)$ (cf. Chapter 3, §1), and let

$$b_i = z(Q_i), \qquad i = 1, 2, \dots, r.$$

Since the set of branch points does not have an accumulation point in the compact space $\mathfrak{R}$, the set $\{Q_i\}$ is a finite set. Let us call these $b_1, \dots, b_r$, and ∞ the exceptional points in $\mathfrak{R}_0^*$ (the same point in $\mathfrak{R}_0^*$ may appear more than once in $b_1, \dots, b_r$ and ∞). Points are called ordinary points if they are not exceptional points.

Let a be an arbitrary ordinary point on $\mathfrak{R}_0^*$, and let

$$z(P) = a.$$

Then a neighborhood of P is mapped by z onto a neighborhood of a in a one-to-one fashion so that the inverse image $z^{-1}(a)$ contains only P in a neighborhood of P. Since $\mathfrak{R}$ is compact, $z^{-1}(a)$ must be a finite set

$$z^{-1}(a) = \{P_1, P_2, \dots, P_n\}.$$

Since none of the P_i is a branch point, one can take a sufficiently small $\varepsilon > 0$ so that the neighborhoods $U_i(\varepsilon)$ of P_i do not intersect each other. Each $U_i(\varepsilon)$, $i = 1, \dots, n$, is mapped conformally by $Z(P)$ to the neighborhood $U(\varepsilon) = \{z; |z-a| < \varepsilon\}$ of a. Next for some even smaller ε_1, $\varepsilon > \varepsilon_1 > 0$, we will prove that the inverse image $z^{-1}(U(\varepsilon_1))$ of $U(\varepsilon_1) = \{z; |z-a| < \varepsilon_1\}$ is contained in the union of $U_1, \dots, U_n$. Suppose $z^{-1}(U(\varepsilon_1))$ is not contained in $U_1 \cup \cdots \cup U_n$ for any small ε_1. Then there exist points P_j^* in $\mathfrak{R}$, not contained in $U_1 \cup \cdots \cup U_n$, such that

$$\lim_{j\to\infty} z(P_j^*) = a. \tag{2.1}$$

Since $\mathfrak{R}$ is compact, taking a subsequence of P_j^* converges to a point P^* on $\mathfrak{R}$. Then by (2.1)

$$z(P^*) = a.$$

Therefore, P^* must coincide with one of $P_1, \dots, P_n$. But the P_j^*'s are not in $U_1 \cup \cdots \cup U_n$, hence P^* does not belong to U_i, $i = 1, \dots, n$, leading to a contradiction.

For an arbitrary ordinary point a, let $n = N(a)$ be the number of points in $z^{-1}(a)$. As we saw in the above, for a' in a sufficiently small neighborhood $U(\varepsilon_1)$ of a, $N(a')$ and $N(a)$ are the same. Set

$$M_n = \{a; N(a) = n \quad \text{for } n = 1, 2, 3, \dots.$$

All the M_n are open sets, $M_n \cap M_m = \emptyset$ for $n \neq m$, and the totality S_0' of ordinary points in $\mathfrak{R}_0^*$ is in the union of these disjoint sets M_n, $n = 1, 2, \ldots$. Since S_0' is the complement of a finite set in the Riemann sphere, S_0' is connected. Therefore, $S_0' = M_n$ must hold for some natural number n. That is, for any ordinary point a, there are always n points $P_1, \ldots, P_n$ in $\mathfrak{R}$ so that $z(P_i) = 0$ and the locally uniformizing variable

$$t' = z - a = z(P) - z(P_i) = t \tag{2.2}$$

maps conformally a neighborhood of P_i to a neighborhood of a.

Next let b be a finite exceptional point. Then as before,

$$z^{-1}(b) = \{P_1', \ldots, P_r'\} \tag{2.3}$$

is a finite set. Then for a sufficiently small $\varepsilon_1 > 0$, there is a locally uniformizing variable t_i at each P_ε' so that

$$t' = z - b = t_i^{e_i}, \qquad i = 1, \ldots, r \tag{2.4}$$

maps the disjoint neighborhoods $U_i = \{P; |t_i(P)| < \varepsilon_1^{1/e_i}\}$ of P_i onto the neighborhood $U(\varepsilon_1) = \{z; |z-b| < \varepsilon_1\}$ e_i-foldedly for each i. Furthermore, as was for an ordinary point, one can show

$$z^{-1}(U(\varepsilon_1)) = U_1 \cup \cdots \cup U_r. \tag{2.5}$$

We can assume that b is the only exceptional point in $U(\varepsilon_1)$. Hence, for any point c in $U(\varepsilon_1)$, $c \neq b$, there are e_i points in each U_i that are mapped to c by z. From (2.5) we have

$$n = e_1 + e_2 + \cdots + e_r. \tag{2.6}$$

In the case when $z = \infty$, one chooses $t' = 1/z$ instead of $t' = z - b$ and one can obtain similar results as above.

With this preparation, we now consider an analytic function $g(P)$ on $\mathfrak{R}$. For an ordinary point z in $\mathfrak{R}_0^*$, let

$$\{P_1(z), \ldots, P_n(z)\}$$

be the inverse image of the point z under the map $z(P)$ from $\mathfrak{R}$ to $\mathfrak{R}_0^*$. Then consider the fundamental symmetric functions of $g(P_i(z))$ $i = 1, 2, \ldots, n$,

$$s_1(z) = \sum_{i=1}^{n} g(P_i(z)), \quad s_2(z) = \sum_{i<j} g(P_i(z))g(P_j(z)), \quad \ldots.$$

Since $g(P)$ is an analytic function, g may be expanded as a power series of a locally uniformizing variable in a neighborhood of each point on $\mathfrak{R}$. If a neighborhood $U(\varepsilon_1)$ of an ordinary point a is chosen as before, then the neighborhood $U_i(\varepsilon_1)$ of each P_i is mapped conformally by $z(P)$ to the neighborhood $U(\varepsilon_1)$. Then from (2.2), $g(P_i(z))$, and hence also

$s_1(z), s_2(z), \ldots$ can be expanded as power series in $z-a$. Moreover, in a neighborhood of an exceptional point, $s_i(z)$ can be expressed as a Puiseux series in either $z-b$ or $1/z$ as we saw earlier. As in the previous section, all the $s_i(z)$, as functions on $\mathfrak{R}_0^*$, are rational functions of z. Let P be a point on $\mathfrak{R}$ such that $a = z(P)$ is an ordinary point. We have

$$g(P)^n - s_1(z(P))g(P)^{n-1} + \cdots + (-1)^n s_n(z(P)) = 0 \tag{2.7}$$

not only in a neighborhood of P, but also (2.7) holds identically on $\mathfrak{R}$. This is because the left-hand side of (2.7) is an analytic function on $\mathfrak{R}$. Therefore, an arbitrary function $g(P)$ in $K(\mathfrak{R})$ is an algebraic element over $K_0 = k(z)$, $z = z(P)$, of degree at most n.

Hence, if one chooses a function $w = w(P)$ in $K(\mathfrak{R})$ so that $[K(w) : K_0]$ is maximal. Then for an arbitrary element $w_1 = w_1(P)$ of $K(\mathfrak{R})$, $K_0(w, w_1)$ is a finite extension of K_0. Noting that the characteristic of $k(z)$ is 0, we obtain

$$K_0(w, w_1) = K_0(w_2) \quad \text{for some } w_2 = w_2(P).$$

Since $K_0(w_2)$ contains $K_0(w)$,

$$[K_0(w) : K_0] \leq [K_0(w_2) : K_0].$$

From our choice of w, we have

$$[K_0(w) : K_0] = [K_0(w_2) : K_0], \qquad K_0(w) = K_0(w_2).$$

That is, w_1 is contained in $K_0(w)$. Since w_1 was an arbitrary function in $K(\mathfrak{R})$,

$$K(\mathfrak{R}) = K_0(w) = k(z, w).$$

We have found that the analytic function field $K(\mathfrak{R})$ of a closed Riemann surface is the algebraic function field generated by z and w over k. Next we will study the relation between $\mathfrak{R}$ and $K(\mathfrak{R})$ more closely. For an ordinary point a, as before, let

$$z^{-1}(a) = \{P_1, \ldots, P_n\}.$$

Then $t = z - a$ is a locally uniformizing variable at each P_i. Hence, $w = w(P)$ can be expanded as a power series in t

$$w = P_i(t) = \sum_j a_{ij} t^j$$

in a neighborhood of P_i. For $i \neq j$, we have two different power series for P_i and P_j. Suppose $P_i(t) = P_j(t)$. Since any function $w_1 = w_1(P)$ in $K(\mathfrak{R})$ is a rational function of z and w, the t-expansions of w_1 at P_i and P_j are always the same. From Theorem 3.11, we can find a function $w_1(P)$ in $K(\mathfrak{R})$ such that

$$w_1(P_i) \neq w_1(P_j),$$

which leads to a contradiction. According to Theorem 2.6, from

$$z = a + t, \qquad w = \sum_j a_{ij} t^j, \qquad i = 1, 2, \ldots, n,$$

we get n distinct places on $K(\mathfrak{R})$. Let $\overline{P}_i$ be the place associated with P_i. Then $\overline{P}_1, \ldots, \overline{P}_n$ are extensions on $K(\mathfrak{R})$ of the place P_a on $k(z)$ determined by the ideal $(z-a)$. But, since we have

$$[K(\mathfrak{R}) : k(z)] = [K_0(w) : K_0] \leq n,$$

Theorem 2.6 implies that P_a can have at most n extensions on $K(\mathfrak{R})$. Therefore, we have

$$[K(\mathfrak{R}) : k(z)] = n,$$

and then $\overline{P}_1, \ldots, \overline{P}_n$ are all the extensions of P_a on $K(\mathfrak{R})$.

Let b be a finite exceptional point, and let P'_i, t_i, e_i be as in (2.3) and (2.4). Then in a neighborhood of P'_i, $w = w(P)$ is expanded as

$$w = \sum_j b_{ij} t_i^j.$$

However, by Theorem 3.11, there is a function in $K(\mathfrak{R})$, i.e., a rational function of z and w, whose expansion begins with a nonzero linear term. Therefore, the greatest common divisor of e_i and all j such that $b_{ij} \neq 0$, must be 1. Again by Theorem 2.6,

$$z = b + t_i^{e_i} \quad \text{and} \quad w = \sum_j b_{ij} t_i^j$$

provide a place on $K(\mathfrak{R})$, denoted by $\overline{P}'_i$. Just as for an ordinary point, $\overline{P}'_i \neq \overline{P}'_j$ holds for $i \neq j$. Once more from Theorem 2.6 and also from (2.6), $\overline{P}'_1, \ldots, \overline{P}'_r$ are all the extensions on $K(\mathfrak{R})$ of the place P_b on $k(z)$. One can treat the case $z = \infty$ in the same manner.

We can summarize what we have obtained as follows: There is a one-to-one correspondence between points P on the Riemann surface $\mathfrak{R}$ and places P on $K(\mathfrak{R})$. A locally uniformizing variable t, or t_i at P gives a prime element of $\overline{P}$ in the algebraic sense. As a consequence, for an arbitrary $u = u(P)$ in $K(\mathfrak{R})$, the analytical expansion of u in the locally uniformizing variable t in a neighborhood of P coincides with the formal t-expansion of u obtained by regarding $K(\mathfrak{R})$ algebraically. Hence, the value $u(P)$ of u at P is exactly the value $\overline{u}(\overline{P})$ at $\overline{P}$ as defined by (1.1) in §1,

$$u(P) = \overline{u}(\overline{P}).$$

Therefore, if $\mathfrak{R}(K(\mathfrak{R}))$ denotes the Riemann surface belonging to $K(\mathfrak{R})$ whose existence was proved in Theorem 4.2, then Theorem 3.14 implies that

$$\mathfrak{R}(K(\mathfrak{R})) = \mathfrak{R}. \tag{2.8}$$

Therefore, we have obtained the following theorem.

THEOREM 4.3. *The analytic function field $K(\mathfrak{R})$ of a closed Riemann surface $\mathfrak{R}$ is an algebraic function field over the field k of complex numbers, and the Riemann surface $\mathfrak{R}(K(\mathfrak{R}))$ belonging to the analytic function field $K(\mathfrak{R})$ is isomorphic to the Riemann surface $\mathfrak{R}$.*

In general, when an arbitrary polynomial $f(z, W)$ with coefficients in the field of complex numbers is given such that the degree in W is n, for any complex number z, there are n values of W for which

$$f(z, W) = 0.$$

Regard the values of W as a function of z, denoted as

$$w = w(z).$$

This multivalued function $w(z)$ of z is said to be an *algebraic function.* In order to study the properties of $w(z)$, it is crucial to consider the Riemann surface $\mathfrak{R}(K)$ belonging to the algebraic function field $K = k(z, w)$ determined by $f(z, w)$ as in the proof of Theorem 4.2. Let $\overline{z}(P)$ and $\overline{w}(P)$ be the analytic functions on $\mathfrak{R}(K)$ corresponding to z and w. Then for any complex value z_0, generally there are n points P on $\mathfrak{R}(K)$ satisfying $z(P) = z_0$, and the values of $\overline{w}(P)$ at these points are the values of the multi-valued function $w(z)$ at $z = z_0$. Therefore, the algebraic function $w(z)$ displays the relation between the single-valued functions $z(P)$ and $w(P)$ on the closed Riemann surface $\mathfrak{R}(K)$. On the other hand, from Theorem 4.3, the analytic function field $K(\mathfrak{R})$ of a closed Riemann surface $\mathfrak{R}$ is an algebraic function field over the field of complex numbers. Hence, for arbitrary nonconstant analytic functions $z(P)$ and $w(P)$ on $\mathfrak{R}$, $z = z(P)$, and $w = w(P)$ are related by an irreducible equation

$$f(z, w) = 0.$$

Then if one regards $w(P)$ as a function of $z(P)$, an algebraic function $w(z)$ is obtained.

Therefore, algebraically speaking, an algebraic function is an algebraic relation between elements z and w in an algebraic function field K over the field of complex numbers. Analytically speaking, an algebraic function is an analytical relation between analytic functions $z(P)$ and $w(P)$ on a closed Riemann surface $\mathfrak{R}$. Consequently, the study of algebraic functions is the study of algebraic function fields over the field of complex numbers and closed Riemann surfaces. This is our approach in this treatise to the theory of algebraic functions. See the introduction.

We will study further the relation between the algebraic function field obtained in Theorem 4.2 and the closed Riemann surface obtained in Theorem 4.3.

THEOREM 4.4. *Let K_1 and K_2 be algebraic function fields over k, and let $\mathfrak{R}(K_1)$ and $\mathfrak{R}(K_2)$ be the closed Riemann surfaces belonging to them. Let*

σ be an isomorphism from K_1 be the subfield $K_1' = \sigma(K_1)$ of K_2 over k. Then for arbitrary $u \in K_1$ and $u' = \sigma(u) \in K_1'$, let $\overline{u}(P)$ and $\overline{u}'(P)$ be the corresponding functions in $K(\mathfrak{R}(K_1))$ and $K(\mathfrak{R}(K_2))$, respectively. Then there exists a unique analytic map f' from $\mathfrak{R}(K_2)$ onto $\mathfrak{R}(K_1)$ satisfying

$$\overline{u}'(P') = \overline{u}(f'(P')) \quad \text{for all } P' \in \mathfrak{R}(K_2). \tag{2.9}$$

Furthermore, the ramification index at an arbitrary place P' on K_2 for K_1' equals the ramification index of P' for the map f'. In particular, if K_1 is a subfield of K_2 and σ is the identity, then the above $P = f'(P')$ is the projection of the place P' on K_2.

Conversely, let $\mathfrak{R}_1$ and $\mathfrak{R}_2$ be closed Riemann surfaces, and let $P = f'(P')$ be an analytic map from $\mathfrak{R}_2$ onto $\mathfrak{R}_1$. For any function $\overline{u}(P)$ in $K(\mathfrak{R}_1)$, define

$$\overline{u}'(P') = \overline{u}(f'(P')), \qquad P' \in \mathfrak{R}_2.$$

Then $\overline{u}'(P')$ belongs to $K(\mathfrak{R}_2)$, and $\overline{\sigma}(\overline{u}) = \overline{u}'$ defines is an isomorphism $\overline{\sigma}$ from $K(\mathfrak{R}_1)$ to the subfield $\overline{\sigma}(K(\mathfrak{R}_1))$ of $K(\mathfrak{R}_2)$ over k. In particular, if $\mathfrak{R}_1$ and $\mathfrak{R}_2$ coincide with $\mathfrak{R}(K_1)$ and $\mathfrak{R}(K_2)$, respectively, and if φ_1 and φ_2 denote the isomorphisms assigning the functions $\overline{u}(P)$ and $\overline{u}'(P')$ for elements u and u' in K_1 and K_2, respectively, then we have $\overline{\sigma}\varphi_1 = \varphi_2\sigma$. Therefore, we obtain

$$K_2/K_1' \cong K(\mathfrak{R}_2)/\overline{\sigma}(K(\mathfrak{R}_1)). \tag{2.10}$$

PROOF. Since $K_1' = \sigma(K_1)$, for an arbitrary place P'' on K_1' there exists a unique place $P = \varphi(P'')$ on K_1 such that

$$v_P(u) = v_{P''}(\sigma(u))$$

holds for all the elements u in K_1. Then φ defines a one-to-one correspondence between all the places on K_1 and all the places on K_1'. Since K_1' is a subfield of K_2, let $P'' = f(P')$ be the projection of any place P' on K_2 to K_1'. Then the map

$$f' = \varphi f$$

is the map stated on the first part of the theorem. We will verify it as follows. It is plain that f' is a map from the set $\mathfrak{R}(K_2)$ of places on K_2 to the set $\mathfrak{R}(K_1)$ of places on K_1. Let e be the ramification index of a place P' on K_2 for K_1'. By the definition, for any element u in K_1 we have

$$v_{P'}(\sigma(u)) = ev_P(u), \qquad P = f'(P'). \tag{2.11}$$

Hence, if $v_{P'}(\sigma(u)) < 0$ then $v_P(u) < 0$. Moreover, if $v_P(u) \geq 0$ and $v_P(u-a) > 0$, we have $v_{P'}(\sigma(u)) \geq 0$ and $v_{P'}(\sigma(u)-a) = v_{P'}(\sigma(u-a)) > 0$. Therefore, from the definitions for $\overline{u}(P)$ and $\overline{u}'(P')$ we have

$$\overline{u}'(P') = \overline{u}(P) = \overline{u}(f'(P')).$$

Hence, f' satisfies (2.9). The fact that f' is an analytical map is an immediate consequence of Theorem 3.13. Next let f'_1 be another analytical map from $\mathfrak{R}(K_2)$ to $\mathfrak{R}(K_1)$ having the same property as f'. Then

$$\overline{u}(f'(P')) = \overline{u}(f'_1(P'))$$

for arbitrary $u \in K_1$ and $P' \in \mathfrak{R}(K_2)$. Hence, $f'(P') = f'_1(P)$, $f' = f'_1$ from Theorem 3.11. The uniqueness has been proved as well. Choose locally uniformizing variables $t = t_P(Q)$ and $t' = t'_{P'}(Q')$ at P' and $P = f'(P')$ as in §1. Then the algebraic expansion of an element u (or u') in K_1 (or K_2) with respect to t (or t') coincides with the analytical expansion of $\overline{u}(P)$ (or $\overline{u}'(P')$) in $t_P(Q)$ (or $t'_{P'}(Q')$). For an element u in K_1 such that $v_P(u) = 1$ in (2.11), let

$$u = \sum_{i=1}^{\infty} c_i t^i \quad \text{and} \quad \sigma(u) = \sum_{i=e}^{\infty} c'_i t'^i, \qquad c_i, c'_i \in k.$$

Then $c_1 \neq 0$, $c'_e \neq 0$, and

$$\overline{u}(Q) = \sum_{i=1}^{\infty} c_i t_P(Q)^i \quad \text{and} \quad \overline{u}'(Q') = \sum_{i=e}^{\infty} c'_i t'_{P'}(Q')^i.$$

Letting $Q = f'(Q')$ in (2.9), we obtain

$$\sum_{i=1}^{\infty} c_i t_P(Q)^i = \sum_{i=e}^{\infty} c'_i t'_{P'}(Q')^i.$$

The left-hand side is a locally uniformizing variable at P, and the right-hand side is the eth power of a suitable locally uniformizing variable at P'. Therefore, the image of a neighborhood of P' under f' is an n-folded neighborhood of P. Therefore, by definition, the ramification index of P' for f' is e. We have completed the first half of the proof.

It is also plain that $\overline{\sigma}$ maps $K(\mathfrak{R}_1)$ isomorphically onto the subfield $\overline{\sigma}(K(\mathfrak{R}_1))$ of $K(\mathfrak{R}_2)$. Let φ_1 and φ_2 be the maps

$$u \mapsto \overline{u}(P) \quad \text{and} \quad u' \mapsto \overline{u}'(P'),$$

respectively. Then (2.9) implies that

$$\overline{\sigma}(\varphi_1(u)) = \varphi_2(\sigma(u))$$

holds for all u in K_1. That is,

$$\overline{\sigma}\varphi_1 = \varphi_2\sigma.$$

Hence, the elements corresponding to the elements in $K'_1 = \sigma(K_1)$ are just functions in $(\overline{\varphi}K(\mathfrak{R}_1))$. Therefore, (2.10) holds.

From the above theorem, in particular, the branch points on $\mathfrak{R}(K_2)$ for f' coincide with the branch points of K_2/K'_1, i.e., the places on K_2 whose ramification indices over K'_1 are greater than 1. Since the set of branch points on $\mathfrak{R}(K_2)$ for f' cannot have an accumulation point in $\mathfrak{R}(K_2)$, the compactness of $\mathfrak{R}(K_2)$ implies that the set of branch points on $\mathfrak{R}(K_2)$ is a finite set. Therefore, K_2/K_1', for a general extension field of algebraic function fields over k, can have only a finite number of branch points in the algebraic sense. However, Riemann's formula (5.23) in Chapter 2 also verifies the above. See §4 of Chapter 5.

In the special case $\varphi^{-1} = f'$ in Theorem 4.4, we have the following results.

THEOREM 4.5. *Let K_1 and K_2 be two algebraic function fields over k, and let σ be an isomorphism from K_1 to K_2 over k. Then there exists a unique isomorphism φ from $\mathfrak{R}(K_1)$ to $\mathfrak{R}(K_2)$ such that for any $u \in K_1$ and $u' = \sigma(u) \in K_2$ we have*

$$\overline{u}(P) = \overline{u}'(\varphi(P)), \qquad P \in \mathfrak{R}(K_1).$$

Conversely, when $P' = \varphi(P)$ is an isomorphism from a closed Riemann surface $\mathfrak{R}_1$ to a closed Riemann surface $\mathfrak{R}_2$, for any $\overline{u}$ in $K(\mathfrak{R}_1)$ define a function $\overline{u}'$ on $\mathfrak{R}_2$ by

$$\overline{u}'(P') = \overline{u}(\varphi^{-1}(P')), \qquad P' \in \mathfrak{R}_2,$$

and denote this function by $\overline{\sigma}(\overline{u})$. Then $K(\mathfrak{R}_1)$ is mapped isomorphically by $\overline{\sigma}$ onto $K(\mathfrak{R}_2)$. If $\mathfrak{R}_1 = \mathfrak{R}(K_1)$ and $\mathfrak{R}_2 = \mathfrak{R}(K_2)$ and if φ_1 and φ_2 denote isomorphisms $K_1 \cong K(\mathfrak{R}_1)$, $K_2 \cong K(\mathfrak{R}_2)$, mapping $u \in K_1$ and $u_2 \in K_2$ to the functions $\overline{u}(P)$ and $\overline{u}'(P)$ in $K(\mathfrak{R}_1)$ and $K(\mathfrak{R}_2)$, respectively, then we have

$$\overline{\sigma}\varphi_1 = \varphi_2\sigma.$$

From this theorem there is a one-to-one correspondence between the isomorphisms σ from K_1 to K_2 and the isomorphisms φ from the closed Riemann surface $\mathfrak{R}_1$ to the closed Riemann surface $\mathfrak{R}_2$ as follows

$$\sigma \mapsto \varphi \mapsto \overline{\sigma} \mapsto \sigma.$$

In particular, if $K = K_1 = K_2$, then σ and φ are automorphisms of K and $\mathfrak{R}(K)$, respectively. Denote the automorphism corresponding to σ by φ_σ. Then

$$\varphi_{\sigma_1}\varphi_{\sigma_2} = \varphi_{\sigma_1\sigma_2}$$

for any automorphisms σ_1 and σ_2 of K. Therefore, we have obtained

THEOREM 4.6. *The automorphism group $A(K)$ of an algebraic function field K over the field k of complex numbers is isomorphic to the automorphism group $A(\mathfrak{R}(K))$ of the closed Riemann surface $\mathfrak{R}(K)$ belonging to K under the above correspondence*

$$\sigma \leftrightarrow \varphi_\sigma, \qquad \sigma \in A(K), \qquad \varphi_\sigma \in A(\mathfrak{R}(K)).$$

By Theorems 4.2, 4.3, and 4.5, the relation between algebraic function fields K over the complex number field k and closed Riemann surfaces

$\mathfrak{R}$ becomes very clear. Let us denote the classes of isomorphic algebraic function fields over k by $\{K\}$, $\{K'\}$, ..., and the classes of isomorphic closed Riemann surfaces by $\{\mathfrak{R}\}$, $\{\mathfrak{R}'\}$, ... Then letting $\{K\}$ correspond to the class $\{\mathfrak{R}(K)\}$ and conversely $\{\mathfrak{R}\}$ to the class $\{K(\mathfrak{R})\}$, we obtain a one-to-one correspondence between all the classes of algebraic function fields and all the classes of closed Riemann surfaces. One can regard two isomorphic algebraic function fields as defining the same abstract algebraic function field, and as we defined in the final paragraph of §1 in Chapter 3, one regards two isomorphic closed Riemann surfaces as defining the same abstract closed Riemann surface. Then one can rephrase the result by saying that abstract algebraic function fields and abstract closed Riemann surfaces are in a one-to-one correspondence. As can be foreseen, abstract properties of the corresponding K and $\mathfrak{R}$ are closely related. We will derive algebraic results on K from the topological and analytical properties of $\mathfrak{R}$, or conversely we will derive some results on $\mathfrak{R}$ from the algebraic properties of K, which are the main themes in what follows.

In the previous section, we proved that the correspondence

$$u \leftrightarrow \overline{u}(P),$$

between an algebraically defined K and the analytic function field $\overline{K} = K(\mathfrak{R}(K))$ of the closed Riemann surface $\mathfrak{R}(K)$ belonging to K, is an isomorphism. As long as one's interest is about the abstract properties of an algebraic function field in the above sense, one need not distinguish between K and $\overline{K}$. The same can be said for $\mathfrak{R}$ and $\mathfrak{R}(K(\mathfrak{R}))$. Hence, in what follows, we set

$$u = \overline{u}(P) \quad \text{and} \quad P = \overline{P},$$

and we will identify K with $K(\mathfrak{R}(K))$ and $\mathfrak{R}$ with $\mathfrak{R}(K(\mathfrak{R}))$. However, one needs to modify the above convention in some cases. For instance, even when K_1 is a subfield of K_2, $K(\mathfrak{R}(K_1))$ is not a subfield of $K(\mathfrak{R}(K_2))$. In this case, $K(\mathfrak{R}(K_2))$ only contains a subfield isomorphic to $K(\mathfrak{R}(K_1))$.

With the above convention, let $K = K(\mathfrak{R}(K))$ and $\mathfrak{R} = \mathfrak{R}(K(\mathfrak{R}))$. Then the correspondences $K \to \mathfrak{R}(K)$ and $\mathfrak{R} \to K(\mathfrak{R})$ are inverse to each other. Table 4.1 gives the corresponding algebraic and analytic notions for K and $\mathfrak{R}$ that have been obtained in this and the previous sections.

We may need to clarify the correspondence between a prime element t for P and the corresponding locally uniformizing variable $t_P(Q)$. That is, for the prime element t in Theorem 4.1 and for the corresponding locally uniformizing variable $t_P(Q)$, the algebraic expansion $u = \sum a_i t^i$ of u in K and the analytical expansion $u(P) = \sum a_i t_P(Q)^i$ of the corresponding analytic function $u(P)$ are expressed as the same power series. From Theorem 3.1 an arbitrary locally uniformizing variable $t'_P(Q)$ at P can be obtained

TABLE 4.1

algebraic function field K	closed Riemann surface $\mathfrak{R}$
element u of K	analytic function $u(Q)$ on R
place P on K	point P on R
prime element t for P	locally uniformizing variable $t_P(Q)$ at P
t-expansion of u in K_P:	expansion of $u(Q)$ in a neighborhood of P:
$u = \sum a_i t^i$	$u(Q) = \sum a_i t_P(Q)^i$
normalized valuation $v_P(u)$ belonging to P	order $v_P(u(Q))$ of $u(P)$ at P
differential $w\,dz$ in K	differential $w\,dz$ on $\mathfrak{R}$
$v_P(w\,dz)$	$v_P(w\,dz)$

from the above $t_P(Q)$ as a converging power series

$$t'_P(Q) = \sum_{i=1}^{\infty} b_i t_P(Q)^i, \qquad b_i \in k, \ b_1 \neq 0.$$

Then let

$$t' = \sum_{i=1}^{\infty} b_i t^i.$$

Then t' is a prime element for P, and t' and $t'_P(Q)$ are related just as t and $t_P(Q)$ are as above. The correspondence between a prime element and a locally uniformizing variable in the chart is the correspondence between this t' and the $t'_P(Q)$. Notice that not every prime element is obtained at t'. For example, a power series with radius of convergence 0

$$t'' = \sum_{i=1}^{\infty} c_i t^i, \qquad c_i \in k, \ c_1 \neq 0$$

is a prime element for P in the algebraic sense. But it is not a prime element coming from a locally uniformizing variable.

We will state some of the results as a theorem, which were obtained in the proof of Theorem 4.3.

THEOREM 4.7. *Let K be an algebraic function field over the field of complex numbers, and let $\mathfrak{R}$ be the closed Riemann surface belonging to K. Then an arbitrary element $z = z(Q)$ of K takes any complex value $n = [K : k(z)]$ times. Of course, we understand that at a zero of $z(Q) - a$ of order l, $z(Q)$ takes the value a l times. (For $a = \infty$, l is the order of the pole of $z(Q)$.)*

PROOF. We have given the analytical proof for this theorem. One can prove it algebraically as follows. The order $v_P(z(Q) - a)$ of the function

$z(Q)-a$ at a zero point P equals the valuation $v_P(z-a)$ of $z-a$ at the place P. Therefore, we only need to show that the order of the numerator of the divisor $(z-a)$ on K is equal to $[K : k(z)]$. Since $[K : k(z-a)] = [K : k(z)]$, Theorem 2.7 implies the above.

NOTE. Conversely, when one assumes the analytical proof of this theorem, one can obtain another proof for Theorem 2.7 for the case where k is the field of complex numbers. The reader should notice that a similar situation occurs when two equivalent statements are proved for K and $\mathfrak{R}$, one by the algebraic method and the other by the analytic method.

§3. Topological properties of closed Riemann surfaces

Using the results in §1 and §2, we can completely characterize the topological structure of a closed Riemann surface $\mathfrak{R}$, i.e., the properties of the underlying space S of $\mathfrak{R}$.

With the convention $\mathfrak{R} = \mathfrak{R}(K(\mathfrak{R}))$ in the last section, we apply the results in §1 to $K = K(\mathfrak{R})$. Therefore, S is homeomorphic to S' in §1 obtained by the identification of n regular $2r$-gons $\overline{T}_i$, $i = 1, \dots, n$. Since each $\overline{T}_i$ can be decomposed into $2r$ two dimensional simplexes with their common vertex at the center of $\overline{T}_i$, S' is decomposed into $2nr$ triangles. Through the homeomorphism, S is also decomposed into $2nr$ two-dimensional simplexes, and S forms a complex in the sense of algebraic topology. Let C denote this complex, and let $C^{(s)}$ be the complex obtained by s barrycentric subdivisions of C. (Precisely speaking, C does not satisfy some of the conditions for a complex, but actually $C^{(s)}$ does. Later we will compute the Euler characteristic of S by means of C. Then C need not be exactly a complex.) We define the notion of a barrycentric subdivision as follows. As we saw in §1, except at the vertices, each triangle Δ in C is mapped conformally by $z(P)$ onto the triangle Δ' on the sphere $\mathfrak{R}_0^*$ in a one-to-one fashion. Fix three points P_1, P_2, P_3 on the three sides of Δ, and then let z_0 be a point inside Δ'. Then connect z_0 by great circles on $\mathfrak{R}_0^*$ with each vertex of Δ' and the images $z(P_1)$, $z(P_2)$, $z(P_3)$. Thus, Δ' is subdivided into six smaller triangles. The inverse images by $z(P)$ provide six subdivided triangles for each triangle of C. We denote this subdivision of C by $C^{(1)}$. Similarly, $C^{(2)}$ is obtained from $C^{(1)}$ by applying the above construction. Then each side (one-dimensional simplex) in $C^{(s)}$ is an analytic curve on $\mathfrak{R}$ in the strict sense, and furthermore for a sufficiently large s, each triangle is contained in a neighborhood $U = \{Q; |t_P(Q) < \varepsilon\}$ of a point P, where $t_P(Q)$ is a locally uniformizing variable at P.

In general, if a two-dimensional complex C satisfies the following conditions, then C is said to be an orientable 2-dimensional closed manifold:

(1) an arbitrary side in C (1-dimensional simplex) belongs to exactly two different triangles adjacent to each other,

(2) for an arbitrary point in C (0-dimensional simplex), the collection

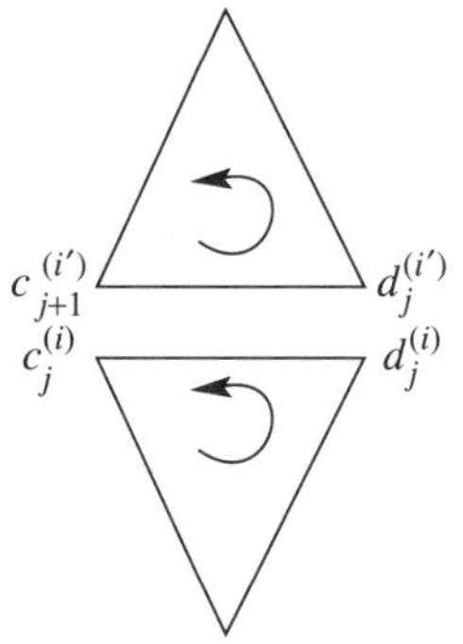

FIGURE 4.2

of all the triangles containing the point forms a chain

$$\Delta_1, \Delta_2, \ldots, \Delta_n \qquad (\Delta_{n+1} = \Delta_1),$$

where Δ_i and Δ_{i+1} share a side containing the point,

(3) for any two triangles Δ and Δ' in C, there is a chain of triangles next to each other connecting Δ and Δ':

$$\Delta = \Delta_1, \Delta_2, \ldots, \Delta_n = \Delta',$$

(4) an orientation of each triangle can be arranged so that any pair of triangles sharing a common side induce opposite orientations on the common side.

When an orientable two-dimensional closed manifold C is considered just as a topological space disregarding the triangulation, we call this manifold a closed surface, denoted by $\overline{C}$. The complexes C and $C^{(s)}$ obtained from the closed Riemann surface $\mathfrak{R}$ are orientable closed manifolds of dimension two, and $S = \overline{C} = \overline{C^{(s)}}$. This is obvious since S is a compact manifold of dimension two on which an R-function system is defined. One can verify this by considering the triangulation of S'. In particular, in order to prove the orientableness of S', hence of C, give the same orientation to every triangle of the regular $2r$-gon $\overline{T}_i$, $i = 1, \ldots, n$, and transfer the orientation to S. The identification induces the opposite orientations on the common side as one can observe in Figure 4.2. Hence, S is an orientable closed surface.

We will describe, without proofs, some of the topological properties of an orientable closed surface S that will be needed for the study of closed Riemann surfaces.

Let $S = \overline{C}$, and let α_0, α_1, α_2 be the numbers of vertices (0-dimensional simplexes), sides (1-dimensional simplexes), triangles (2-dimensional simplexes) of C, respectively. Let

$$N = -\alpha_0 + \alpha_1 - \alpha_2 \tag{3.1}$$

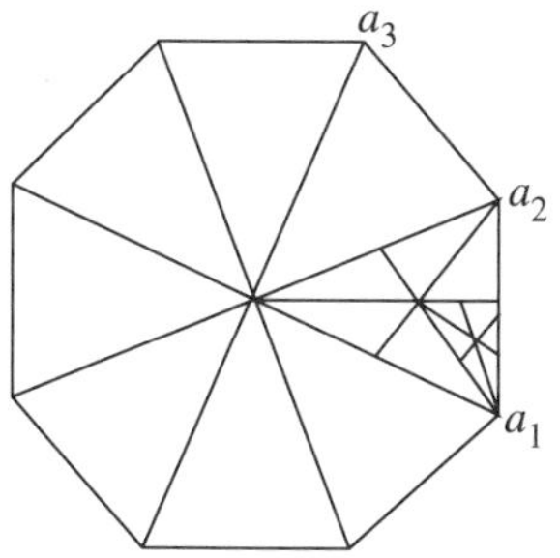

FIGURE 4.3

be the Euler characteristic of C. Then the genus g of S is defined by

$$N = 2g - 2, \qquad \left(g = \frac{N}{2} - 1\right). \tag{3.2}$$

Therefore, the genus is a nonnegative rational integer. One can prove that a necessary and sufficient condition for orientable closed surfaces $\overline{C}_1$ and $\overline{C}_2$ to be homeomorphic is that their genuses g_1 and g_2 coincide. Hence, in particular, the genus of $S = \overline{C}$ is independent of the choice of a triangulation. If $g_1 = g_2$ holds, then the complex $C_1^{(s)}$ obtained by a series of barrycentric subdivisions is isomorphic to a subdivision C_2' of C_2. Note that the above isomorphism means a one-to-one correspondence between the simplexes of $C_1^{(s)}$ and the simplexes of C_2' preserving the incident relation.

From the above remark, for an arbitrary rational integer $g \geq 0$, there exists, up to homeomorphisms, at most one orientable closed surface of genus g. Next we will show that for each g, there exists such an orientable closed surface of genus g. For $g = 0$, one can take a sphere. Let $g \geq 1$. Then consider a regular $4g$-gon $\overline{V}$ on the plane. Let $a_1, a_2, \ldots, a_{4g}$ $(a_{4g+1} = a_1)$ be the vertices as in Figure 4.3. Let S_0 be the topological space obtained by the following identifications of sides of $\overline{V}$ in the indicated order:

$$\overline{a_{4i-3}a_{4i-2}} \quad \text{and} \quad \overline{a_{4i}a_{4i-1}}, \tag{3.3}$$

$$\overline{a_{4i-2}a_{4i-1}} \quad \text{and} \quad \overline{a_{4i+1}a_{4i}}, \qquad i = 1, \ldots, g. \tag{3.4}$$

Divide $\overline{V}$ into $4g$ triangles with the common vertex at the center of $\overline{V}$. Let V be the complex obtained from $\overline{V}$ by applying the barrycentric subdivision twice. Then denote by C_0 the complex obtained by the corresponding subdivision on S_0. Then C_0 is an orientable closed manifold of dimension two having genus g. Hence, $S_0 = \overline{C}_0$ is a closed surface of genus g. (The reader is encouraged to verify that C_0 satisfies the four conditions for a closed manifold of dimension two.) The above C_0 is called the canonical form of an orientiable two-dimensional closed manifold of genus g. For an arbitrary C of genus g, its barrycentric subdivision $C^{(s)}$ is homeomorphic to a subdivision C_0' of C_0. Hence, $S_0 = \overline{C}_0$ and $\overline{C}$ are homeomorphic.

Therefore, to study the topological properties of the closed surface $\overline{C}$, we may use the simpler complex C_0, instead of C.

The $4g$ vertices a_i, $i = 1, 2, \ldots, 4g$, of $\overline{V}$ determine a unique vertex a_0 of C_0, and (3.3) and (3.4) determine closed curves α_i' and β_i', $i = 1, \ldots, g$, on $\overline{C}_0$ going through a_0. Here α_i' and β_i' are parametrized by the natural parameters in $[0, 1]$ on the straight lines $\overline{a_{4i-3}a_{4i-2}}$ and $\overline{a_{4i-2}a_{4i-1}}$, respectively. Then the fundamental group $G(a_0)$ for $\overline{C}_0$, with base point a_0, is generated by the homotopy classes $A_i' = \{\alpha_i'\}$ and $B_i' = \{\beta_i'\}$, and those generators satisfy a unique relation

$$A_1'B_1'A_1'^{-1}B_1'^{-1}\cdots A_g'B_g'A_g'^{-1}B_g'^{-1} = 1.$$

Let $G(a_0)'$ be the commutator subgroup of $G(a_0)$. Then the quotient group $G(a_0)/G(a_0)'$ is a free Abelian group generated by the residue classes $\overline{A}_i'$ and $\overline{B}_i'$ of A_i' and B_i'. The quotient group $G(a_0)/G(a_0)'$ has the following topological interpretation. For an arbitrary closed path α' with base point a_0, let A' be the homotopy class to which α' belongs, and let $\overline{A}'$ be the residue class of A' in $G(a)/G(a)'$. Regarding α' as a one-dimensional continuous cycle, let $\overline{\alpha}'$ be the homology class of α'. Then the correspondence

$$\overline{A}' \leftrightarrow \overline{\alpha}'$$

defines an isomorphism from the quotient group $G(a_0)/G(a_0)'$ to the first homology group $B_1(\overline{C}_0)$ of $\overline{C}_0$ with integer coefficients. (Note that such a relation between the fundamental group and the first homology group exists for higher dimensional manifolds.) Hence, $B_1(\overline{C}_0)$ is a free Abelian group with basis $\overline{\alpha}_i'$ and $\overline{\beta}_i'$, $i = 1, 2, \ldots, g$. Therefore, the one-dimensional Betti number P_1 of the complex C_0 is $2g$. Since C_0 is an orientable closed manifold of dimension two, the 0-dimensional and two-dimensional Betti numbers, p_0 and p_2, of C_0 are both 1. Hence, the Euler-Poincaré formula

$$N(= -\alpha_0 + \alpha_1 - \alpha_2) = -p_0 + p_1 - p_2 = 2g - 2.$$

Since $S_0 = \overline{C}_0$ is an orientable closed surface, the intersection number (α', β') is defined for any 1-dimensional continuous cycles α' and β' on S_0, and such intersection numbers (α', β') satisfy the following properties:

(1) for $\alpha' \sim \alpha''$, i.e., α' and α'' are homologous, $(\alpha', \beta') = (\alpha'', \beta')$;
(2) $(\alpha' + \alpha'', \beta') = (\alpha', \beta') + (\alpha'', \beta')$ and $(\alpha', \beta') = -(\beta', \alpha')$;
(3) for the α_i' and β_i' above

$$(\alpha_i', \alpha_j') = 0, \quad (\beta_i', \beta_j') = 0, \quad (\alpha_i', \beta_j') = \delta_{ij}, \qquad i, j = 1, \ldots, g. \quad (3.5)$$

From (1), (α', β') depends only upon the homology classes to which α' and β' belong. Hence, we can define a function by

$$(\overline{\alpha}', \overline{\beta}') = (\alpha', \beta')$$

for the elements $\overline{\alpha}'$ and $\overline{\beta}'$ in the homology group $B_1(\overline{C}_0)$. Since $\overline{\alpha}'_i$ and $\overline{\beta}'_i$ generate $B_1(\overline{C}_0)$, the above function is uniquely determined by (2) and (3). In particular, the value of $(\overline{\alpha}', \overline{\beta}')$ is always a rational integer. We will need to consider one-dimensional continuous cycles with coefficients in the real numbers and to also consider the one-dimensional homology group $B_1^*(\overline{C}_0)$ with coefficient group equal to the real numbers. It is easy to see that we can uniquely extend the notion of intersection numbers for such general cycles, requiring the condition

(4) for any real number ξ

$$(\xi\alpha', \beta') = \xi(\alpha', \beta').$$

We will use only these four properties of intersection numbers. We will not go into detail about the geometric properties of intersection numbers, in particular, their invariance under homeomorphisms. However, note that for a one-dimensional continuous cycle γ' with coefficients in the real numbers, let

$$\gamma' \sim \xi_1\alpha'_1 + \cdots + \xi_g\alpha'_g + \xi_{g+1}\beta'_1 + \cdots + \xi_{2g}\beta'_g, \qquad \xi_i = \text{a real number}.$$

Then we have

$$\xi_i = (\gamma', \beta'_i) \quad \text{and} \quad \xi_{g+i} = -(\gamma', \alpha'_i), \qquad i \leq i \leq g.$$

To the homology class $\overline{\gamma}'$ of γ', assign the vector of dimension $2g$ with coordinates

$$(\gamma', \alpha'_1), (\gamma', \alpha'_2), \ldots, (\gamma', \alpha'_g), (\gamma', \beta'_1), \ldots, (\gamma', \beta'_g).$$

Then $B_1^*(\overline{C}_0)$ and the $2g$-dimensional vector space $V_{2g}(k_0)$ over the field k_0 of real numbers become isomorphic as k_0-modules. The same result holds for any choice of basis elements $\gamma'_1, \ldots, \gamma'_{2g}$ for $B_1^*(\overline{C}_0)$ over k_0 instead of α'_i and β'_i. Let

$$\xi'_i = (\gamma', \gamma'_i), \qquad i = 1, \ldots, 2g.$$

Then the correspondence

$$\overline{\gamma}' \leftrightarrow (\xi'_1, \ldots, \xi'_{2g})$$

is an isomorphism between $B_1^*(\overline{C}_0)$ and $V_{2g}(k_0)$. In particular, $\overline{\gamma}' = 0$ holds if $\xi'_i = 0$ for $i = 1, \ldots, 2g$.

We return to the case of the closed Riemann surface $\mathfrak{R}$ and consider the complex C or the barrycentric subdivision $C^{(s)}$ obtained by triangulating the underlying space S. As we described above, the topological nature of S is completely determined by the genus. We will first examine how it is related to the algebraic function field $K = K(\mathfrak{R})$. For this, we will compute the genus using the triangulation of S' introduced in §1. Since the number of triangles on S' is $2rn$, we have $\alpha_2 = 2rn$. Next each regular $2r$-gon has $4r$

one-dimensional simplexes, but each simplex on the perimeter is identified with another such simplex. Therefore,

$$\alpha_1 = 4rn - \frac{1}{2} \cdot 2rn = 3rn .$$

As for α_0, there are n vertices on S' coming from $a_0^{(i)}$, and $c_j^{(i)}$, $i = 1, \dots, n$, $j = 1, \dots, r$, also provide the total of n vertices on S', each coming from r points among $C_j^{(i)}$. Let $Q_1', \dots, Q_s'$ be the points on S' obtained from rn $d_j^{(i)}$, $i = 1, \dots, n$, $j = 1, \dots, r$. Let e_l be the number of $d_j^{(i)}$ that provide Q_l'. We obtain

$$\alpha_0 = 2n + s, \qquad rn = \sum_{l=1}^{s} e_l .$$

As in §1, let Q_l be the point on S that corresponds to Q_l'. Then e_l coincides with the ramification index of the place Q_l on K with respect to $k(z)$. The other points on S which do not correspond to Q_l', $l = 1, \dots, s$, are unramified for $K/k(z)$. Therefore, letting e_P denote the ramification index of a place P on K for $k(z)$, we have

$$rn - s = \sum_{l=1}^{s} (e_l - 1) = \sum_P (e_P - 1), \qquad \alpha_0 = 2n + rn - \sum_P (e_P - 1).$$

Substitute the above values of α_2, α_1, and α_0 into (3.1) and (3.2) to obtain

$$\begin{aligned} 2g - 2 &= -\alpha_0 + \alpha_1 - \alpha_2 = -2n - rn + \sum_P (e_P - 1) + 3rn - 2rn \\ &= \sum_P (e_P - 1) - 2n . \end{aligned}$$

Compare this with Riemann's formula (5.23) in Chapter 2. Then the above g coincides with the genus of the algebraic function field K.

THEOREM 4.8. *The underlying space S of a closed Riemann surface $\mathfrak{R}$ is an orientable closed surface, and its genus g coincides with the genus of the analytic function field $K = K(\mathfrak{R})$ of $\mathfrak{R}$. We call g simply the genus of $\mathfrak{R}$.*

Recall that Theorem 2.32 states that for any nonnegative rational integer g there exists an algebraic function field of genus g over the field of complex numbers. On the other hand, an orientable closed surface is completely determined topologically by the genius only. The above theorem then implies the following result.

THEOREM 4.9. *A compact topological space S is the underlying space of a Riemann surface if and only if S is an orientable closed surface.*

Thus, the underlying space of a compact Riemann surface is completely characterized topologically.

Riemann's sphere $\mathfrak{R}_0^*$ is a closed surface of genus 0, and its analytic function field is the field $k(z)$ of rational functions. For an arbitrary closed Riemann surface $\mathfrak{R}$ of genus 0, $K = K(\mathfrak{R})$ is an algebraic function field of genus 0 by Theorem 4.8. Hence, Theorem 2.28 implies $K \cong k(z)$. Then we obtain $\mathfrak{R} \cong \mathfrak{R}_0^*$ by Theorem 4.5. That is, up to isomorphic closed Riemann surfaces, $\mathfrak{R}_0^*$ is the only closed Riemann surface of genus 0. Contrary to the genus zero case, there are infinitely many nonisomorphic algebraic function fields of genus 1 by Theorem 2.30. Then Theorem 4.5 implies that there are infinitely many nonisomorphic closed Riemann surfaces of genus 1 as well. (The same can be proved also for the case of genus $g \geq 2$.) This fact indicates that a closed Riemann surface cannot be determined by its topological properties alone. That is, the concept of a Riemann surface contains something more than the topology of the underlying space.

Theorem 4.8 accords a remarkable relationship between the topological property of the closed Riemann surface $\mathfrak{R}$ and the algebraic property of the corresponding algebraic function field K. We will find more relations of this sort. We will return to those results in the next chapter. In this chapter, we will make a few more remarks on triangulations of $\mathfrak{R}$. As before let C and $C^{(s)}$ be the triangulations of the underlying space S of $\mathfrak{R}$ as stated earlier, and let C_0 be the canonical form of the orientable closed manifolds of the same genus g as $\mathfrak{R}$. We will not consider $\mathfrak{R}$ of genus 0 at this moment. As we observed earlier, a certain barrycentric subdivision $C^{(s)}$ is isomorphic as a complex to a subdivision C_0' of C_0. Replacing, if necessary, $C^{(s)}$ by $C^{(t)}$ for a sufficiently large $t \geq s$, we may suppose that each triangle of $C^{(s)}$ is contained in a neighborhood $U_P(Q; |t_P(Q)| < \varepsilon\}$ of a properly chosen point P on $\mathfrak{R}$. Here, $t_P(Q)$ is a locally uniformizing variable at P. For the sake of simplicity, we denote this complex $C^{(s)}$ by C again, called a *standard subdivision* of the closed Riemann surface $\mathfrak{R}$. To be explicit, a standard subdivision C of $\mathfrak{R}$ is a complex satisfying the following conditions:

(1) C is a complex obtained by a triangulation of the underlying space S such that C is isomorphic to a subdivision of the canonical form of the orientable 2-dimensional closed manifolds of the same genus g as $\mathfrak{R}$.

(2) Each triangle belonging to C is contained in an analytic neighborhood $U_P = \{Q; |t_P(Q)| < \varepsilon\}$ of a point P, and each side of the triangle is an analytic curve in U_P in the strict sense.

Since the standard closed manifold C_0 of genus g was obtained by identifications from a regular $4g$-gon in the plane, there exists a map F for the above C as follows:

(1′) F is a continuous map from a regular $4g$-gon $\overline{V}$ onto $\mathfrak{R}$.

(2′) Let $a_1, a_2, \ldots, a_{4g}$ $(a_{4g+1} = a_1)$ be the vertices of $\overline{V}$ indexed in the positive orientation. Then F defines $2g$ closed paths on $\mathfrak{R}$, going through

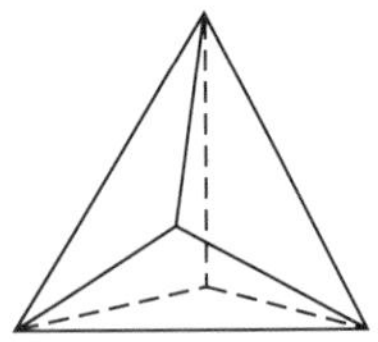

FIGURE 4.4

$P_0 = F(a_1) = \cdots = F(a_{4g})$, by

$$F(\overline{a_{4i-3}a_{4i-2}}) = F(\overline{a_{4i}a_{4i-1}}) = \alpha_i,$$
$$F(\overline{a_{4i-2}a_{4i-1}}) = F(\overline{a_{4i+1}a_{4i}}) = \beta_i, \qquad i = 1, 2, \ldots, g.$$

(3′) The interior $V^{(0)}$ of $\overline{V}$, i.e., the open regular $4g$-gon, is mapped homeomorphically onto the complementary domain U of α_i and β_i, $i = 1, \ldots, g$, in $\mathfrak{R}$.

(4′) For a complex V_1 on the plane obtained from a suitable triangulation of $\overline{V}$, F induces a simplicial map from V_1 to C.

Let us call α_i and β_i, $i = 1, \ldots, g$, *subdivision paths* of the standard subdivision C. In what follows, for a standard subdivision C, the notations F, α_i, β_i, P_0, etc., always mean the above. Since α_i and β_i are composed of sides (one-dimensional simplexes) of C, condition (2) implies that α_i and β_i are analytic paths on $\mathfrak{R}$. Moreover, any results about the curves α_i' and β_i' on $\overline{C}_0$ should be true for α_i and β_i through the homeomorphism from $\overline{C}_0$ to $\overline{C} = S$. For example, the fundamental group $G(P_0)$ with base point P_0 is generated by $A_i = \{\alpha_i\}$ and $B_i = \{\beta_i\}$ with the unique relation

$$A_1 B_1 A_1^{-1} B_1^{-1} \cdots A_g B_g A_g^{-1} B_g^{-1} = 1. \tag{3.6}$$

The first homology group $B_1(\mathfrak{R}) = B_1(S)$ of $\mathfrak{R}$ is the free Abelian group generated by the homology classes $\overline{\alpha}_i$ and $\overline{\beta}_i$ of α_i and β_i, $i = 1, \ldots, g$. Then we have

$$G(P_0)/G(P_0)' \cong B_1(\mathfrak{R}). \tag{3.7}$$

Similar results can be obtained also for the intersection numbers.

Finally, notice that for a given standard subdivision C, as above, one may deform the sides of C as shown in Figure 4.4 so that F does not change the correspondence between triangles in $\overline{V}$ and C. Hence, for a finitely many given points on $\mathfrak{R}$, it is always possible to construct a standard subdivision C so that none of those points lies on the sides of C. We will use this remark later.

CHAPTER 5

Analytic Theory of Algebraic Function Fields

§1. Abelian integrals

In the previous chapter we saw the profound connections between an algebraic function field K over the field of complex numbers and a closed Riemann surface $\mathfrak{R}$. In this chapter we will find more profound results on K through the topological and analytic methods for the closed Riemann surface $\mathfrak{R}$ which the algebraic method in Chapter 2 could not attain. Theorems in the classical algebraic function theory by the analytical method often provide algebraic theorems on algebraic function fields over a general coefficient field, in particular, of characteristic zero.

Already in §3 of Chapter 3, we defined the integral of the differential $w\,dz$ along the analytic curve γ when $w\,dz$ on the Riemann surface $\mathfrak{R}$ is regular on γ. When $\mathfrak{R}$ is the closed Riemann surface belonging to the algebraic function field K, the integral

$$\int_\gamma w\,dz$$

is called an Abelian integral in K or on $\mathfrak{R}$. If the differential $w\,dz$ is of the first kind, the second kind, or the third kind, then the Abelian integral is said to be the first kind, the second kind, or the third kind, respectively.

Let P_0 and P_1 be the initial point and the end point of γ, respectively. For another analytic curve γ', $\int_{\gamma'} w\,dz$ generally differs from $\int_\gamma w\,dz$. The difference is given by

$$\int_{\gamma_0} w\,dz,$$

where $\gamma_0 = \gamma'\gamma^{-1}$. That is, if one knows the value of the integral along the closed analytic curve, then one knows the variation of the integral along different paths. In this sense it is important to know the value of $\int w\,dz$ along a closed analytic curve γ. The value of the integral is called the *period* of $w\,dz$ or $\int w\,dz$ with respect to γ as follows.

The period of the differential $w\,dz$ depends upon the closed analytic curve γ as follows.

THEOREM 5.1. *If the residue of $w\,dz$ is 0 at every point on $\mathfrak{R}$, i.e., $w\,dz$ is a sum of differentials of the first kind and the second kind; then the period* *of $w\,dz$ depends only upon the homotopy class to which γ belongs. Here γ is regarded as a continuous cycle on the underlying space S of $\mathfrak{R}$. That is, for closed analytic curves γ_1 and γ_2 such that $\gamma_1 \sim \gamma_2$, we have*

$$\int_{\gamma_1} w\,dz = \int_{\gamma_2} w\,dz, \tag{1.1}$$

where $w\,dz$ has no poles on γ_1 and γ_2.

PROOF. Let the genus g of $\mathfrak{R}$ be $g \geq 1$, and let C be an arbitrary subdivision of $\mathfrak{R}$. Let γ be an arbitrary closed analytic curve on $\mathfrak{R}$. As we noted earlier, one may assume that there are no poles of $w\,dz$ on the sides of C. After a further subdivision, one can assume the initial (= end) point P_1 is a vertex of C. Choose the intersecting points, $P_1, \dots, P_l, P_{l+1} = P_1$, of γ with the sides of C so that the part γ_i of γ from P_i to P_{i+1} is contained in a triangle Δ_i. Let γ_i' be the segment connecting P_i to P_{i+1} along the sides of C, and let $\gamma' = \gamma_2' \cdots \gamma_1'$. (See Figure 5.1.) Clearly, γ_i and γ_i' are homotopic. Hence, γ and γ' are homotopic. Then Theorem 3.6 implies

$$\int_{\gamma} w\,dz = \int_{\gamma'} w\,dz. \tag{1.2}$$

Since $\gamma \approx \gamma'$ holds, as 1-dimensional continuous cycles, γ and γ' belong to the same homologous class. For a properly chosen standard subdivision C, it is sufficient to prove (1.1) for the case where γ_1 and γ_2 are composed of sides of C. Let $\mu_1, \dots, \mu_{\alpha_1}$ be the sides of C that are analytic curves and also 1-dimensional continuous simplexes of S, and then let

$$\gamma_1 = \mu_{i_l}^{\pm 1} \cdots \mu_{i_1}^{\pm 1} \quad \text{and} \quad \gamma_2 = \mu_{j_m}^{\pm 1} \cdots \mu_{j_1}^{\pm 1}.$$

From (3.8) and (3.9) in Chapter 3 we obtain

$$\int_{\gamma_1} w\,dz = \sum_{s=1}^{l} \pm \int_{\mu_{i_s}} w\,dz, \qquad \int_{\gamma_2} w\,dz = \sum_{s=1}^{m} \pm \int_{\mu_{j_s}} w\,dz.$$

Regarding γ_1 and γ_2 as 1-dimensional cycles, let

$$\gamma_1 = \sum_{i=1}^{\alpha_1} a_i \mu_i \text{ and } \gamma_2 = \sum_{i=1}^{\alpha_1} b_i \mu_i, \qquad a_i \text{ and } b_i = \text{ rational integers},$$

as subcomplexes of C. Then a_i, for example, is the number of μ_i, in the algebraic sense, contained in $\mu_{i_1}^{\pm 1}, \dots, \mu_{i_l}^{\pm 1}$. Hence

$$\int_{\gamma_1} w\,dz = \sum_{i=1}^{\alpha_1} a_i \int_{\mu_i} w\,dz \quad \text{and} \quad \int_{\gamma_2} w\,dz = \sum_{i=1}^{\alpha_1} b_i \int_{\mu_i} w\,dz.$$

Therefore, in order to prove (1.1), it is sufficient to prove, for example, if $\mu_1 + \mu_2 + \mu_3$ is the boundary of a triangle Δ of C, then

$$\int_{\mu_1} + \int_{\mu_2} + \int_{\mu_3} w\,dz = 0.$$

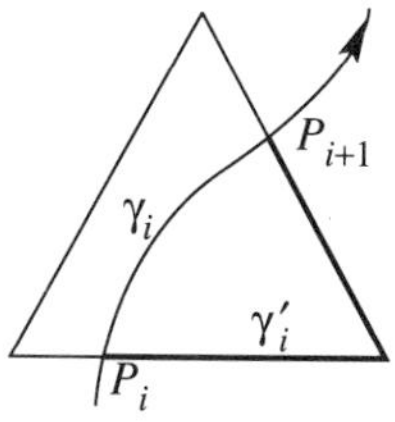

FIGURE 5.1

But Δ is contained in an analytical neighborhood $U = \{Q; |t_P(Q)| < \varepsilon\}$ of a point P. Then Cauchy's theorem implies the above. That is, we have proved (1.1) for the case $g \geq 1$. For the case $g = 0$, $\mathfrak{R}$ is the Riemann sphere. From a suitable triangulation C, one can prove (1.1) as above. Q.E.D.

The above proof also implies

$$\int_\gamma w\,dz = \int_{\gamma'} w\,dz + \int_{\gamma''} w\,dz \tag{1.3}$$

for $\gamma \sim \gamma' + \gamma''$.

Let the genus of $\mathfrak{R}$ be greater than or equal to 1, and let $w\,dz$ be a differential of the first kind in K. Then let $B_1(\mathfrak{R})$ be the first homology group with coefficients in rational integers, i.e., the integral first homology group $B_1(\overline{C})$ of $S = \overline{C}$. Let $\gamma_1, \ldots, \gamma_{2g}$ be closed analytic curves forming a base for $B_1(\mathfrak{R})$. Then any closed analytic curve γ can be expressed as

$$\gamma \sim \sum_{i=1}^{2g} c_i\gamma_i, \qquad c_i = \text{ rational integer.}$$

If we let

$$\zeta_i = \int_{\gamma_i} w\,dz,$$

then from (1.3) we have

$$\int_\gamma w\,dz = \sum_{i=1}^{2g} c_i\zeta_i. \tag{1.4}$$

That is, the period of $w\,dz$ with respect to any closed analytic curve can be written as a linear combination of ζ_i with coefficients in rational integers. In this sense

$$(\zeta_1, \ldots, \zeta_{2g})$$

is said to be the *fundamental system* of *periods* of $w\,dz$ with respect to γ_i, $i = 1, \ldots, 2g$.

For the base α_i and β_i, $i = 1, \ldots, g$, for $B_1(\mathfrak{R})$ obtained from a standard subdivision C of $\mathfrak{R}$, we have the following theorem.

THEOREM 5.2. *For differentials $w\,dz$ and $w'\,dz$ of the first kind in an algebraic function field K of genus $g \geq 1$, define the fundamental system of* *periods with respect to* α_i *and* β_i *as*

$$\begin{aligned} \omega_i &= \int_{\alpha_i} w\,dz\,, & \omega_{g+i} &= \int_{\beta_i} w\,dz\,, \\ \omega'_i &= \int_{\alpha_i} w'\,dz\,, & \omega'_{g+i} &= \int_{\beta_i} w'\,dz\,, \end{aligned} \qquad i = 1\,,\,\dots\,,\,g\,.$$

Then

$$\sum_{i=1}^{g}(\omega_i\omega'_{g+i} - \omega_{g+i}\omega'_i) = 0\,. \tag{1.5}$$

Separate ω_i *into the real part and the imaginary part*

$$\omega_i = \xi_i + \sqrt{-1}\eta_i\,, \qquad i = 1\,,\,\dots\,,\,2g\,.$$

Provided $w\,dz \neq 0$, *we have*

$$\sum_{i=1}^{g}(\xi_i\eta_{g+i} - \xi_{g+i}\eta_i) > 0\,. \tag{1.6}$$

Then (1.5) *and* (1.6) *are called the Riemann relation and the Riemann inequality, respectively.*

PROOF. For a standard subdivision C of $\mathfrak{R}$, let $\overline{V}$ and F be as in §3 of Chapter 4, and let U be the image of the interior $V^{(0)}$ of $\overline{V}$ under the map F. Since U is simply connected as $V^{(0)}$ is, from Theorem 3.7

$$w_1 = w_1(P) = \int_{P_1}^{P} w\,dz$$

is a single-valued regular analytic function in U. Here P_1 is a fixed point in U. Similarly set

$$w_2 = w_2(P) = \int_{P_1}^{P} w'\,dz\,.$$

Let us denote all the triangles in $\overline{V}$ by $\Delta_1\,,\,\dots\,,\,\Delta_l$. Then all the triangles in C are given by $F(\Delta_i)$, $i = 1\,,\,\dots\,,\,l$. We will compute the integral of $w_1\,dw_2$ along the sides of δ_i of $F(\Delta_i)$ in the positive direction once around. When Δ_i does not have a common point with the circumference $\Gamma = \overline{a_1a_2} + \overline{a_3a_4} + \cdots + \overline{a_{4g}a_1}$ of $\overline{V}$, we have $\Delta_i \subset V^{(0)}$. Hence, $F(\Delta_i) \subset U$. Since w_1 and w_2 are regular functions in U, clearly

$$\int_{\delta_i} w_1\,dw_2 = 0\,. \tag{1.7}$$

When Δ_i intersects with Γ, $F(\Delta_i)$ is not contained in U. Then w_1 and w_2 have points on δ_i where they are not defined. We will then interpret $w_1\,dw_2$ as follows. By the definition of a standard subdivision, $F(\Delta_i)$ is contained in an analytical neighborhood $U_P = \{Q; |t_P(Q)| < \varepsilon\}$ of a point P. Then for a point P_2 in $U \cap U_P$, consider the function

$$f(Q) = w_1(P_2) + \int_{P_2}^{Q} w\,dz$$

in U_P. The function $f(Q)$ is single valued and regular, and clearly $f(Q)$ coincides with $w_1(Q)$ in $U \cap U_P$. That is, $w_1(Q)$ is extended by $f(Q)$ to a domain containing $F(\Delta_i)$, i.e., it is an analytic continuation of $w_1(P)$. Similarly, $w_2(P)$ is also extended to U_P. If one considers the integral along the circumference δ_i of $F(\Delta_i)$ of the differential $w_1\,dw_2$ for the extended w_1 and w_2, then (1.7) is again secured. Consequently,

$$\sum_{i=1}^{l} \int_{\delta_i} w_1\,dw_2 = 0. \tag{1.8}$$

For Δ_i contained in $V^{(0)}$, along the side shared by triangles next to each other, the integration is computed twice in opposite directions. Hence the left-hand side of (1.8) becomes the integral over $F(\Gamma)$:

$$\int_{F(\Gamma)} w_1\,dw_2.$$

Next we will compute the above integral. First, parameterize $\overline{a_1a_2}$ and $\overline{a_4a_3}$ as we mentioned in $(2')$, §3, Chapter 4. For a parameter s, let $a(s)$ and $b(s)$ be the points $0 \le s \le 1$. From our assumption we have

$$F(a(s)) = F(b(s)). \tag{1.9}$$

The definition of $w_1(P)$ implies

$$\begin{aligned} w_1(F(b(s))) - w_1(F(a(s))) &= \{w_1(F(b(s))) - w_1(F(a_3))\} \\ &\quad + \{w_1(F(a_3)) - w_1(F(a_2))\} \\ &\quad + \{w_1(F(a_2)) - w_1(F(a(s)))\} \\ &= \int_{F(a_3)}^{F(b(s))} w\,dz + \int_{F(a_2)}^{F(a_3)} w\,dz + \int_{F(a(s))}^{F(a_2)} w\,dz, \end{aligned}$$

where the integrals are over the images of $\overline{a_3b(s)}$, etc. under F. See Figure 5.2 (next page). Then from (1.9) the sum of the first and the third integrals is 0 since they are identical, but the paths are opposite. The second term is exactly equal to

$$\int_{\beta_1} w\,dz = \omega_{g+1}.$$

Therefore, we have

$$w_1(F(b(s))) = w_1(F(a(s))) + \omega_{g+1}. \tag{1.10}$$

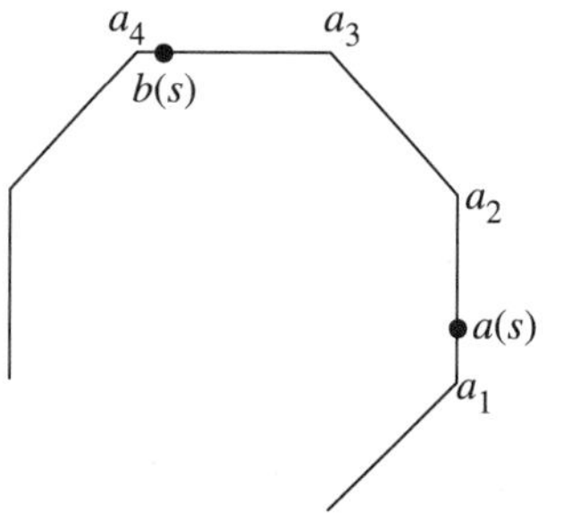

FIGURE 5.2

On the other hand, $dw_2 = w'\,dz$ is invariant on $F(b(s)) = F(a(s))$. Hence,

$$\begin{aligned}\int_{F(a_3)}^{F(a_4)} w_1\,dw_2 &= -\int_0^1 w_1(F(b(s)))w'\frac{dz}{ds}\,ds \\ &= -\int_0^1 w_1(F(a(s)))w'\frac{dz}{ds}\,ds - \omega_{g+1}\int_0^1 w'\frac{dz}{ds}\,ds \\ &= -\int_{F(a_1)}^{F(a_2)} w_1\,dw_2 - \omega_{g+1}\int_{\alpha_1} w'\,dz \\ &= -\int_{F(a_1)}^{F(a_2)} w_1\,dw_2 - \omega_{g+1}\omega_1'.\end{aligned}$$

That is,

$$\int_{F(a_1)}^{F(a_2)} + \int_{F(a_3)}^{F(a_4)} w_1\,dw_2 = -\omega_{g+1}\omega_1'.$$

Just as above, for $F(\overline{a_2a_3})$ and $F(\overline{a_4a_5})$, we have

$$\int_{F(a_2)}^{F(a_3)} + \int_{F(a_4)}^{F(a_5)} w_1\,dw_2 = \omega_1\omega_{g+1}'.$$

For other $F(\overline{a_{4i-3}a_{4i-2}})$, $F(\overline{a_{4i-1}a_{4i}})$, $F(\overline{a_{4i-2}a_{4i-1}})$, $F(\overline{a_{4i}a_{4i-1}})$, we can compute exactly as above. Adding all of those, we get

$$\int_{F(\Gamma)} w_1\,dw_2 = \sum_{i=1}^{g}(\omega_i\omega_{g+i}' - \omega_{g+i}\omega_i').$$

Since the left-hand side is 0 from (1.8), we have proved (1.5).

Next we will prove (1.6). First, separate $w_1(P)$ into the real part and the imaginary part as

$$w_1(P) = \xi(P) + \sqrt{-1}\eta(P).$$

As above, for each δ_i we will compute

$$\int_{\delta_i}\xi\,d\eta = \int_{\delta_i}\xi\frac{d\eta}{ds}\,ds, \qquad i = 1, \dots, l$$

and then add. From (1.10)

$$\xi(F(b(s))) = \xi(F(a(s))) + \xi_1.$$

Since $d\eta = \mathrm{Im}(w\,dz)$ is invariant on $F(b(s)) = F(a(s))$, and for example,

$$\int_{\alpha_1} d\eta = \mathrm{Im}\left(\int_{\alpha_1} w\,dz\right) = \mathrm{Im}(\omega_1) = \eta_1$$

holds, we obtain

$$\sum_{i=1}^{l} \int_{\delta_i} \xi\,d\eta = \sum_{i=1}^{g} (\xi_i \eta_{g+i} - \xi_{g+1}\eta_i). \tag{1.11}$$

Since each Δ_i is contained in a $U_P = \{Q;\, |t_P(Q)| < \varepsilon\}$, one computes $\int_{\delta_i} \xi\,d\eta$ as follows. Let $t_P = x + \sqrt{-1}y$. Then

$$\xi \frac{d\eta}{ds} = \xi \left(\frac{\partial \eta}{\partial x}\frac{dx}{ds} + \frac{\partial \eta}{\partial y}\frac{dy}{ds}\right).$$

Notice $\mathrm{div}^* v = 0$ for $v = \mathrm{grad}\,\eta$. From formula (4.13) in Lemma 3.3 and the above remarks, we have

$$\int_{\delta_i} \xi\,d\eta = \iint_{F(\Delta_i)} \left(\frac{\partial \eta}{\partial y}\frac{\partial \xi}{\partial x} - \frac{\partial \eta}{\partial x}\frac{\partial \xi}{\partial y}\right) dx\,dy.$$

Since $w_1(P)$ is regular in t_P, $\partial\xi/\partial x = \partial\eta/\partial y$ and $\partial\xi/\partial y = -\partial\eta/\partial x$. Then the right-hand side becomes

$$\iint_{F(\Delta_i)} \left[\left(\frac{\partial \xi}{\partial x}\right)^2 + \left(\frac{\partial \eta}{\partial x}\right)^2\right] dx\,dy = \iint_{F(\Delta_i)} \left|\frac{dw_1}{dt_P}\right|^2 dx\,dy$$
$$= \iint_{F(\Delta_i)} \left|w \frac{dz}{dt_P}\right|^2 dx\,dy.$$

Hence, for nonzero differential $w\,dz$, the right-hand side of (1.11) is certainly > 0. Therefore, (1.6) is proved. Q.E.D.

From Theorem 5.2 we obtain various results on the Abelian integrals of the first kind. When the periods of a differential $w\,dz$ of the first kind with respect to all $\alpha_1, \dots, \alpha_g$ are zero, i.e., $\xi_i = \eta_i = 0$, $i = 1, \dots, g$, in the above notation, we have $\sum_{i=1}^{g}(\xi_i\eta_{g+i} - \xi_{g+i}\eta_i) = 0$. Then (1.6) implies $w\,dz = 0$. Hence, for an arbitrary differential $w\,dz$ in K of the first kind, assign the periods $(\omega_1, \dots, \omega_g)$ for α_i, $i = 1, \dots, g$:

$$w\,dz \mapsto (\omega_1, \dots, \omega_g). \tag{1.12}$$

Then this map is an isomorphism as k-modules from the set $\mathfrak{L}_0$ of differentials of the first kind in K into the g-dimensional vector space $V_g(k)$ over the field of complex numbers. However, $\mathfrak{L}_0$ and $V_g(k)$ are both g-dimensional k-modules. Hence, the map (1.12) is an isomorphism from $\mathfrak{L}_0$ onto $V_g(k)$, i.e.,

$$\mathfrak{L}_0 \cong V_g(k). \tag{1.13}$$

Therefore, for arbitrary g complex numbers $\omega_1, \ldots, \omega_g$, there exists a unique differential $w\,dz$ in K of the first kind such that

$$\int_{\alpha_i} w\,dz = \omega_i, \qquad i = 1, \ldots, g.$$

In general, for any differential $v\,dz$ in K which is regular on α_i, $i = 1, \ldots, g$, let

$$\int_{\alpha_i} v\,dz = \omega_i, \qquad i = 1, \ldots, g.$$

For the differential $w\,dz$ above, let $u\,dz = v\,dz - w\,dz$. Then

$$\int_{\alpha_i} u\,dz = 0, \qquad i = 1, \ldots, g.$$

That is, for an arbitrary differential, there is a differential of the first kind so that the sum of these differentials has period zero with respect to α_i, $i = 1, \ldots, g$. This result will be useful later.

In the above one can replace α_i with β_i, $i = 1, \ldots, g$.

Next we will examine the real part ξ_i, $i = 1, \ldots, 2g$, of the period ω_i of $w\,dz$ of the first kind for α_i and β_i. For this case as well, if $\xi_1 = \xi_2 = \cdots = \xi_{2g} = 0$, then the left-hand side of (1.6) becomes zero. Just as above, the map over the field k_0 of real numbers

$$w\,dz \mapsto (\xi_1, \ldots, \xi_{2g}) \tag{1.14}$$

is an isomorphism from $\mathfrak{L}_0$ onto the $2g$-dimensional vector space $V_{2g}(k_0)$ as k_0-modules. That is, we have

$$\mathfrak{L}_0 \cong V_{2g}(k_0). \tag{1.15}$$

Therefore, for given real numbers ξ_i, $i = 1, \ldots, 2g$, there is a unique differential $w\,dz$ of the first kind in K satisfying

$$\operatorname{Re}\left(\int_{\alpha_i} w\,dz\right) = \xi_i \quad \text{and} \quad \operatorname{Re}\left(\int_{\beta_i} w\,dz\right) = \xi_{g+i}, \qquad i = 1, \ldots, g.$$

Moreover, for an arbitrary differential in K, regular on α_i and β_i, there exists a differential of the first kind so that their sum has the real part of the period zero with respect to α_i and β_i.

Let $\gamma_1, \ldots, \gamma_{2g}$ be closed analytic curves forming basis elements for $B_1(\mathfrak{R})$, and let γ be an arbitrary closed analytic curve such that as real coefficient one-dimensional cycles

$$\gamma \sim \sum_{i=1}^{2g} \lambda_i \gamma_k, \qquad \lambda_i = \text{real number}. \tag{1.16}$$

Then we define the period of an arbitrary differential $w\,dz$ for γ as follows.

$$\int_\gamma w\,dz = \sum_{i=1}^{2g} \lambda_i \int_{\gamma_i} w\,dz.$$

It should be clear that this definition is independent of the choice of the base $\gamma_1, \dots, \gamma_{2g}$. When γ is a closed analytic curve, the above definition coincides with the usual definition. See (1.4). Needless to say, $\int_\gamma w\,dz$ depends only upon the homology class of γ.

For this extended definition of $\int_\gamma w\,dz$, we have the following.

THEOREM 5.3. *Let γ be an arbitrary real coefficient cycle of dimension one on a closed Riemann surface of genus $g \geq 1$. Then there exists a unique differential $w\,dz$ of the first kind in K such that*

$$\operatorname{Re}\left(\int_{\gamma'} w\,dz\right) = (\gamma, \gamma') \tag{1.17}$$

for all real coefficient one-dimensional cycles γ', where (γ, γ') is the intersection number of γ and γ'. Let us denote this uniquely determined differential by $w_\gamma\,dz$. Then $w_{\gamma_1}\,dz = w_{\gamma_2}\,dz$ holds if and only if $\gamma_1 \sim \gamma_2$ holds. Furthermore, for the homology class $\overline{\gamma}$ of γ, the correspondence

$$w_\gamma\,dz \leftrightarrow \overline{\gamma} \tag{1.18}$$

is an isomorphism as k_0-modules between the set $\mathfrak{L}_0$ of the differentials of the first kind in K and the first homology group $B_1^*(\mathfrak{R})$ with real coefficients, i.e.,

$$\mathfrak{L}_0 \cong B_1^*(\mathfrak{R}). \tag{1.19}$$

PROOF. Let

$$\gamma' \sim \sum_{i=1}^{g} \mu_i \alpha_i + \sum_{i=1}^{g} \mu_{g+i} \beta_i, \qquad \mu_i = \text{ a real number}.$$

Then we have

$$\begin{aligned}\operatorname{Re}\left(\int_{\alpha_i} w\,dz\right) &= \sum_{i=1}^{g} \mu_i \operatorname{Re}\left(\int_{\alpha_i} w\,dz\right) \\ &\quad + \sum_{i=1}^{g} \mu_{g+i} \operatorname{Re}\left(\int_{\beta_i} w\,dz\right)\end{aligned}$$

and

$$(\gamma, \gamma') = \sum_{i=1}^{g} \mu_i (\gamma, \alpha_i) + \sum_{i=1}^{g} \mu_{g+i} (\gamma, \beta_i).$$

Hence, for (1.17) to hold for any γ' it is necessary and sufficient that

$$\operatorname{Re}\left(\int_{\alpha_i} w\,dz\right) = (\gamma, \alpha_i) \quad \text{and} \quad \operatorname{Re}\left(\int_{\beta_i} w\,dz\right) = (\gamma, \beta_i), \qquad i = 1, \dots, g.$$

We have already proved the existence of such a unique $w\,dz$ for the given (γ, α_i) and (γ, β_i). We will prove the later half next.

If we have $w_{\gamma_i}\,dz = w_{\gamma_2}\,dz$, then for any γ' we have $(\gamma_1, \gamma') = (\gamma_2, \gamma')$ and $(\gamma_1 - \gamma_2, \gamma') = 0$. By the remark in the first paragraph following the definition of the intersection number, we have $\gamma_1 - \gamma_2 \sim 0$, i.e., $\gamma_1 \sim \gamma_2$. Conversely, for $\gamma_1 \sim \gamma_2$ we have $(\gamma_1, \gamma') = (\gamma_2, \gamma')$ for all γ'. Then by the above uniqueness, we have $w_{\gamma_1}\,dz = w_{\gamma_2}\,dz$.

Lastly, for any real numbers λ and μ

$$\begin{aligned}
\operatorname{Re}\left(\int_{\gamma'} (\lambda w_{\gamma_1}\,dz + \mu w_{\gamma_2}\,dz)\right) &= \lambda \operatorname{Re}\left(\int_{\gamma'} w_{\gamma_1}\,dz\right) + \mu \operatorname{Re}\left(\int_{\gamma'} w_{\gamma_2}\,dz\right) \\
&= \lambda(\gamma_1, \gamma') + \mu(\gamma_2, \gamma') \\
&= (\lambda\gamma_1 + \mu\gamma_2, \gamma') = \operatorname{Re}\left(\int_{\gamma'} w_{\lambda\gamma_1+\mu\gamma_2}\,dz\right).
\end{aligned}$$

Hence,

$$\lambda w_{\gamma_1}\,dz + \mu w_{\gamma_2}\,dz = w_{\lambda\gamma_1+\mu\gamma_2}\,dz.$$

That is, the correspondence is an isomorphism over k_0. Q.E.D.

Next we will consider a function defined on $\mathfrak{L}_0 \times B_1^*(\mathfrak{R})$ defined by

$$(w\,dz, \overline{\gamma}) = \exp\left(2\pi\sqrt{-1}\operatorname{Re}\left(\int_{\gamma} w\,dz\right)\right). \tag{1.20}$$

For arbitrary $w\,dz$ and $w'\,dz$ in $\mathfrak{L}_0$ and arbitrary $\overline{\gamma}$ and $\overline{\gamma}'$ in $B_1^*(\mathfrak{R})$ we clearly have

$$\begin{aligned}
(w\,dz + w'\,dz, \overline{\gamma}) &= (w\,dz, \overline{\gamma})\cdot(w'\,dz, \overline{\gamma}), \\
(w\,dz, \overline{\gamma} + \overline{\gamma}') &= (w\,dz, \overline{\gamma})\cdot(w\,dz, \overline{\gamma}').
\end{aligned}$$

Hence, for a fixed $\overline{\gamma}$, $(w\,dz, \overline{\gamma})$ defines a character of the Abelian group $\mathfrak{L}_0$, and for a fixed $w\,dz$ it defines a character of the Abelian group $B_1^*(\mathfrak{R})$. For subdivision curves α_i and β_i of a standard subdivision C of $\mathfrak{R}$, we have by (1.19) that the corresponding differentials of the first kind $w_{\alpha_i}\,dz$ and $w_{\beta_i}\,dz$, $i = 1, \dots, g$, form a base for $\mathfrak{L}_0$. For arbitrary $w\,dz$ and γ, let

$$\begin{aligned}
w\,dz &= \sum_{i=1}^{g} \lambda_i w_{\alpha_i}\,dz + \sum_{i=1}^{g} \lambda_{g+i} w_{\beta_i}\,dz, \\
\gamma &\sim \sum_{i=1}^{g} \mu_i \alpha_i + \sum_{i=1}^{g} \mu_{g+i}\beta_i.
\end{aligned}$$

Then we have the following

$$\begin{aligned}\operatorname{Re}\left(\int_\gamma w\,dz\right) &= \sum_{i,j=1}^{g}\left[\lambda_i\mu_i \operatorname{Re}\left(\int_{\alpha_j} w_{\alpha_i}\,dz\right) + \lambda_i\mu_{g+j}\operatorname{Re}\left(\int_{\beta_j} w_{\alpha_i}\,dz\right)\right.\\ &\qquad + \lambda_{g+i}\mu_j \operatorname{Re}\left(\int_{\alpha_j} w_{\beta_i}\,dz\right)\\ &\qquad\qquad \left. + \lambda_{g+i}\mu_{g+j}\operatorname{Re}\left(\int_{\beta_j} w_{\beta_i}\,dz\right)\right]\\ &= \sum_{i,j=1}^{g}[\lambda_i\mu_j(\alpha_i\,,\,\alpha_j) + \lambda_i\mu_{g+j}(\alpha_i\,,\,\beta_j) + \lambda_{g+i}\mu_j(\beta_i\,,\,\alpha_j)\\ &\qquad + \lambda_{g+i}\mu_{g+j}(\beta_i\,,\,\beta_j)],\end{aligned}$$

which equals

$$\sum_{i=1}^{g}(\lambda_i\mu_{g+i} - \lambda_{g+i}\mu_i)$$

by condition (3) in the definition of the intersection number. Consequently, we obtain

$$(w\,dz\,,\,\overline{\gamma}) = \exp\left(2\pi\sqrt{-1}\sum_{i=1}^{g}(\lambda_i\mu_{g+i} - \lambda_{g+i}\mu_i)\right).$$

Consider λ_i and μ_i, $i = 1, \ldots, 2g$, as coordinates for $w\,dz$ and $\overline{\gamma}$, respectively, and introduce topologies on $\mathfrak{L}_0$ and $B_1^*(\mathfrak{R})$. The topologies on $\mathfrak{L}_0$ and $B_1^*(\mathfrak{R})$ are the same for other sets of basis elements for $\mathfrak{L}_0$ and $B_1^*(\mathfrak{R})$. That is, those topologies are particular topologies for $\mathfrak{L}_0$ and $B_1^*(\mathfrak{R})$. Both topological Abelian groups are isomorphic to $V_{2g}(k_0)$. When $\overline{\gamma}$ runs through all the elements in $B_1^*(\mathfrak{R})$, $\chi_{\overline{\gamma}}(w\,dz) = (w\,dz\,,\,\overline{\gamma})$ provide all the continuous characters of $\mathfrak{L}_0$. The corresponding statement for $B_1^*(\mathfrak{R})$ is also true for the characters $\chi_{w\,dz}(\overline{\gamma}) = (w\,dz\,,\,\overline{\gamma})$. Therefore, we have the following theorem.

THEOREM 5.4. *The topological Abelian groups* $\mathfrak{L}_0$ *and* $B_1^*(\mathfrak{R})$, *isomorphic to the real vector space* $V_{2g}(k_0)$ *of dimension* $2g$, *are dual to each other for the character* (1.20) *in the sense of Pontrjagin.*

For topological Abelian groups, see Pontrjagin, *Topological Groups*, Princeton University Press, Princeton, NJ, 1939.

Theorem 5.3 and 5.4 are sharpened results of Theorem 4.8 asserting that the algebraic genus of the algebraic function field K coincides with the topological genus of the closed Riemann surface $\mathfrak{R}$ belonging to K. The fact that $\mathfrak{L}_0$ and $B_1^*(\mathfrak{R})$ are isomorphic and also dual to each other is based on the underlying space S of $\mathfrak{R}$ being dual to itself. In general, an n-dimensional orientable closed manifold is dual to itself. This is one of the underlying facts for many remarkable theorems on closed manifolds. See the beginning of §3, Chapter 4.

We will come back to the topic of the Abelian integrals of the first kind in a later section. Next we will generalize Theorem 5.2 as follows.

THEOREM 5.5. *Let C be a standard subdivision of a closed Riemann surface $\mathfrak{R}$ of genus $g \geq 1$, and let α_i and β_i, $i = 1, \dots, g$, be subdivision curves going through a fixed point P_0. Let $P_1, \dots, P_l, Q_1, \dots, Q_l$ be points on $\mathfrak{R}$ that are not on α_i and β_i, and let $w\,dz$ and $w'\,dz$ be differentials in $K(\mathfrak{R})$ possibly having poles on $P_1, \dots, P_l$ and $Q_1, \dots, Q_l$, respectively. Let also ω_i and ω_{g+i}, or ω'_i and ω'_{g+i} be the periods of $w\,dz$ for α_i and β_i, or $w'\,dz$ for α_i and β_i, respectively, where $i = 1, \dots, g$. Then let, respectively,*

$$w\frac{dz}{dt_i} = \sum_s a_s^{(i)} t_i^s \quad \text{and} \quad w'\frac{dz}{dt_i} = \sum_s a'_s t_i^s$$

be the expansions of $w\,dz$ and $w'\,dz$ with respect to a locally uniformizing variable t_i at P_i. Similarly, for a locally uniformizing variable t'_j at Q_j let

$$w\frac{dz}{dt'_j} = \sum_s b_s^{(j)} t'^s_j \quad \text{and} \quad w'\frac{dz}{dt'_j} = \sum_s b'^{(j)}_s t'^s_j$$

be the t'_j-expansions of $w\,dz$ and $w'\,dz$, respectively. Then we have

$$\frac{1}{2\pi\sqrt{-1}}\sum_{i=1}^{g}(\omega_i\omega'_{g+i} - \omega_{g+i}\omega'_i) + \sum_{i=1}^{l}\left(a_{-1}^{(i)}\int_{P_0}^{P_i} w'\,dz + \sum_{s=0}^{\infty}\frac{a_{-s-2}^{(i)}a'^{(i)}_s}{s+1}\right)$$
$$-\sum_{j=1}^{m}\left(b'^{(j)}_{-1}\int_{P_0}^{Q_j} w\,dz + \sum_{s=0}^{\infty}\frac{b_s^{(j)}b'^{(j)}_{-s-2}}{s+1}\right) = 0, \tag{1.21}$$

where the integral paths from P_0 to P_i and Q_i are properly chosen, independent of $w\,dz$ and $w'\,dz$, depending only upon P_0, P_i, and Q_i.

PROOF. For C, let $\overline{V}$, $V^{(0)}$ and F be as in §3 of the preceding chapter. Since P_i and Q_j are not on α_i and β_i, there uniquely exist p_i and q_i, respectively, in $V^{(0)}$ such that

$$F(p_i) = P_i \quad \text{and} \quad F(q_j) = Q_j, \qquad i = 1, \dots, l, \; j = 1, \dots, m.$$

Connect p_i and q_i with a_1 by continuous curves L_i and L'_j in $V^{(0)}$ such that L_i and L'_j do not intersect in $V^{(0)}$, and let Γ_i and Γ'_j be small simple closed curves around p_i and q_i, respectively. For a sufficiently large r, the images of $L_i, L'_j, \Gamma_i, \Gamma'_j$ by the map F can be approximated as closely as one desires by the sides of the rth time barycentric subdivision $C^{(r)}$ of C. Rewrite the inverse images of those approximating curves as the same $L_i, L'_j, \Gamma_i, \Gamma'_j$, and rewrite $C^{(r)}$ as C. Then $F(L_i)$, $F(L'_j)$, $F(\Gamma_i)$, $F(\Gamma'_j)$ are sides of C, i.e., one-dimensional subcomplexes of C. Let $W^{(0)}$ be the

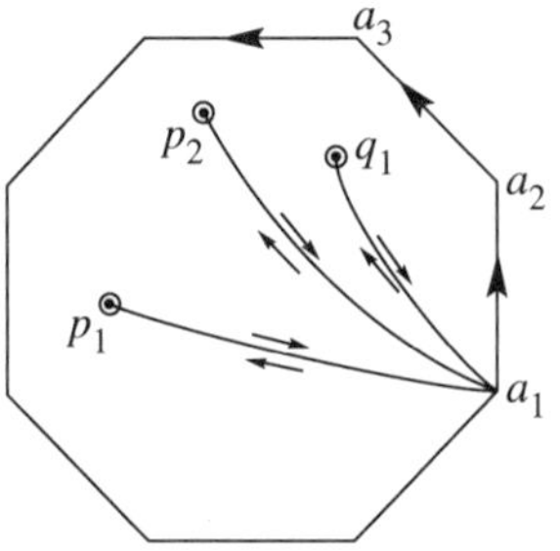

FIGURE 5.3

complement in $V^{(0)}$ of L_i, L'_j, and the small regions bounded by Γ_i and Γ'_j, and let $U' = F(W^{(0)})$. Then $W^{(0)}$ is clearly simply connected, and hence so is U'. Let Q_0 be any point in U', and define

$$w_1(P) = \int_{Q_0}^{P} w\,dz \quad \text{and} \quad w_2(P) = \int_{Q_0}^{P} w'\,dz.$$

Since $w\,dz$ and $w'\,dz$ are regular in U', we obtain single-valued regular functions $w_1(P)$ and $w_2(P)$. Extend $w_1(P)$ and $w_2(P)$ from U' to its boundary of U' as we did in the proof of Theorem 5.2. Then the integral of $w_1\,dw_2$ along the boundaries δ of all the triangles of C contained in $\overline{U}'$ becomes zero, i.e.,

$$\sum{}' \int_{\delta} w_1\,dw_2 = 0.$$

As we saw in the proof of Theorem 5.2, the left-hand side is the sum of the integrals along the boundary of U':

$$F(L_i),\ F(L'_j),\ F(\Gamma_i),\ F(\Gamma'_j),\ F(\Gamma),$$

where $\Gamma = \overline{a_1a_2} + \overline{a_2a_3} + \cdots + \overline{a_{4g}a_1}$. Since we can make Γ_i and Γ'_j as small as possible, letting Γ_i and Γ'_j converge to P_i and Q_j, we obtain

$$\int_{F(\Gamma)} + \sum_{i=1}^{l}\left(\int_{P_0}^{P_i} + \int_{P_i}^{P_0} + \lim \int_{F(\Gamma_i)}\right) + \sum_{j=1}^{m}\left(\int_{P_0}^{Q_j} + \int_{Q_j}^{P_0} + \lim \int_{F(\Gamma'_j)}\right) = 0, \tag{1.22}$$

where the integrand is $w_1\,dw_2$, and see Figure 5.3 for the integral paths. Next we will compute each term.

As for $\int_{F(\Gamma)}$, we have, just as before,

$$\int_{F(\Gamma)} w_1\,dw_2 = \sum_{i=1}^{g} (\omega_i\omega'_{g+i} - \omega_{g+i}\omega'_i).$$

Next we will compute $\int_{P_0}^{P_i} + \int_{P_i}^{P_0}$. The integral

$$w_1(P) = \int_{Q_0}^{P} w\,dz$$

increases by $-2\pi\sqrt{-1}a_{-1}^{(i)}$ as P goes once around P_i in the negative direction. Therefore, the difference of the values of $w_1(P)$ at the same point, between the path from P_0 to P_i, i.e., the integral along the side of the triangle on the left of $\overline{P_0P_1}$, and the path from P_i to P_0, i.e., the integral along the side of the triangle on the right of $\overline{P_0P_1}$, is $-2\pi\sqrt{-1}a_{-1}^{(i)}$. Hence,

$$\int_{P_i}^{P_0} + \int_{P_0}^{P_i} = -2\pi\sqrt{-1}a_{-1}^{(i)} \int_{P_i}^{P_0} dw_2 = 2\pi\sqrt{-1}a_{-1}^{(i)} \int_{P_0}^{P_i} w'\,dz.$$

We have

$$\int_{F(\Gamma_i)} w_1\,dw_2 = \int_{F(\Gamma_i)} w_1 \frac{dw_2}{dt_i}\,dt_i = \int_{F(\Gamma_i)} \frac{d(w_1w_2)}{dt_i} - \int_{F(\Gamma_i)} w_2 \frac{dw_1}{dt_i}\,dt_i.$$

Then the first term on the right represents the increase of the function $w_1(P)w_2(P)$ as the point P goes around near by P_i. Meanwhile, $w_1(P)$ increases by $-2\pi\sqrt{-1}a_{-1}^{(i)}$, and $w_2(P)$ is invariant since $w'\,dz$ is regular at P_i. Hence, by taking the limit, the value approaches

$$-2\pi\sqrt{-1}a_{-1}^{(i)} w_2(P_i).$$

The second term equals $-2\pi\sqrt{-1}$ times the residue of the function $w_2(dw_1/dt_i)$ at $t_i = 0$. Since $w'\,dz$ is regular at P_i, in a neighborhood of P_i we have

$$w_2(P) = w_2(P_i) + \int_{P_i}^{P} w' \frac{dz}{dt_i}\,dt_i = w_2(P_i) + \sum_{s=0}^{\infty} \frac{a_s'^{(i)}}{s+1} t_i^{s+1}.$$

Hence, we find the residue of $w_2 dw_1/dt_i$ from the above and from

$$\frac{dw_1}{dt_i} = w\frac{dz}{dt_i} = \sum_s a_s^{(i)} t_i^s,$$

and we get

$$a_{-1}^{(i)} w_2(P_i) + \sum_{s=0}^{\infty} \frac{a_{-s-2}^{(i)} a_s'^{(i)}}{s+1}.$$

Hence,

$$\lim \int_{F(\Gamma_i)} w_1\,dw_2 = 2\pi\sqrt{-1} \sum_{s=0}^{\infty} \frac{a_{-s-2}^{(i)} a_s'^{(i)}}{s+1}.$$

As for the terms $\int_{P_0}^{Q_j} + \int_{Q_j}^{P_0}$ involving Q_j, note that $w\,dz$ is regular at Q_j and that the value of $w_1(P) = \int_{Q_0}^{P} w\,dz$ is unchanged as P goes around Q_j. Hence, the values of $w_1(P)$, at the same point, are the same whether P moves from P_0 to Q_j, or from Q_j to P_0. Of course, $dw_2 = w'\,dz$ takes the same value at a point. Therefore, we obtain

$$\int_{P_0}^{Q_j} + \int_{Q_j}^{P_0} = 0.$$

Lastly, we can compute the limit of $\int_{F(\Gamma'_j)}$ in the same way as above. That is, it is equal to $2\pi\sqrt{-1}$ times the residue at $t'_j = 0$ of

$$w_1 \frac{dw_2}{dt'_j} = \left(\int_{Q_0}^{Q_j} w\,dz + \sum_{s=0}^{\infty} \frac{b_s^{(j)}}{s+1} t'^{s+1}_j\right)\left(\sum_s b'^{(j)}_s t'^s_j\right),$$

i.e.,

$$-2\pi\sqrt{-1}\left(b'^{(j)}_{-1} \int_{Q_0}^{Q_j} w\,dz + \sum_{s=0}^{\infty} \frac{b_s^{(j)} b'^{(j)}_{-s-2}}{s+1}\right).$$

Substitute all those results back into (1.22), and divide (1.22) by $2\pi\sqrt{-1}$ to obtain

$$\begin{aligned}\frac{1}{2\pi\sqrt{-1}} &\sum_{i=1}^{g}(\omega_i\omega'_{g+i} - \omega_{g+i}\omega'_i)\\ &+\sum_{i=1}^{l}\left(a_{-1}^{(i)}\int_{P_0}^{P_i} w'\,dz + \sum_{s=0}^{\infty}\frac{a_{-s-2}^{(i)} a'^{(i)}_s}{s+1}\right)\\ &-\sum_{j=1}^{m}\left(b'^{(j)}_{-1}\int_{Q_0}^{Q_j} w\,dz + \sum_{s=0}^{\infty}\frac{b_s^{(j)} b'^{(j)}_{-s-2}}{s+1}\right) = 0.\end{aligned} \tag{1.23}$$

Since Q_0 was arbitrary in U', by letting Q_0 converge to P_0 we obtain formula (1.21). Q.E.D.

The term in (1.23) that is dependent on Q_0 is

$$\sum_{j=1}^{m} b'^{(j)}_{-1} \int_{Q_0}^{Q_j} w\,dz. \tag{1.24}$$

According to the above proof, (1.24) should not depend on Q_0. However, for another Q'_0 we consider the difference between (1.24) and

$$\sum_{j=1}^{m} b'^{(j)}_{-1} \int_{Q'_0}^{Q_j} w\,dz$$

to find

$$\left(\sum_{j=1}^{m} b'^{(j)}_{-1} \int_{Q_0}^{Q'_0} w\,dz\right). \tag{1.25}$$

Notice then that $\sum_{j=1}^{m} b'^{(j)}_{-1}$ is the sum of all the residues of the differential $w'\,dz$. Hence, by Theorem 2.19, the sum $\sum_{j=1}^{m} b'^{(j)}_{-1} = 0$. That is, the theorem is consistent with the above. The proof of Theorem 2.19 was entirely algebraic. Since (1.25) is zero, conversely we obtain an analytical proof of Theorem 2.19. Note that the coefficient field k is the field of complex numbers. See the note at the end of §2, Chapter 4.

For given differentials $w\,dz$ and $w'\,dz$ whose poles are different from each other, there is a standard subdivision C so that the poles of these differentials are not on the subdivision curves of C. See the last paragraph of §3, Chapter 4. Then (1.21) provides a relationship between the differentials $w\,dz$ and $w'\,dz$. Next we will treat special differentials so that (1.21) gives a simple relation between the differentials.

Let P be an arbitrary point on $\mathfrak{R}$ and let t be a locally uniformizing variable at P. Let $u\,dz$ be a differential of the second kind which is regular except at the point P. Let its expansion in a neighborhood of P be given by

$$u\frac{dz}{dt} = -\frac{1}{t^2} + a_0 + a_1 t + \cdots .$$

From Theorem 2.25, such a differential $u\,dz$ exists. For such a differential $u\,dz$, the sum of any differential of the first kind and $u\,dz$ has the same property as $u\,dz$. In particular, there exists a differential $\overline{w}^{\,t}_P\,dz$ whose periods for all α_i, $i = 1, \dots, g$, are zero, where α_i and β_i are chosen not to contain P. See the third paragraph preceding Theorem 5.3. Then $\overline{w}^{\,t}_P\,dz$, as the notation suggests, is uniquely determined by P and the locally uniformizing variable t to satisfy the above condition. The reason for uniqueness is as follows. The difference between such differentials is a differential of the first kind whose periods for α_i, $i = 1, \dots, g$, are all zero. Then the isomorphism of (1.12) and (1.13) implies that the difference must be zero.

Similarly to the above, one can show that there exists a differential $\overline{w}_{P,Q}\,dz$ of the third kind whose residues at P and Q are 1 and -1, respectively, and the differential is regular except at these points P and Q. Moreover, the periods of $\overline{w}_{P,Q}\,dz$ for α_i, $i = 1, \dots, g$, are all zero. Using the isomorphism in (1.14) and (1.15) instead of (1.12) and (1.13), one can prove that there exist uniquely determined $w^t_P\,dz$ and $w_{P,Q}\,dz$ of the second or third kind with the same singularities as $\overline{w}^{\,t}_P\,dz$ and $\overline{w}_{P,Q}\,dz$, respectively, so that their periods for α_i and β_i, $i = 1, \dots, g$, are all purely imaginary.

First apply (1.21) for

$$w\,dz = \overline{w}^{\,t}_P\,dz \quad \text{and} \quad w'\,dz = \overline{w}^{\,t'}_Q\,dz, \qquad P \neq Q,$$

and $P_1 = P$ and $Q_1 = Q$. Then the term involving the periods and the terms involving integrals vanish. From the terms with $\sum_{s=0}^{\infty}$, we are left with $a^{(1)}_{-2}a'^{(1)}_0$ and $b^{(1)}_0 b'^{(1)}_{-2}$. Hence, we obtain

$$\overline{w}^{\,t}_P\frac{dz}{dt'} = \overline{w}^{\,t'}_Q\frac{dz}{dt}. \tag{1.26}$$

Here note that $\overline{w}^{\,t}_P dz/dt'$ and $\overline{w}^{\,t'}_Q dz/dt$ are evaluated at $t' = 0$ and $t = 0$.

Let P, Q, and Q' be distinct points, and let

$$w\,dz = \overline{w}^{\,t}_P\,dz \quad \text{and} \quad w'\,dz = \overline{w}_{Q,Q'}\,dz.$$

As above, e.g., letting $P_1 = P$, $Q_1 = Q$, and $Q_2 = Q'$, we obtain from (1.21)

$$-\overline{w}_{Q,Q'}\frac{dz}{dt} - \left(\int_{P_0}^{Q} \overline{w}^{\,t}_{P}\, dz - \int_{P_0}^{Q'} \overline{w}^{\,t}_{P}\, dz\right) = 0,$$

i.e.,

$$\overline{w}_{Q,Q'}\frac{dz}{dt} = -\int_{Q'}^{Q} \overline{w}^{\,t}_{P}\, dz. \tag{1.27}$$

Furthermore, take four distinct points P, P', Q, and Q', and let

$$w\, dz = \overline{w}_{P,P'}\, dz \quad \text{and} \quad w'\, dz = \overline{w}_{Q,Q'}\, dz.$$

Similarly, we obtain

$$\int_{Q'}^{Q} \overline{w}_{P,P'}\, dz = \int_{P'}^{P} \overline{w}_{Q,Q'}\, dz. \tag{1.28}$$

Notice that (1.26), (1.27), and (1.28) are relations among special differentials of the second or third kind and their Abelian integrals, which are special cases of (1.21). As we saw in the proof of Theorem 2.27, any differential in K can be written as a sum of such $\overline{w}^{\,t}_{P}\, dz$ and $\overline{w}_{P,Q}\, dz$, a differential of the first kind and a differential of the second kind with a pole of higher order. Hence, those relations are indeed fundamental. In particular, (1.28) is referred to as the *exchange law of variable and parameter*.

In order to obtain the relation between a differential of the first kind and a differential of the second or third kind, it is better to consider $w^t_P\, dz$ and $w_{P,Q}\, dz$ rather than $\overline{w}^{\,t}_{P}\, dz$ and $\overline{w}_{P,Q}\, dz$. From Theorem 5.3,

$$w_{\alpha_i}\, dz \quad \text{and} \quad w_{\beta_i}\, dz, \qquad i = 1, \dots, g,$$

form a base for the set $\mathfrak{L}_0$ of differentials of the first kind in K over the real field k_0. Let us apply (1.21) for those differentials and $w^t_P\, dz$ and $w_{P,Q}\, dz$. First set

$$w\, dz = w_{\alpha_i}\, dz \quad \text{and} \quad w'\, dz = w^t_P\, dz.$$

By definition, the periods w'_j of $w^t_P\, dz$ for α_i and β_i are all purely imaginary. Therefore, the real part of

$$\frac{1}{2\pi\sqrt{-1}}\sum_{j=1}^{g}(\omega_j\omega'_{g+j} - \omega_{g+j}\omega'_j) \tag{1.29}$$

equals the value of (1.29) obtained by replacing ω_j and ω_{g+j} by their real parts. From the definition we have

$$\operatorname{Re}(\omega_j) = \operatorname{Re}\left(\int_{\alpha_j} w_{\alpha_i}\, dz\right) = (\alpha_i, \alpha_j) = 0,$$

$$\operatorname{Re}(\omega_{g+j}) = \operatorname{Re}\left(\int_{\beta_j} w_{\alpha_i}\, dz\right) = (\alpha_i, \beta_j) = \delta_{ij}.$$

As a consequence, the real part of (1.29) is

$$\frac{-\omega_i'}{2\pi\sqrt{-1}} = \frac{-1}{2\pi\sqrt{-1}}\int_{\alpha_i} w_P^t\,dz\,.$$

Considering only the real parts of the other terms in (1.21), we get

$$\frac{-1}{2\pi\sqrt{-1}}\int_{\alpha_i} w_P^t\,dz + \operatorname{Re}\left(w_{\alpha_i}\frac{dz}{dt}\right) = 0\,,$$

i.e.,

$$\int_{\alpha_i} w_P^t\,dz = 2\pi\sqrt{-1}\operatorname{Re}\left(w_{\alpha_i}\frac{dz}{dt}\right)\,,$$

and similarly

$$\int_{\beta_i} w_P^t\,dz = 2\pi\sqrt{-1}\operatorname{Re}\left(w_{\beta_i}\frac{dz}{dt}\right)\,.$$

Since α_i and β_i are basis elements for $B_1^*(\mathfrak{R})$, the above two equalities imply

$$\int_{\gamma} w_P^t\,dz = 2\pi\sqrt{-1}\operatorname{Re}\left(w_{\gamma}\frac{dz}{dt}\right) \tag{1.30}$$

for an arbitrary cycle γ with real coefficients. The left-hand side is the extended period of $w_P^t\,dz$ for γ as we defined the extended period for a differential of the first kind in the paragraph above Theorem 5.3.

Lastly, let

$$w\,dz = w_{\alpha_i}\,dz \quad \text{and} \quad w'\,dz = w_{P,Q}\,dz\,.$$

Considering the real parts of (1.21) as above, we have

$$\int_{\alpha_i} w_{P,Q}\,dz = 2\pi\sqrt{-1}\operatorname{Re}\left(-\int_Q^P w_{\alpha_i}\,dz\right)\,,$$

and similarly

$$\int_{\beta_i} w_{P,Q}\,dz = 2\pi\sqrt{-1}\operatorname{Re}\left(-\int_Q^P w_{\beta_i}\,dz\right)\,.$$

Hence, for a general one-dimensional cycle γ we have

$$\int_{\gamma} w_{P,Q}\,dz = 2\pi\sqrt{-1}\operatorname{Re}\left(-\int_Q^R w_{\gamma}\,dz\right)\,. \tag{1.31}$$

Note also that the left-hand side is the extended period of $w_{P,Q}\,dz$. The integral path on the right is properly chosen from Q to P, and as we noted in Theorem 5.5, the integral does not depend upon γ. Formula (1.31) will play a significant role later.

For $w_P^t\,dz$ and $w_{P,Q}\,dz$, one can obtain the relations among the real parts like (1.26) from (1.21). When one chooses a differential of the second or third kind whose periods for α_i and β_i are all real, one obtains a formula similar to (1.26) for the imaginary parts. We will leave the formula for the reader to establish.

§2. Additive functions and multiplicative functions

In this section we will consider an Abelian integral from a different view. From Theorem 3.19, the covering Riemann surface of an arbitrary closed Riemann surface $\mathfrak{R}$ is either the Riemann sphere, a finite complex t-plane, or the interior $\{t; |t| < 1\}$ of a unit circle. Let us denote such a simply connected covering Riemann surface by $\mathfrak{R}^*$, and let f^* be the covering map from $\mathfrak{R}^*$ to $\mathfrak{R}$. For either case above, we can express a point on $\mathfrak{R}^*$ by a complex number t. We write the image of t simply as

$$P_t = f^*(t).$$

Then for an arbitrary function $z = z(P)$ in the analytic function field $K = K(R)$ of $\mathfrak{R}$, define

$$z'(t) = z(P_t) = z(f^*(t)).$$

Clearly, $z'(t)$ belongs to the analytic function field $\widetilde{K}$ of $\mathfrak{R}^*$. Furthermore, the correspondence

$$z(P) \leftrightarrow z'(t)$$

is an isomorphism from K into $\widetilde{K}$. We will regard K as a subfield of $\widetilde{K}$, and we will not distinguish $z(P)$ from $z'(t)$. When an element z in K is considered as a function on $\mathfrak{R}$, we write $z = z(P)$, and we write $z = z(t)$ as a function on $\mathfrak{R}^*$. Similarly, for elements $z = z(P) = z(t)$ and $w = w(P) = w(t)$ in K, we do not distinguish the differential $w(P)\,dz(P)$ on $\mathfrak{R}$ from the differential $w(t)\,dz(t)$ on $\mathfrak{R}^*$. We denote the differential by $w\,dz$. The map f^* is a one-to-one and conformal map from a neighborhood of each point t on $\mathfrak{R}^*$ onto a neighborhood of P_t on $\mathfrak{R}$. Hence, we can regard a locally uniformizing variable t^* at t as a locally uniformizing variable at P_t as well. Therefore the t^*-expansion

$$w\frac{dz}{dt^*} = \sum a_i t^{*^i}$$

of $w\,dz$ may be considered as the expansion of $w\,dz$ in K and also as the expansion of $w\,dz$ in $\widetilde{K}$. Hence, the order and the residue of $w\,dz$ at t coincide with those at P_t.

Let $G = A(\mathfrak{R}^*, f^*; \mathfrak{R})$ be the automorphism group of the covering Riemann surface $\{\mathfrak{R}^*, f^*; \mathfrak{R}\}$, i.e., the fundamental group of the underlying space S of R for the underlying space S^*. Since G is a subgroup of the automorphism group $A(\mathfrak{R}^*)$ of $\mathfrak{R}^*$, define

$$u'(t) = u(\varphi^{-1}(t))$$

for an arbitrary φ in G and an arbitrary function $u(t)$ in $\widetilde{K}$. Then $u'(t)$ also belongs to $\widetilde{K}$. Let

$$u' = \sigma(u).$$

Then σ defines an automorphism of $\widetilde{K}$. Let $\mathfrak{G}$ be the totality of those σ obtained from the maps φ in G. Then $\mathfrak{G}$ is a group of automorphisms of $\widetilde{K}$. The map

$$\varphi \mapsto \sigma \tag{2.1}$$

is a homomorphism from G onto $\mathfrak{G}$. Let σ and σ' be the elements in $\mathfrak{G}$ corresponding to φ and φ', $\varphi \neq \varphi'$, in G. Then there is a point t such that $t_1 = \varphi^{-1}(t)$ differs from $t_2 = \varphi'^{-1}(t)$. Hence, for an analytic function u on $\mathfrak{R}^*$ satisfying $u(t_1) \neq u(t_2)$ (e.g., take t itself), we have $\sigma(u) \neq \sigma'(u)$ from the definition, i.e., $\sigma \neq \sigma'$. Therefore, the map (2.1) is one-to-one, and consequently we obtain

$$G \cong \mathfrak{G}.$$

We denote this isomorphism from $\mathfrak{G}$ to G by

$$\varphi = \Phi(\sigma), \qquad \varphi \in G, \ \sigma \in \mathfrak{G}.$$

On the other hand, for an arbitrary point P_0 on $\mathfrak{R}$, G is isomorphic to the homotopy group $G(P_0)$ of $\mathfrak{R}$ with base point P_0 (Lemma 3.11). Hence, we have

$$\mathfrak{G} \cong G(P_0). \tag{2.2}$$

The commutator subgroup $\mathfrak{G}'$ of $\mathfrak{G}$ corresponds to the commutator subgroup $G(P_0)'$ of $G(P_0)$ through the above isomorphism. From (2.2) we have the isomorphism

$$\mathfrak{G}/\mathfrak{G}' \cong G(P_0)/G(P_0)'.$$

For the P_0 in $(2')$ of the definition of a standard subdivision C, §3, Chapter 4, the isomorphism (3.7) in that section implies

$$G(P_0)/G(P_0)' \cong B_1(\mathfrak{R}),$$

where $B_1(\mathfrak{R})$ is the first integral homology group of $\mathfrak{R}$. If we let $\overline{\mathfrak{G}} = \mathfrak{G}/\mathfrak{G}'$, then

$$\overline{\mathfrak{G}} \cong B_1(\mathfrak{R}). \tag{2.3}$$

We denote the above isomorphism by

$$\overline{\gamma} = \Psi(\overline{\sigma}), \qquad \overline{\gamma} \in B_1(\mathfrak{R}), \ \overline{\sigma} \in \overline{\mathfrak{G}}. \tag{2.4}$$

By following the steps to get the isomorphism (2.3), we can specify the map Ψ explicitly as follows. Let σ be any element in $\mathfrak{G}$ which belongs to the residue class $\overline{\sigma}$ in $\overline{\mathfrak{G}} = \mathfrak{G}/\mathfrak{G}'$. Then let $\varphi = \Phi(\sigma)$. Next fix a point t_0 on $\mathfrak{R}^*$ such that $P_0 = f^*(t_0)$. For a path γ^* on $\mathfrak{R}^*$ connecting t_0 to $\varphi(t_0)$, let $\gamma = f^*(\gamma^*)$. Then Ψ assigns to $\overline{\sigma}$ the homology class of $\overline{\gamma}$ of γ, i.e., $\overline{\gamma} = \Psi(\overline{\sigma})$.

The above description of the map Ψ appears to depend upon P_0 and t_0. However, as we will show below, Ψ is independent of P_0 and t_0. Let $\overline{\sigma}$, σ, and φ be as above, and let t_0' be an arbitrary point on $\mathfrak{R}^*$ other than t_0.

For a path δ^* connecting t_0' and $\varphi(t_0')$, let $\delta = f^*(\delta^*)$ be the image of δ^*. Then δ is homologous to γ, i.e.,

$$\overline{\delta} = \overline{\gamma} = \Psi(\overline{\sigma}).$$

We will prove this as follows. Let δ_1^* be a path from t_0 to t_0', and then let $\delta_2^* = \varphi(\delta_1^*)$. Then δ_2^* is the path from $\varphi(t_0)$ to $\varphi(t_0')$. The path $\delta_1^* \delta^{*^{-1}} \delta_2^{*^{-1}} \gamma^*$ is a closed curve with base point t_0. Since $\mathfrak{R}^*$ is simply connected, the above path is homotopic to the constant path. Hence, $f^*(\delta_1^* \delta^{*^{-1}} \delta_2^{*^{-1}} \gamma^*)$ is also homotopic to the constant path. Therefore, as a one-dimensional cycle we have

$$f^*(\delta_1^* \delta^{*^{-1}} \delta_2^{*^{-1}} \gamma^*) = -f^*(\delta_1^*) - f^*(\delta^*) + f^*(\delta_2^*) + f^*(\gamma^*) \sim 0.$$

Since $f^*(\delta_2^*) = f^*(\varphi(\delta_1^*)) = f^*(\delta_1^*)$, we obtain

$$-f^*(\delta^*) + f^*(\gamma^*) \sim 0.$$

That is,

$$\gamma \sim \delta, \quad \text{i.e.,}\ \overline{\gamma} = \overline{\delta}.$$

Therefore, the map Ψ in (2.4) can be obtained from any point t_0'.

From the definition of $G = A(\mathfrak{R}^*, f^*; \mathfrak{R})$, any map φ in G satisfies

$$f^*(\varphi(t)) = f^*(t) \quad \text{and} \quad P_{\varphi(t)} = P_t.$$

Hence, for a function $z = z(P) = z(t)$ in K, we have

$$z(\varphi(t)) = z(P_{\varphi(t)}) = z(P_t) = z(t).$$

Therefore, for all the elements σ in $\mathfrak{G}$, we have

$$\sigma(z) = z. \tag{2.5}$$

Conversely, if a function $z(t)$ in $\widetilde{K}$ satisfies (2.5) for all σ in $\mathfrak{G}$, then

$$z(\varphi(t)) = z(t) \quad \text{for } \varphi \in G.$$

Then the function defined by

$$z_1(P) = z(f^{*^{-1}}(P)), \qquad P \in \mathfrak{R}$$

is a single-valued function on $\mathfrak{R}$. Since f^* is a one-to-one and conformal map from a neighborhood of t in $\mathfrak{R}^*$ onto a neighborhood of P_t, the above function $z_1(P)$ is analytic on $\mathfrak{R}$ and

$$z(t) = z_1(P_t).$$

By our notation, we may write

$$z(t) = z_1(t), \quad \text{i.e.,}\ z = z', \quad z \in K.$$

Consequently, the set of elements in $\widetilde{K}$ that are invariant under all the automorphisms in $\mathfrak{G}$ are exactly the elements in K. In the above sense, $\mathfrak{G}$

is called the Galois group of $\widetilde{K}/K$. We will show later that for the extension $\widetilde{K}/K$, $\mathfrak{G}$ plays the role of the Galois group for a finite Galois extension. However, in this section we need $\mathfrak{G}$ only to define the following.

DEFINITION 5.1. When a function u in $\widetilde{K}$ is transformed by an arbitrary automorphism σ in $\mathfrak{G}$ of the form

$$\sigma(u) = u + \omega_\sigma, \qquad \omega_\sigma \in k, \tag{2.6}$$

then u is called an *additive function* in $\widetilde{K}$ with respect to $\mathfrak{G}$, and ω_σ is called the *period* of u for σ.

Note that all the functions in K are additive functions of period 0. For K of genus zero, $\mathfrak{R} = \mathfrak{R}^*$ and $K = \widetilde{K}$ hold. Hence, all the functions are additive functions of period 0. We will show next that for $g \geq 1$ there is a function in $\widetilde{K}$, but not in K, which is an additive function.

Generally for an additive function $u = u(t)$ satisfying (2.6) we have

$$\omega_{\sigma\tau} = \omega_\sigma + \omega_\tau$$

for any σ and τ in $\mathfrak{G}$. That is, the map $\sigma \mapsto \omega_\sigma$ is a homomorphism from $\mathfrak{G}$ into the additive group of complex numbers. Then the value ω_σ depends only upon the residue class of σ in $\overline{\mathfrak{G}} = \mathfrak{G}/\mathfrak{G}'$. In particular, for $\sigma \in \mathfrak{G}'$ we have

$$\omega_\sigma = 0. \tag{2.7}$$

Next let $z = z(t)$ be a nonconstant function in K, and let

$$w(t) = \frac{du(t)}{dt} \Big/ \frac{dz(t)}{dt}.$$

Then $w(t)$ is clearly an analytic function in $\widetilde{K}$. Furthermore, for any map φ in $\mathfrak{G}$ we have

$$w(\varphi(t)) = \frac{du(\varphi(t))}{d\varphi(t)} \Big/ \frac{dz(\varphi(t))}{d\varphi(t)} = \frac{du(\varphi(t))}{dt} \Big/ \frac{dz(\varphi(t))}{dt}.$$

Letting $\sigma = \Phi^{-1}(\varphi)$, we obtain

$$\sigma(u) = u + \omega_\sigma, \quad u(\varphi(t)) + \omega_\sigma = u(t), \quad \frac{du(\varphi(t))}{dt} = \frac{du(t)}{dt}.$$

On the other hand, since z is a function in K, we have

$$\sigma(z) = z, \quad z(\varphi(t)) = z(t), \quad \frac{du(\varphi(t))}{dt} = \frac{du(t)}{dt}.$$

Therefore,

$$w(\varphi(t)) = w(t), \qquad \varphi \in G.$$

That is, $w = w(t)$ belongs to K. For any point t_1 on $\mathfrak{R}^*$, $t - t_1$ is a locally uniformizing variable at t_1. Hence, we have

$$\frac{du}{d(t-t_1)} = \frac{du}{dt} = w\frac{dz}{dt} = w\frac{dz}{d(t-t_1)}.$$

Therefore, the differential du in $\widetilde{K}$ coincides with the differential $w\,dz$ in K:

$$du = w\,dz.$$

Furthermore, for the expansion of u at t_1,

$$u(t) = \sum a_i(t-t_1)^i,$$

we have

$$w\frac{dz}{d(t-t_1)} = \frac{du}{d(t-t_1)} = \sum a_i i(t-t_1)^{i-1}.$$

Then the residue of $w\,dz$ is 0 either on $\mathfrak{R}^*$ or on $\mathfrak{R}$.

Next let $w\,dz$ be an arbitrary differential in K whose residue is always 0 on $\mathfrak{R}$. Then as a differential on $\mathfrak{R}^*$ the residue of $w\,dz$ is always 0. Since $\mathfrak{R}^*$ is simply connected, Theorem 3.7 implies that there is a function $u = u(t)$ in $\widetilde{K}$, unique up to an added constant, such that

$$du = w\,dz.$$

Next we will prove that this function u in $\widetilde{K}$ is additive for $\mathfrak{G}$. First, note that for an arbitrary analytic curve γ^* on $\mathfrak{R}^*$

$$\int_{\gamma^*} w\,dz = \int_{f^*(\gamma^*)} w\,dz. \tag{2.8}$$

The differential on the left-hand side is considered to be the differential on $\mathfrak{R}^*$, and the right-hand side integral is regarded as the differential on $\mathfrak{R}$. When γ^* is contained in a neighborhood $U^* = \{t^*; |t^*| < \varepsilon\}$ of t which is injectively and conformally mapped by f^*, then the locally uniformizing variable t^* at t may also be considered to be a locally uniformizing variable at P_t. Then (2.8) holds since the expansion of $w\,dz/dt^*$ in U^* coincides with the expansion of $w\,dz/dt^*$ in $U = f^*(U^*)$. The general case can be proved by subdividing γ^* into sufficiently small analytic curves γ_i^*, and then taking the sum of the integrals over γ_i^*.

Let t be any point on $\mathfrak{R}^*$ which is not a pole of $w\,dz$ and let φ be an arbitrary map in G. Then t and $\varphi(t)$ can be connected by a path containing no poles of $w\,dz$ since $\varphi(t)$ is not a pole of $w\,dz$. We obtain from (2.8)

$$u(\varphi(t)) - u(t) = \int_{\gamma^*} du = \int_{\gamma^*} w\,dz = \int_{\gamma} w\,dz, \tag{2.9}$$

where $\gamma = f^*(\gamma^*)$. By the earlier remark, the homology class $\overline{\gamma}$ of γ satisfies

$$\overline{\gamma} = \Psi(\overline{\sigma}), \qquad \sigma = \Phi^{-1}(\varphi)$$

independently of t. From Theorem 5.1, $\int_\gamma w\,dz$ depends only upon the homology class $\overline{\gamma}$ of γ. Hence, the right-hand side of (2.9) does not depend on t. Therefore, if we set

$$\omega_\sigma = -\int_\gamma w\,dz, \qquad \overline{\gamma} = \psi(\overline{\sigma}),$$

then for all points where $w\,dz$ is regular, we obtain

$$u(\varphi(t)) + w_\sigma = u(t), \qquad \omega_\sigma \in k.$$

The functions in the above are analytic functions in t. Hence, the equality holds for all t on $\mathfrak{R}^*$, namely,

$$\sigma(u) = u + \omega_\sigma$$

holds, i.e., u is an additive function with period ω_σ. We have established the following theorem.

THEOREM 5.6. *Let K, $\widetilde{K}$, $\mathfrak{R}$, $\mathfrak{R}^*$, and $\mathfrak{G}$ be as before. Then the totality of additive functions in $\widetilde{K}$ with respect to $\mathfrak{G}$ coincides with the set of integrating functions on $\mathfrak{R}^*$ of differentials in K whose residues are 0. Let u be an integrating function of such a differential $w\,dz$. Then the period of u for σ is given by*

$$\omega_\sigma = -\int_\gamma w\,dz, \qquad \bar{\gamma} = \Psi(\bar{\sigma}). \tag{2.10}$$

From Theorem 5.6, we know that the integral of a differential of residue zero, i.e., essentially the Abelian integral of a differential of the first or second kind, is nothing but an additive function in $\widetilde{K}$. The Abelian integral of the third kind in $\widetilde{K}$ also has a profound relationship to a special function in $\widetilde{K}$. A multiplicative function in $\widetilde{K}$ for $\mathfrak{G}$ is such a function.

DEFINITION 5.2. When a function $v \neq 0$ in $\widetilde{K}$ is transformed by an arbitrary automorphism σ in $\mathfrak{G}$ in the following manner

$$\sigma(v) = \chi_\sigma v, \qquad \chi_\sigma \in k\ (\chi_\sigma \neq 0), \tag{2.11}$$

v is called a *multiplicative function* in $\widetilde{K}$ with respect to $\mathfrak{G}$. We call χ_σ the *multiplier* of v for σ.

Just as for an additive function, we have

$$\chi_{\sigma\tau} = \chi_\sigma \chi_\tau$$

for any σ and τ in $\mathfrak{G}$. That is, $\sigma \mapsto \chi_\sigma$ is a homomorphism from $\mathfrak{G}$ into the multiplicative group of complex numbers. Hence, χ_σ, like ω_σ, depends only on the residue class of σ in $\overline{\mathfrak{G}} = \mathfrak{G}/\mathfrak{G}'$. In particular, for $\sigma \in \mathfrak{G}'$, we have

$$\chi_\sigma = 1. \tag{2.12}$$

Let $v = v(t)$ be a multiplicative function in $\widetilde{K}$, and let t_1 be a zero or a pole of order m. Then from (2.11), $v(t)$ has a zero or a pole of the same order m at all the points $\varphi(t_1)$ for $\varphi \in G$. Then we say that v has a zero or a pole of order m at the point $P = P_{t_1}$. Since $\mathfrak{R}$ is compact, there are only finitely many zeros or poles as one can easily recognize. Therefore, the product of those points

$$A = \prod P^{\pm m} \tag{2.13}$$

is a divisor of K, which is called the divisor of the multiplicative function v. We denote this divisor by $A = (v)$. Note that $\pm m$ on the right-hand side of (2.13) indicates $+m$ for a zero P of order m, and $-m$ for a pole P of order m. If v_1 and v_2 are multiplicative functions in $\widetilde{K}$, then $v_1 v_2$ is clearly a multiplicative function in $\widetilde{K}$ satisfying

$$(v_1 v_2) = (v_1)(v_2). \tag{2.14}$$

As in the case for an additive function, consider

$$w'(t) = \frac{1}{v}\frac{dv}{dt} \Big/ \frac{dz}{dt}$$

for a nonconstant function z in K. Then $w'(t)$ is an analytic function belonging to K, and the differential dv/v coincides with $w'\,dz$, i.e.,

$$\frac{dv}{v} = w'\,dz.$$

Since we also have $dv/v = d\log v$, the expansion of v shows that for a zero or a pole of v at t_1 of order m the differential $w'\,dz$ has a pole of order 1 with residue m or a pole of order m, respectively, at $P = P_{t_1}$. By Theorem 2.19 the sum of those residues must be 0. Then from (2.13), we obtain

$$n((v)) = n(A) = \sum \pm m = 0,$$

i.e., the degree of the divisor (v) of the multiplicative function v is always zero.

Fix a point t_0 on $\mathfrak{R}^*$ such that $v(t_0) \neq 0, \infty$. Let γ^* be an analytic curve containing neither zeros nor poles of $v(t)$. Since we have $d\log v = dv/v = w'\,dz$, the integration of $w'\,dz$ along γ^* is given by

$$\log v(t) - \log v(t_0) = \int_{\gamma^*} d\log v = \int_{\gamma^*} w'\,dz = \int_{t_0}^{t} w'\,dz,$$

i.e.,

$$v(t) = v(t_0)\exp\left(\int_{t_0}^{t} w'\,dz\right). \tag{2.15}$$

Conversely, let $w'\,dz$ be a differential on $\mathfrak{R}$ with poles of order at most 1 such that all the residues at the poles are rational integers. Then consider the function

$$v(t) = \exp\left(c + \int_{t_0}^{t} w'\,dz\right), \qquad c \in k$$

on $\mathfrak{R}^*$. The points t_0 and t in $\mathfrak{R}^*$ are chosen not to be the poles of $w'\,dz$, and the path from t_0 to t avoids the poles of $w'\,dz$ as well. First we will show that the above $v(t)$ is a single-valued function defined for t where $w'\,dz$ is regular, i.e., the value $v(t)$ is independent of the choice of path from t_0 to t. As one can see from the proof of Theorem 3.7, it is sufficient to show that the value of $v(t)$ is unchanged as t goes around once near t_0 in the simply connected $\mathfrak{R}^*$. When $w'\,dz$ is regular at t_1, $v(t)$ is also a regular single-valued function in a neighborhood of t_1. If t_1 is a pole of $w'\,dz$ whose residue is m, let

$$w'\frac{dz}{dt} = w'\frac{dz}{d(t-t_1)} = \frac{m}{t-t_1} + \sum_{i=0}^{\infty} a_i(t-t_1)^i, \qquad 0 < |t-t_1| < \varepsilon. \quad (2.16)$$

When t goes around once near t_1 in the positive direction, we have

$$\int w'\,dz = \int w'\frac{dz}{dt}\,dt = 2\pi\sqrt{-1}m.$$

Taking the exponential of this integral, the value of $v(t)$ does not change. Hence, $v(t)$ is single valued. Moreover, in a neighborhood of t_1 at which m is the residue of $w'\,dz$, we get the following from (2.16):

$$\int_{t_0}^{t} w'\,dz = c' + m\log(t-t_1) + \sum_{i=0}^{\infty}\frac{b_i}{i+1}(t-t_1)^{i+1}.$$

Hence,

$$v(t) = (t-t_1)^m \exp\left(c + c' + \sum_{i=0}^{\infty}\frac{b_i}{i+1}(t-t_1)^{i+1}\right).$$

Therefore, $v(t)$ is analytic at every point in $\mathfrak{R}^*$, and $v(t)$ is regular and $v(t) \neq 0$, where $w'\,dz$ is regular. Furthermore, at a pole of $w'\,dz$ with residue m, $v(t)$ has a zero of order m if $m > 0$, and $v(t)$ has a pole of order $-m$ if $m < 0$.

Let t be a point where $w'\,dz$ is regular, and let φ be an arbitrary map in G. Then

$$\int_{t_0}^{\varphi(t)} w'\,dz = \int_{t_0}^{t} w'\,dz + \int_{t}^{\varphi(t)} w'\,dz$$

and

$$v(\varphi(t)) = \exp\left(\int_{t}^{\varphi(t)} w'\,dz\right) v(t).$$

Hence, as we observed for the case of an additive function, for a path γ^* from t to $\varphi(t)$ and $\gamma = f^*(\gamma^*)$

$$\overline{\gamma} = \Psi(\overline{\sigma}), \qquad \sigma = \Phi^{-1}(\varphi)$$

is independent of t. Therefore,

$$\chi_\sigma = \exp\left(-\int_{t}^{\varphi(t)} w'\,dz\right) = \exp\left(-\int_{\gamma^*} w'\,dz\right) = \exp\left(-\int_{\gamma} w'\,dz\right)$$

also does not depend upon t. That is, we have

$$v(\varphi(t)) = \chi_\sigma^{-1} v(t), \qquad \chi_\sigma \in k.$$

The above equation is obtained at the point where $w'\,dz$ is regular. However, since both sides are analytic functions on $\mathfrak{R}^*$, the equation is valid for all the points in $\mathfrak{R}^*$. Hence,

$$\sigma(v) = \chi_\sigma v\,, \qquad \chi_\sigma \in k\,.$$

We have proved that $v(t)$ is a multiplicative function.

THEOREM 5.7. *Let K, $\widetilde{K}$, $\mathfrak{R}$, $\mathfrak{R}^*$, and $\mathfrak{G}$ be as before. For any differential $w'\,dz$ in K which has at most poles of order 1 on R whose residues are rational numbers,*

$$v(t) = \exp\left(c + \int_{t_0}^{t} w'\,dz\right), \qquad c \in k \tag{2.17}$$

is a multiplicative function in $\widetilde{K}$ with respect to $\mathfrak{G}$. The multiplier is given by

$$\chi_\sigma = \exp\left(-\int_\gamma w'\,dz\right), \qquad \overline{\gamma} = \Psi(\overline{\sigma})\,. \tag{2.18}$$

Conversely, any multiplicative function $v(t)$ in $\widetilde{K}$ for $\mathfrak{G}$ is obtained uniquely in the form of (2.17) *for some $w'\,dz$.*

As was shown earlier, the degree of the divisor (v) of a multiplicative function v is 0, and $v_P((v))$ equals the residue of $dv/v = w'\,dz$ at P. Let

$$(v) = \frac{P_1 \cdots P_r}{Q_1 \cdots Q_r}\,.$$

Choose a standard subdivision C of $\mathfrak{R}$ so that P_i and Q_j are not on α_i and β_i, $i = 1, \dots, g$. Then let $w_{P_iQ_i}\,dz$, $i = 1, \dots, r$, be differentials of the third kind in K such as defined in the paragraph preceding equation (1.26). Since $\sum_{i=1}^r w_{P_iQ_i}\,dz$ and $w'\,dz$ have the same singularities,

$$w\,dz = w'\,dz - \sum_{i=1}^{r} w_{P_iQ_i}\,dz$$

is a differential of the first kind in K. Then, from (2.18) and (1.31), we obtain

$$\chi_\sigma = \exp\left(2\pi\sqrt{-1}\sum_{i=1}^{r}\operatorname{Re}\left(\int_{Q_i}^{P_i} w_\gamma\,dz - \int_\gamma w\,dz\right)\right), \qquad \overline{\gamma} = \Psi(\overline{\sigma})\,.$$

That is, we have proved the following lemma.

LEMMA 5.1. *The degree of the divisor (v) of any multiplicative function v is always zero. Conversely, for an arbitrary divisor*

$$A = \frac{P_1 \cdots P_r}{Q_1 \cdots Q_r} \tag{2.19}$$

on K of degree 0, there exists a multiplicative function $v(t)$ such that $(v) = A$. All such multiplicative functions are given by

$$v(t) = \exp\left(c + \sum_{i=1}^{r} \int_{t_0}^{t} w_{P_iQ_i}\, dz + \int_{t_0}^{t} w\, dz\right), \tag{2.20}$$

where c is an arbitrary constant and $w_{P_iQ_i}\, dz$ are differentials of the third kind in K as described in the paragraph preceding (2.16), and $w\, dz$ is an arbitrary differential of the first kind in K. Moreover, the multiplier of $v(t)$ is given by

$$\chi_\sigma = \exp\left(2\pi\sqrt{-1}\sum_{i=1}^{r} \mathrm{Re}\left(\int_{Q_i}^{P_i} w_\gamma\, dz - \int_\gamma w\, dz\right)\right), \qquad \overline{\gamma} = \Psi(\overline{\sigma}). \tag{2.21}$$

In particular, if $w\, dz = 0$ in (2.21), then we have

$$|\chi_\sigma| = 1, \qquad \sigma \in \mathfrak{G}. \tag{2.22}$$

Conversely, if (2.22) holds for all the σ in $\mathfrak{G}$, then for all the γ in (2.21) we have

$$\mathrm{Re}\left(\int_\gamma w\, dz\right) = 0.$$

Therefore, from (1.14) and (1.15), we have

$$w\, dz = 0.$$

Definition 5.3. When the absolute value of the multiplier χ_σ of a multiplicative function $v(t)$ for $\mathfrak{G}$ is one, $v(t)$ is said to be a *multiplicative function in the strict sense.*

As we explained above, for a given divisor, as in (2.19), of degree 0, there is a multiplicative function v in the strict sense satisfying $(v) = A$. Then $v(t)$ is given by

$$v(t) = \exp\left(c + \sum_{i=0}^{r} \int_{t_0}^{t} w_{P_iQ_i}\, dz\right). \tag{2.23}$$

Therefore, for an arbitrary function $z = z(t)$ in K, if the divisor (z) is known, then one can express $z(t)$ explicitly through an Abelian integral of the third kind in K. From (2.23), in general, a multiplicative function is uniquely determined by its divisor up to a constant multiplication. The following is a group theoretic interpretation of the above. Let $\mathfrak{M}$ be the set of multiplicative functions in $\widetilde{K}$ for $\mathfrak{G}$, and let $\mathfrak{M}_0$ be the set of multiplicative functions in the strict sense. Then since $v_1/v_2 \in \mathfrak{M}$ for $v_1, v_2 \in \mathfrak{M}$, $\mathfrak{M}$ is a multiplicative group, and in particular $\mathfrak{M}_0$ is a subgroup of $\mathfrak{M}$. By (2.14) and the above remark, the correspondence from an element v in $\mathfrak{M}_0$ to its divisor

$$v \mapsto (v)$$

is a homomorphism from $\mathfrak{M}_0$ onto the divisor group $\mathfrak{D}_0$ of degree 0 on K. The elements which correspond to the identity element E in $\mathfrak{D}_0$ are all the elements a in k satisfying $a \neq 0$, i.e., the multiplicative group k^* of k. Hence, the group isomorphism theorem implies

$$\mathfrak{M}_0/k^* \cong \mathfrak{D}_0. \tag{2.24}$$

From (2.12) the multiplier χ_σ of a multiplicative function depends only on the class $\overline{\sigma}$ in $\overline{\mathfrak{G}} = \mathfrak{G}/\mathfrak{G}'$ to which σ belongs. Hence, we may write

$$\chi_{\overline{\sigma}} = \chi_\sigma, \qquad \sigma \in \mathfrak{G}, \ \overline{\sigma} \in \overline{\mathfrak{G}}.$$

Then condition (2.22) simply means that $\chi_{\overline{\sigma}}$ is a character of the free Abelian group $\overline{\mathfrak{G}}$. That is, for a multiplicative function in the strict sense there corresponds a character (its multiplier) $\chi_{\overline{\sigma}}$

$$v \mapsto \chi_{\overline{\sigma}}. \tag{2.25}$$

Furthermore, for $v_1 \mapsto \chi_{\overline{\sigma}}^{(1)}$ and $v_2 \mapsto \chi_{\overline{\sigma}}^{(2)}$, we clearly have $v_1 v_2 \mapsto \chi_{\overline{\sigma}}^{(1)} \chi_{\overline{\sigma}}^{(2)}$. Therefore, the map (2.25) is a homomorphism from $\mathfrak{M}_0$ onto a subgroup X' of the character group $X(\overline{\mathfrak{G}})$ of $\overline{\mathfrak{G}}$. The multiplicative functions v_0, in the strict sense, that are mapped onto the identity element of X', i.e., the identity character, are those which satisfy $\sigma(v_0) = v_0$ for all $\sigma \in \mathfrak{G}$. That is, those v_0 are all the functions in K satisfying $v_0 \neq 0$. Again by the isomorphism theorem of groups,

$$\mathfrak{M}_0/K^* \cong X', \tag{2.26}$$

where K^* is the multiplicative group of K.

The subgroup of $\mathfrak{D}_0$ corresponding to K^*/k^* by isomorphism (2.24) is the principal divisor group $\mathfrak{D}_H$ of K. Hence, we have

$$\mathfrak{M}_0/K^* \cong \mathfrak{M}_0/k^*/K^*/k^* \cong \mathfrak{D}_0/\mathfrak{D}_H.$$

Combining this with (2.26), we obtain

$$\mathfrak{D}_0/\mathfrak{D}_H \cong X'. \tag{2.27}$$

Next we will express the isomorphism from $\overline{\mathfrak{D}}_0 = \mathfrak{D}_0/\mathfrak{D}_H$ to X' explicitly. Let $\overline{A}$ be an arbitrary class in $\overline{\mathfrak{D}}_0$. Choose a multiplicative function v, in the strict sense, so that (v) belongs to $\overline{A}$. Then let the character $\chi_{\overline{\sigma}}$ of $\overline{\mathfrak{G}}$ determined by v be denoted by $(\overline{\sigma}, \overline{A})$, i.e.,

$$\sigma(v) = \chi_\sigma v, \qquad \chi_\sigma = (\overline{\sigma}, \overline{A}), \qquad (v) \in \overline{A}.$$

Let $A = P_1 \cdots P_r / Q_1 \cdots Q_r$ be a divisor belonging to $\overline{A}$. Then $(\overline{\sigma}, \overline{A})$ is given by

$$(\overline{\sigma}, \overline{A}) = \exp\left(2\pi\sqrt{-1}\sum_{i=1}^{r} \mathrm{Re}\left(\left(\int_{Q_i}^{P_i} w_\gamma \, dz\right)\right)\right), \qquad \overline{\gamma} = \Psi(\overline{\sigma}). \tag{2.28}$$

For $\overline{\sigma}, \overline{\tau} \in \overline{\mathfrak{G}}$ and $\overline{A}, \overline{B} \in \overline{\mathfrak{D}}_0$, it is clear that

$$\begin{aligned} (\overline{\sigma\tau}, \overline{A}) &= (\overline{\sigma}, \overline{A})(\overline{\tau}, \overline{A}), \\ (\overline{\sigma}, \overline{AB}) &= (\overline{\sigma}, \overline{A})(\overline{\sigma}, \overline{B}). \end{aligned} \tag{2.29}$$

The first equation in the above indicates that for fixed $\overline{A}$, $(\overline{\sigma}, \overline{A})$ is a character of $\overline{\mathfrak{G}}$. The second equation indicates that the correspondence from $\overline{A}$ to that character is a homomorphism from $\overline{\mathfrak{D}}_0$ into the character group $X(\overline{\mathfrak{G}})$ of $\overline{\mathfrak{G}}$. Then (2.27) implies that this homomorphism is an isomorphism from $\overline{\mathfrak{D}}_0$ to a subgroup X' of $X(\overline{\mathfrak{G}})$.

As the character group of the discrete Abelian group $\overline{\mathfrak{G}}$, $X(\overline{\mathfrak{G}})$ is a compact Abelian group. (For properties of the character group of a topological Abelian group, we refer to the book by Pontrjagin cited earlier.) Regarding $X(\mathfrak{G})$ as a topological group, we will prove that X' is closed in $X(\overline{\mathfrak{G}})$. Fix any g points $Q_1, \dots, Q_g$ on $\mathfrak{R}$. Then from Theorem 2.23, an arbitrary class $\overline{A}$ in $\overline{\mathfrak{D}}_0$ contains a divisor of the form

$$A = \frac{P_1 \cdots P_g}{Q_1 \cdots Q_g}.$$

For a point $(P_1, \dots, P_g)$ in the product space $\mathfrak{R}^{(g)} = \mathfrak{R} \times \cdots \times \mathfrak{R}$, ($g$ copies of $\mathfrak{R}$) assign the character $(\overline{\sigma}, \overline{A})$ of $\overline{\mathfrak{G}}$ determined by $\overline{A}$. Then this assignment is a map from $\mathfrak{R}^{(g)}$ onto X'. Furthermore, by (2.28) and the definition of the topology on the character group, the above map $A \mapsto (\overline{\sigma}, \overline{A})$ is a continuous map from $\mathfrak{R}^{(g)}$ to $X(\overline{\mathfrak{G}})$. Therefore, the image X' of the continuous map of the compact space $\mathfrak{R}^{(g)}$ must be compact. That is, X' is a closed subgroup of $X(\overline{\mathfrak{G}})$.

Let $\overline{\sigma}$ be any element of $\overline{\mathfrak{G}}$ different from the identity element in $\overline{\mathfrak{G}}$, and let $\overline{\gamma} = \Psi(\overline{\sigma})$. Since Ψ is an isomorphism, $\overline{\gamma} \neq 0$ holds, i.e, γ is not homologous to 0. As was pointed out in the paragraph following the definition of the intersection number, there exists a closed analytic curve δ such that

$$\operatorname{Re}\left(\int_\delta w_\gamma \, dz\right) = (\gamma, \delta) \neq 0.$$

Let Q_1 be the base point of δ. When P_1 moves along δ from the base point Q_1, and back to Q_1, the value of

$$\operatorname{Re}\left(\int_{Q_1}^{P_1} w_\gamma \, dz\right)$$

changes continuously from 0 to $(\gamma, \delta) \neq 0$. Therefore for some P_1 the value is not a rational integer. For these P_1 and Q_1, let $A = P_1/Q_1$. Then from (2.28), we have $(\overline{\sigma}, \overline{A}) \neq 1$. That is, we have proved that there exists $\overline{A}$ in $\mathfrak{D}_0$ satisfying $(\overline{\sigma}, \overline{A}) \neq 1$ for $\overline{\sigma} \neq 1$. Namely, X' is dense in $X(\mathfrak{G})$. On the other hand, we have proved that X' is a closed subgroup of $X(\mathfrak{G})$. Consequently, $X' = X(\mathfrak{G})$. Hence, we obtain

$$\overline{\mathfrak{D}}_0 \cong X(\mathfrak{G}). \tag{2.30}$$

The following theorem summarizes what we have obtained.

THEOREM 5.8. *For a given divisor A of K with degree zero, there exists a multiplicative function v in $\widetilde{K}$ for $\mathfrak{G}$ such that $(v) = A$. Furthermore, such a v is uniquely determined up to multiplication by an element of k. For an arbitrary character $\chi_{\overline{\sigma}}$ of $\overline{\mathfrak{G}} = \mathfrak{G}/\mathfrak{G}'$, there exists a multiplicative function v, in the strict sense, having $\chi_{\overline{\sigma}}$ as its multiplier. The multipliers of v_1 and v_2 give the same character of $\overline{\mathfrak{G}}$ if and only if the divisors (v_1) and (v_2) of v_1 and v_2 are equivalent with respect to K. Thus,*

$$\mathfrak{M}_0/k^* \cong \mathfrak{D}_0, \qquad X(\overline{\mathfrak{G}}) \cong \overline{\mathfrak{D}}_0 = \mathfrak{D}_0/\mathfrak{D}_H. \tag{2.31}$$

For the topology on $\overline{\mathfrak{D}}_0$ induced from $X(\overline{\mathfrak{G}})$ by the isomorphism in (2.30), $\overline{\mathfrak{D}}_0$ is also a compact Abelian group. Then the function $(\overline{\sigma}, \overline{A})$ is continuous for $\overline{\sigma} \in \overline{\mathfrak{G}}$ and $\overline{A} \in \overline{\mathfrak{D}}_0$. Moreover, as we proved in the above, if $(\overline{\sigma}, \overline{A}) = 1$ for all $\overline{A}$, then $\overline{\sigma} = 1$, and if $(\overline{\sigma}, \overline{A}) = 1$ for all $\overline{\sigma}$, then clearly $\overline{A} = 1$ holds. This is what the isomorphism means in (2.27). Hence, considering (2.29), we see that the discrete Abelian group $\overline{\mathfrak{G}}$ and the compact Abelian group $\overline{\mathfrak{D}}_0$ are a dual pair through $(\overline{\sigma}, \overline{A})$. Since the groups

$$\overline{\mathfrak{G}} = \mathfrak{G}/\mathfrak{G}' \cong G(P_0)/G(P_0)'$$

are free Abelian groups generated by $2g$ elements, the dual $\overline{\mathfrak{D}}_0$ must be isomorphic to the direct sum of $2g$ copies of $\mathbb{R}^+/\mathbb{Z}^+$

$$\overline{\mathfrak{D}}_0 \cong (\mathbb{R}^+/\mathbb{Z}^+) + \cdots + (\mathbb{R}^+/\mathbb{Z}^+),$$

where $\mathbb{R}^+$ and $\mathbb{Z}^+$ are the additive groups of real numbers and rational integers, respectively. We have proved the following.

THEOREM 5.9. *Let K, $\widetilde{K}$, $\mathfrak{G}$, and $\overline{\mathfrak{G}}$ be as before, and let the genus of K be g. Then the divisor class group $\overline{\mathfrak{D}}_0 = \mathfrak{D}_0/\mathfrak{D}_H$ of degree 0 of K is a compact Abelian group in a certain topology, isomorphic to the direct sum of $2g$ copies of the quotient group $\mathbb{R}^+/\mathbb{Z}^+$. Moreover, $\overline{\mathfrak{D}}_0$ and the discrete Abelian group $\overline{\mathfrak{G}} = \mathfrak{G}/\mathfrak{G}'$ form a dual pair of topological Abelian groups in the sense of Pontrjagin, with respect to the character*

$$(\overline{\sigma}, \overline{A}) = \exp\left(2\pi\sqrt{-1}\sum_{i=1}^{r} \operatorname{Re}\left(\left(\int_{Q_i}^{P_i} w_\gamma\, dz\right)\right)\right), \tag{2.32}$$

where $\overline{\gamma} = \Psi(\overline{\sigma})$ and $A = P_1 \cdots P_r/Q_1 \cdots Q_r$. Equation (2.32) is said to be an integral character of $\overline{\mathfrak{G}}$ and $\overline{\mathfrak{D}}_0$.

The dependency of the topology of $\overline{\mathfrak{D}}_0$ upon the point $\overline{A}$ of $\overline{\mathfrak{D}}_0$, and P_i, Q_i will be explained in detail in the next section.

Theorem 5.9 indicates the relationship between the divisor class group $\overline{\mathfrak{D}}_0$ of degree 0 and the Galois group $\mathfrak{G}$ of $\widetilde{K}/K$. For the study of Abelian extensions of K, this theorem is particularly important. However, if one is interested only in K and its Riemann surface $\mathfrak{R}$, then using (2.3) the following is better suited.

THEOREM 5.10. *The divisor class group* $\overline{\mathfrak{D}}_0$ *of degree* 0 *for the algebraic function field* K *and the first homology group* $B_1(\mathfrak{R})$ *of the closed Riemann surface belonging to* K *form a dual pair of topological groups in the sense of Pontrjagin with respect to the character*

$$(\overline{\gamma}, \overline{A}) = \exp\left(2\pi\sqrt{-1}\sum_{i=1}^{r}\operatorname{Re}\left(\int_{Q_i}^{P_i} w_\gamma\, dz\right)\right), \qquad A = \frac{P_1\cdots P_r}{Q_1\cdots Q_r}.$$

Theorem 5.10 also expresses, as does Theorem 5.4, a profound relationship between the algebraic property of K and the topological structure of $\mathfrak{R}$. It was J. Igusa who in, *Zur klassischen Theorie der algebraischen Funktionen*, J. Math. Soc. Japan, **1** (1948), pointed out for the first time the duality between $\overline{\mathfrak{D}}_0$ and $B_1(\mathfrak{R})$. It is remarkable that these theorems are expressed as duality theorems through the character defined by the Abelian integral. Note that a differential $w\,dz$ in K is defined not only algebraically but also analytically as a differential on $\mathfrak{R}$, and this is the essential fact behind Theorems 5.9 and 5.10.

As we explained, an additive function and a multiplicative function in $\widetilde{K}$ play a significant role for the algebraic function field K. We take this opportunity to generalize notions of those functions as follows.

DEFINITION 5.4. Let $\mathfrak{G}$ be the Galois group of $\widetilde{K}/K$ as before. Let

$$\sigma \mapsto M_\sigma, \qquad M_\sigma = (\alpha_{ij}^\sigma), \qquad \alpha_{ij}^\sigma \in k, \qquad i, j = 1, \ldots, n$$

be an arbitrary representation of $\mathfrak{G}$ in the field of complex numbers. Then n functions $u_i = u_i(t)$, $i = 1, \ldots, n$, in $\widetilde{K}$ are called *representation functions* belonging to the representation $\{M_\sigma\}$ if for all $\sigma \in \mathfrak{G}$ we have

$$\sigma(u_j) = \sum_{i=1}^{n} \alpha_{ij}^\sigma u_i, \quad j = 1, \ldots, n,$$

or in matrix notation,

$$(\sigma(u_1), \ldots, \sigma(u_n)) = (u_1, \ldots, u_n)M_\sigma. \tag{2.33}$$

It is clear that the notion of representation functions generalizes that of a multiplicative function. In fact, any multiplicative function v is a representation function belonging to the representation of $\mathfrak{G}$

$$\sigma \mapsto \chi_\sigma.$$

Moreover, for an arbitrary additive function u in $\widetilde{K}$, let

$$\sigma(u) = u + \omega_\sigma, \qquad \sigma \in \mathfrak{G}.$$

Then the functions $u_1 = 1$ and $u_2 = u$ are representation functions belonging to the representation of $\mathfrak{G}$

$$\sigma \mapsto \begin{pmatrix} 1 & \omega_\sigma \\ 0 & 1 \end{pmatrix}.$$

We ask whether there always exist representation functions belonging to a given representation $\{M_\sigma\}$ of $\mathfrak{G}$. Let $u_1 = u_2 = \cdots = u_n = 0$. Then (2.33) clearly holds. We have the following theorem concerning a nontrivial solution set $\{u_i\}$ for (2.33).

THEOREM 5.11. *Let $\{M_\sigma\}$ be an arbitrary representation of dimension n of $\mathfrak{G}$. Then there exists a square matrix of size n*

$$Z^* = Z^*(t) = (\zeta^*_{ij}(t)), \qquad \zeta^*_{ij}(t) \in \widetilde{K}, \qquad i, j = 1, \dots, n$$

such that

$$\sigma(Z^*) = Z^* M_\sigma$$

and the determinant of Z^* is not zero. In particular, each row of Z^* gives a nontrivial solution for (2.33).

PROOF. When the genus of K is zero, $\mathfrak{G}$ is the identity group. Then there is nothing to prove. In §5 we will treat the case $g = 1$. We will prove the theorem for $g \geq 2$. Then the simply connected covering Riemann surface $\mathfrak{R}^*$ of $\mathfrak{R} = \mathfrak{R}(K)$ coincides with the interior $\mathfrak{R}^*_1 = \{t;\, |t| < 1\}$ of the unit disc, and $G = \Phi(\mathfrak{G})$ is the discontinuous group of motions. (For the proofs of the next lemma and the above theorem see H. Poincaré, *Sur les fonctions fuchsiennes*, Acta Math. **1** (1882) (as we mentioned in the note following Lemma 3.12, §6, Chapter 3) and also H. Poincaré, *Mémoire sur les fonctions zétafuchsiennes*, Acta Math. **5** (1884).)

For a matrix with complex entries

$$M = (\alpha_{i,j}), \qquad \alpha_{ij} \in k, \qquad i, j = 1, \dots, n,$$

$$\|M\| = \left(\sum_{i,j=1}^{n} |\alpha_{ij}|^2 \right)^{1/2} \tag{2.34}$$

is called the norm (or the absolute value) of the matrix M. The norm has the following properties.

$$\begin{cases} |\alpha_{ij}| \leq \|M\|, \\ \|M_1 + M_2\| \leq \|M_1\| + \|M_2\|, \\ \|M_1 \cdot M_2\| \leq \|M_1\| \cdot \|M_2\|, \\ \|aM\| = |a| \cdot \|M\|, \qquad a \in k. \end{cases} \tag{2.35}$$

LEMMA 5.2. *Let $\{M_\sigma\}$ be a given representation of $\mathfrak{G}$. For any point t_1 on $\mathfrak{R}^*$ and an arbitrary positive real number ε, there are a real number λ_1 independent of t_1 and ε and a positive real number c_1 depending on t_1 and ε with the following property. For all the points t in $\overline{U}_\varepsilon = \{t;\, \rho(t, t_1) \leq \varepsilon\}$ and any element σ of $\mathfrak{G}$ we have*

$$\|M_\sigma\| \leq c_1 \left| \frac{d\varphi(t)}{dt} \right|^{\lambda_1}, \qquad \varphi = \Phi(\sigma), \tag{2.36}$$

where ρ is the non-Euclidean distance in $\mathfrak{R}^*$.

PROOF. Let $\varphi_1, \dots, \varphi_m$ be a set of generators for G, as in Lemma 3.12, then let

$$\sigma_i = \Phi^{-1}(\varphi_i), \qquad M_{\sigma_i} = M_i, \qquad i = 1, \dots, m$$
$$\operatorname{Max}(\|M_i^{\pm 1}\|, i = 1, \dots, m) = e^{\lambda_2}.$$

By Lemma 3.12, for any σ in $\mathfrak{G}$ we have

$$\varphi = \Phi(\sigma), \qquad \varphi = \varphi_{i_1}^{\pm 1} \cdots \varphi_{i_r}^{\pm 1}, \qquad r \le \lambda \rho(t, \varphi(t)) + \mu.$$

Hence,

$$\sigma = \sigma_{i_1}^{\pm 1} \cdots \sigma_{i_r}^{\pm 1}, \qquad M_\sigma = M_{i_1}^{\pm 1} \cdots M_{i_r}^{\pm 1},$$
$$\|M_\sigma\| \le \|M_{i_1}^{\pm 1}\| \cdots \|M_{i_r}^{\pm 1}\| \le e^{\lambda_2 r} \le e^{\lambda_2 \lambda \rho(t, \varphi(t)) + \lambda_2 \mu}.$$

For $\rho(t, t_1) \le \varepsilon$, we have

$$\rho(t, \varphi(t)) \le \rho(t, t_1) + \rho(t_1, 0) + \rho(0, \varphi(t)) \le \varepsilon + \rho(0, t_1) + \rho(0, \varphi(t)).$$

Therefore, for some $c_2 > 0$, we have

$$\|M_\sigma\| \le c_2 e^{\lambda_2 \lambda \rho(0, \varphi(t))}. \tag{2.37}$$

On the other hand, by (6.11) in Chapter 3

$$\left| \frac{d\varphi(t)}{dt} \right| = \frac{1 - |\varphi(t)|^2}{1 - |t|^2},$$

and from (6.12) in Chapter 3

$$\rho(0, t) = \frac{1}{2} \log \frac{1 + |t|}{1 - |t|}, \qquad |t| = \frac{e^{2\rho(0, t)} - 1}{e^{2\rho(0, t)} + 1}.$$

Hence, we have

$$\left| \frac{d\varphi(t)}{dt} \right| = \frac{e^{2\rho(0, t)} + 2 + e^{-2\rho(0, t)}}{e^{2\rho(0, \varphi(t))} + 2 + e^{-2\rho(0, \varphi(t))}}.$$

Noting that $\rho(0, t)$ is bounded for $\rho(t, t_1) \le \varepsilon$, we obtain

$$\left| \frac{d\varphi(t)}{dt} \right| \le c_3 e^{-2\rho(0, \varphi(t))}, \qquad e^{2\rho(0, \varphi(t))} \le c_3 \left| \frac{d\varphi(t)}{dt} \right|^{-1}.$$

By setting

$$\lambda_1 = -\frac{\lambda_2 \lambda}{2} \quad \text{and} \quad c_1 = c_2 c_3^{\lambda_2 \lambda / 2},$$

we obtain (2.36). Notice that λ and λ_2 do not depend upon t_1 and ε, and hence t_1 is also independent of t_1 and ε. Q.E.D.

LEMMA 5.3. *Let μ_1 be any real number satisfying $\mu_1 \geq 2$. For an arbitrary point t^* on $\mathfrak{R}^*$ there is a sufficiently small positive ε so that the series*

$$\sum_{\varphi} \left| \frac{d\varphi(t)}{dt} \right|^{\mu_1} \tag{2.38}$$

converges uniformly on $\overline{U}_\varepsilon = \{t;\, \rho(t, t_1) \leq \varepsilon\}$, where φ runs through all the elements in G.

PROOF. Note first that the non-Euclidean distance $\rho(t, t')$ and the Euclidean distance $|t - t'|$ are equivalent. It suffices to prove this lemma for the Euclidean neighborhood $\overline{V}_\varepsilon = \{t;\, |t - t_1| \leq \varepsilon\}$. By the discontinuous nature of G, one can choose ε small enough so that the images $\varphi(\overline{V}_{2\varepsilon})$ of $\overline{V}_{2\varepsilon} = \{t;\, |t - t_1| \leq 2\varepsilon\}$ are mutually disjoint. Then let t_2 be an arbitrary point in $\overline{V}_{2\varepsilon}$, and let $\overline{V}'_\varepsilon = \{t;\, |t - t_2| \leq \varepsilon\}$. Since $f(t) = (d\varphi(t)/dt)^2$ is clearly a regular function in $\mathfrak{R}^*$, we have

$$f(t_2) = \frac{1}{2\pi i} \int_{|t - t_2| = r} \frac{f(t)}{t - t_2}\, dt = \frac{1}{2\pi} \int_0^{2\pi} f(t_2 + re^{i\theta})\, d\theta, \qquad 0 \leq r \leq \varepsilon.$$

Hence, we get

$$|f(t_2)| \leq \frac{1}{2\pi} \int_0^{2\pi} |f(t_2 + re^{i\theta})|\, d\theta.$$

Multiply both sides of the above by r/dr to obtain

$$\frac{\varepsilon^2}{2} |f(t_2)| \leq \frac{1}{2\pi} \int_0^{\varepsilon} r\, dr \int_0^{2\pi} |f(t_2 + re^{i\theta})|\, d\theta = \frac{1}{2\pi} \iint_{V'_\varepsilon} |f(t)|\, dx\, dy,$$

where $t = x + iy$. That is,

$$\left| \frac{d\varphi(t)}{dt} \right|^2_{t = t_2} \leq \frac{1}{\pi\varepsilon^2} \iint_{\overline{V}'_\varepsilon} \left| \frac{d\varphi(t)}{dt} \right|^2 dx\, dy. \tag{2.39}$$

Letting $\varphi(t) = u(x, y) + iv(x, y)$, we have

$$\left| \frac{d\varphi(t)}{dt} \right|^2 dx\, dy = \begin{vmatrix} u_x & u_y \\ v_x & v_y \end{vmatrix} dx\, dy = du\, dv.$$

Therefore, the right-hand side integral of (2.39) is precisely the Euclidean area of $\varphi(\overline{V}'_\varepsilon)$, which is denoted by $I(\varphi)$. Since $\overline{V}'_\varepsilon \subseteq \overline{V}_{2\varepsilon}$ holds, the $\varphi(\overline{V}'_\varepsilon)$ are mutually disjoint. Hence, the sum $\sum_\varphi I(\varphi)$ is not greater than the area π of the unit circle. Then (2.39) implies

$$\sum_{\varphi} \left| \frac{d\varphi(t)}{dt} \right|^2_{t = t_2} \leq \sum_{\varphi} \frac{1}{\pi\varepsilon^2} I(\varphi) \leq \frac{1}{\pi\varepsilon^2}\pi = \frac{1}{\varepsilon^2}. \tag{2.40}$$

Since t_2 was arbitrary in $\overline{V}_{2\varepsilon}$, for $\mu_1 = 2$ we have proved the uniform convergence of (2.38) in $\overline{V}_\varepsilon$. Furthermore, the above inequality implies

$$\left| \frac{d\varphi(t)}{dt} \right| \leq \frac{1}{\varepsilon}.$$

For $\varepsilon_1 \geq 2$, the inequality

$$\sum_{\varphi} \left| \frac{d\varphi(t)}{dt} \right|^{\mu_1} \leq \sum_{\varphi} \left| \frac{d\varphi(t)}{dt} \right|^2 \frac{1}{\varepsilon^{\mu_1 - 2}}$$

completes the proof. Q.E.D.

We are ready to prove Theorem 5.11. First, let $\alpha_{ij}(t)$ be any analytic function having only finitely many poles in $|t| < 1$ and regular on $|t| = 1$, $i, j = 1, \ldots, n$. Let $M(t) = (\alpha_{ij}(t))$ be the matrix having $\alpha_{ij}(t)$ as its entries. For this $M(t)$ and the given $\{M_\sigma\}$, define

$$Z(t) = \sum_{\sigma} M(\varphi(t)) M_\sigma \left(\frac{d\varphi(t)}{dt} \right)^m, \qquad \varphi = \Phi(\sigma), \tag{2.41}$$

where m is an arbitrary natural number with

$$m \geq 2 - \lambda_1$$

for the λ_1 in Lemma 5.2 and $\sum_\sigma$ indicates the sum over all the elements of $\mathfrak{G}$. Let t_1 be an arbitrary point in $\mathfrak{R}^*$ and let ε be a sufficiently small positive real number as in Lemma 5.3. Since $\alpha_{ij}(t)$ is regular on $|t| = 1$, $|\alpha_{ij}(t)|$ and hence $\|M(t)\|$ are bounded for $\rho(0, t) \geq \rho_1$, where $\rho_1 > 0$ is sufficiently large. There are only a finitely many $\varphi(t_1)$ for $\varphi \in G$ that are contained in the bounded region $\{t; \rho(0, t) \leq \rho_1 + \varepsilon\}$. (See the paragraph following Theorem 3.20.) Hence, there are only finitely many images $\varphi(\overline{U}_\varepsilon)$ of $\overline{U}_\varepsilon = \{t; \rho(t, t_1) \leq \varepsilon\}$ that contain t satisfying $\rho(0, t) \leq \rho_1$. Excluding these finitely many φ', for some $c_4 > 0$ we have

$$\|M(\varphi(t))\| \leq c_4, \qquad t \in \overline{U}_\varepsilon. \tag{2.42}$$

Then we will prove that each entry of the series, excluding those finitely many φ' from the series (2.41),

$$\sum_{\sigma}{}' M(\varphi(t)) M_\sigma \left(\frac{d\varphi(t)}{dt} \right)^m \tag{2.43}$$

converges absolutely and uniformly on $\overline{U}_\varepsilon$. From (2.35), it is sufficient to prove that on $\overline{U}_\varepsilon$

$$\sum_{\sigma}{}' \|M(\varphi(t))\| \, \|M_\sigma\| \left| \frac{d\varphi(t)}{dt} \right|^m$$

converges uniformly. From Lemma 5.2 and the inequality (2.42), we obtain

$$\sum_{\sigma}{}' \|M(\varphi(t))\| \, \|M_\sigma\| \left| \frac{d\varphi(t)}{dt} \right|^m \leq \sum_{\sigma}{}' c_4 c_1 \left| \frac{d\varphi(t)}{dt} \right|^{m+\lambda_1} \leq c_4 c_1 \sum_{\sigma} \left| \frac{d\varphi(t)}{dt} \right|^{m+\lambda_1}.$$

Since $m + \lambda_1 \geq 2$, the right-hand side converges uniformly on $\overline{U}_\varepsilon$ by Lemma 5.3. Therefore, the series (2.43) converges absolutely and uniformly on $\overline{U}_\varepsilon$, and each entry is a regular function of t in $U_\varepsilon = \{t; \rho(t, t_1) < \varepsilon\}$. Hence,

$Z(t)$ in equation (2.41), obtained by adding finitely many terms to (2.43), is a matrix whose entries are analytic functions of t with possible poles in U_ε. Since t_1 was taken to be arbitrary in $\mathfrak{R}^*$, each entry of $Z(t)$ is a function belonging to $\widetilde{K}$.

Next let τ be any element of $\mathfrak{G}$, and let $\psi = \Phi(\tau)$. Using $M_\sigma = M_{\sigma\tau^{-1}}M_\tau$, from (2.41) we have

$$\begin{aligned} Z(\psi^{-1}(t)) &= \sum_\sigma M(\varphi\psi^{-1}(t))M_\sigma \left(\frac{d\varphi\psi^{-1}(t)}{d\psi^{-1}(t)}\right)^m \\ &= \sum_\sigma M(\varphi\psi^{-1}(t))M_{\sigma\tau^{-1}} \left(\frac{d\varphi\psi^{-1}(t)}{dt}\right)^m M_\tau \left(\frac{d\psi^{-1}(t)}{dt}\right)^{-m}. \end{aligned}$$

When σ runs through all the elements of $\mathfrak{G}$, so does $\sigma\tau^{-1}$. Hence, the right-hand side of the above equation equals

$$\sum_\sigma M(\varphi(t))M_\sigma \left(\frac{d\varphi(t)}{dt}\right)^m M_\tau \left(\frac{d\psi^{-1}(t)}{dt}\right)^{-m} = Z(t)M_\tau \left(\frac{d\psi^{-1}(t)}{dt}\right)^{-m}.$$

That is,

$$\tau(Z) = ZM_\tau \left(\frac{d\psi^{-1}(t)}{dt}\right)^{-m}. \tag{2.44}$$

Let us consider the case where $n = 1$, and let $\{M_\sigma\}$ be the identity representation, $M_\sigma = 1$, of $\mathfrak{G}$. Then we can take $\lambda_1 = 0$ in Lemma 5.2. For an arbitrary analytic function $f(t)$, which has only a finitely many poles in $|t| < 1$ and is regular on $|t| = 1$, define

$$\Theta(t) = \sum_\sigma f(\varphi(t)) \left(\frac{d\varphi(t)}{dt}\right)^m, \qquad \varphi = \Phi(\sigma). \tag{2.45}$$

For $m \geq 2$, $\Theta(t)$ is a function in $\widetilde{K}$ satisfying

$$\tau(\Theta) = \Theta \cdot \left(\frac{d\psi^{-1}(t)}{dt}\right)^{-m}, \qquad \psi = \Phi(\tau). \tag{2.46}$$

The right-hand sides of (2.45) and (2.46) are called a Zeta-Fuchsian series and a Theta-Fuchsian series of Poincaré, respectively. See the papers of H. Poincaré mentioned above.

In particular, for

$$M(t) = \begin{pmatrix} \frac{1}{t} & & & 0 \\ & \frac{1}{t} & & \\ & & \ddots & \\ 0 & & & \frac{1}{t} \end{pmatrix} = \frac{1}{t}E_n, \qquad E_n = \text{ the identity matrix},$$

each entry of the matrix $Z(t) - M(t)$ is a regular function in a neighborhood of $t = 0$. Therefore, the determinant $|Z(t)|$ of $Z(t)$ has a pole of order n

at $t = 0$. Similarly, for $f(t) = 1/t$, $\Theta(t)$ has a pole of order 1 at $t = 0$. Neither $|Z(t)|$ nor $\Theta(t)$ is identically 0. For m large enough to satisfy

$$m \geq 2 - \lambda_1 \quad \text{and} \quad m \geq 2,$$

the quotient of $Z(t)$ and $\Theta(t)$

$$Z^*(t) = \frac{Z(t)}{\Theta(t)}$$

is a matrix whose entries are clearly functions in $\widetilde{K}$. Moreover, from (2.44) and (2.46), we have

$$\tau(Z^*) = Z^* M_\tau, \qquad \tau \in \mathfrak{G}.$$

Then the determinant of Z^* is given by

$$|Z^*(t)| = \Theta(t)^{-n} |Z(t)| \neq 0,$$

which completes the proof of Theorem 5.11.

We used Lemma 3.12 to prove Lemma 5.2. If the representation $\{M_\sigma\}$ is bounded, i.e., there is a real number $c_5 > 0$ satisfying

$$\|M_\sigma\| \leq c_5, \qquad \sigma \in \mathfrak{G},$$

then directly from Lemma 5.3 we can prove that each entry of $Z(t)$ is a function in $\widetilde{K}$. Hence, in this case, $\mathfrak{R}$ need not be compact. For the fundamental group $G(= \Phi(\mathfrak{G}))$ of any Riemann surface $\mathfrak{R}$ having the interior of a unit circle as its simply connected covering Riemann surface and having the bounded representation $\{M_\sigma\}$, we obtain the representation functions of $\{M_\sigma\}$ from the Zeta-Fuchsian series (2.41) and the Theta-Fuchsian series (2.45). In particular, for $\Theta(t)$ obtained from $f(t) = 1/t$ in the above and the Theta-Fuchsian series $\Theta_1(t)$ obtained from $f_1(t) = 1/t^2$, define

$$z(t) = \frac{\Theta_1(t)}{\Theta(t)}.$$

Since $t = 0$ is a pole of $\Theta_1(t)$ of order 2, $z(t)$ has a pole of order 1 at $t = 0$, and $z(t)$ also satisfies

$$z(\varphi(t)) = z(t)$$

for all the maps φ in G. That is, $z(t) = z(P)$ is an analytic function on the Riemann surface $\mathfrak{R}$, which gives another proof of the existence of a nonconstant analytic function on $\mathfrak{R}$. See Theorem 3.11.

Since representation functions contain additive functions and multiplicative functions, Theorem 5.11 implies that there exist special additive functions and multiplicative functions satisfying a certain condition. For instance, the previously proved assertion $X' = X(\overline{\mathfrak{G}})$ can be directly obtained from this theorem. Also, for subdivision curves α_i and β_i of a standard subdivision of $\mathfrak{R}$, let σ_i and τ_i be the elements in $\mathfrak{G}$ corresponding to α_i

and β_i by the isomorphism in (2.2). Then σ_i and τ_i, $i = 1, \ldots, g$, are generators for $\mathfrak{G}$ with only one relation

$$\sigma_1\tau_1\sigma_1^{-1}\tau_1^{-1}\cdots\sigma_g\tau_g\sigma_g^{-1}\tau_g^{-1} = 1.$$

(See the next to the last paragraph in §3, Chapter 4.) Hence, for any complex numbers $\zeta_1, \ldots, \zeta_{2g}$, there exists a representation $\{M_\sigma\}$ of dimension 2 satisfying

$$M_{\sigma_i} = \begin{pmatrix} 1 & -\zeta_i \\ 0 & 1 \end{pmatrix}, \qquad M_{\tau_i} = \begin{pmatrix} 1 & -\zeta_{g+i} \\ 0 & 1 \end{pmatrix}, \qquad i = 1, \ldots, g.$$

From Theorem 5.11, let

$$Z^*(t) = \begin{pmatrix} z_{11}(t) & z_{12}(t) \\ z_{21}(t) & z_{22}(t) \end{pmatrix}, \qquad z_{ij} \in \widetilde{K}$$

be the matrix in $\widetilde{K}$ corresponding to $\{M_\sigma\}$. Then one of $z_{11}(t)$ and $z_{21}(t)$ is not zero since $|Z^*(t)| \neq 0$. Suppose $z_{11}(t) \neq 0$, and let

$$u_1(t) = 1 \quad \text{and} \quad u_2(t) = u(t) = \frac{z_{12}(t)}{z_{11}(t)}.$$

Then we have

$$(\sigma(u_1), \sigma(u_2)) = (u_1, u_2)M_\sigma, \qquad \sigma \in \mathfrak{G}.$$

By rewriting the above for $\sigma = \sigma_i$ or $\sigma = \tau_i$ we obtain

$$\sigma_i(u) = u - \zeta_i \quad \text{and} \quad \tau_i(u) = u - \zeta_{g+i}, \qquad i = 1, \ldots, g.$$

Therefore, from Theorem 5.6, the periods for α_i and β_i of the differential $du = w\,dz$ in K obtained from the above additive function u are given by ζ_i and ζ_{g+i}. Namely, the above shows that for any given periods there exists a differential $w\,dz$ in K having these periods for α_i and β_i. On the other hand, the isomorphism in (1.12) and (1.13) implies that a differential of the first kind in K is determined by the periods alone for α_i, $i = 1, \ldots, g$. Therefore, differentials of the first kind alone do not work for the above argument, but together with differentials of the second kind we can find a differential whose periods coincide with arbitrarily given periods for the base α_i and β_i of $B_1(\mathfrak{R})$. Other applications of representation functions will appear in §4.

§3. Theorems of Abel-Jacobi and Abelian functions

One can immediately obtain Abel's theorem from a slight modification of Theorem 5.10.

As above, K is an algebraic function field over the field of complex numbers. Let $\mathfrak{R}$ be the associated closed Riemann surface, i.e., $\mathfrak{R} = \mathfrak{R}(K)$ and $K = K(\mathfrak{R})$, and let $\mathfrak{L}_0$ be the set of differentials of the first kind in K. Let P_0 be a fixed point on $\mathfrak{R}$ and let P be an arbitrary point on $\mathfrak{R}$. Further, let

$$w_1\,dz, \ldots, w_g\,dz$$

be a basis for $\mathfrak{L}_0$ over k, and let

$$\int_{\delta} w_i\,dz, \qquad i = 1, \dots, g \tag{3.1}$$

be the ith component of a vector in the complex g-dimensional vector space $V_g(k)$ over k, where δ is an analytic curve from P_0 to P. Note that δ can be arbitrary, but the same path must be used for all components. We denote such a vector in $V_g(k)$ by

$$\mathfrak{v}(P, \delta) = \left(\int_{\delta} w_1\,dz, \dots, \int_{\delta} w_g\,dz\right). \tag{3.2}$$

We will observe the effect on $\mathfrak{v}(P, \delta)$ as one replaces δ by another curve δ' connecting P_0 and P. Let $\gamma = \delta^{-1}\delta'$. Then γ is a closed curve with the initial point P_0, and we obviously have

$$\mathfrak{v}(P, \delta') - \mathfrak{v}(P, \delta) = \left(\int_{\gamma} w_1\,dz, \dots, \int_{\gamma} w_g\,dz\right). \tag{3.3}$$

Conversely, for an arbitrary closed analytic curve γ with initial point P_0 it is clear that equation (3.3) holds for this $\delta' = \delta\gamma$. That is, the set of all the differences of two vectors for paths from P_0 to P coincides with the set of vectors

$$\mathfrak{v}_{\gamma} = \left(\int_{\gamma} w_1\,dz, \dots, \int_{\gamma} w_g\,dz\right)$$

for all closed analytic curves γ. By Theorem 5.1, each integral in the above depends only upon the homology class of γ, and any first homology class on $\mathfrak{R}$ contains a closed analytic curve with initial point P_0. Consequently, the set of all the differences between the two vectors coincides with the set of such vectors $\mathfrak{v}_{\gamma}$ for all closed analytic curves γ. We denote this set $\{\mathfrak{v}_{\gamma}\}$ of vectors by

$$\mathrm{P}(w_1\,dz, \dots, w_g\,dz) \quad \text{or} \quad \mathrm{P}(w_i\,dz),$$

which is called the *periodic vector group* or the *period group* for the base $w_1\,dz, \dots, w_g\,dz$ for $\mathfrak{L}_0$. Since for $\gamma \sim \gamma' + \gamma''$,

$$\int_{\gamma} w_i\,dz = \int_{\gamma'} w_i\,dz + \int_{\gamma''} w_i\,dz, \qquad i = 1, \dots, g,$$

the periodic vector group is a subgroup of $V_g(k)$. Given a set of closed analytic curves forming a basis $\{\gamma_1, \dots, \gamma_{2g}\}$ for the first homology group $B_1(\mathfrak{R})$, let

$$\Omega = \begin{pmatrix} w_{11} & \cdots & w_{1,2g} \\ \vdots & & \vdots \\ w_{g,1} & \cdots & w_{g,2g} \end{pmatrix}, \qquad w_{ij} = \int_{\gamma_j} w_i\,dz.$$

Then $\mathrm{P}(w_i\,dz)$ in the above coincides with the totality of linear combinations of the column vectors of Ω with coefficients in the rational integers. See

(1.4). This g by $2g$ matrix Ω is called the *period matrix* or *Riemann matrix* of $w_1\,dz, \ldots, w_g\,dz$ with respect to $\gamma_1, \ldots, \gamma_{2g}$.

From what we have described, as one varies δ for a fixed P, $\mathfrak{v}(P, \delta)$ fulfills a residue class of $V_g(k)$ modulo $\mathrm{P}(w_i\,dz)$. We denote this class, uniquely determined by P, by

$$\bar{\mathfrak{v}}(P).$$

In general, for the decomposition of an arbitrary divisor A on K

$$A = \frac{P_1 \cdots P_r}{Q_1 \cdots Q_s}$$

into prime divisors, we set

$$\bar{\mathfrak{v}}(A) = \sum_{l=1}^{r} \bar{\mathfrak{v}}(P_l) - \sum_{m=1}^{s} \bar{\mathfrak{v}}(Q_m), \tag{3.4}$$

where addition and subtraction are carried out in the quotient group $V_g(k)/\mathrm{P}(w_i\,dz)$. Hence $\bar{\mathfrak{v}}(A)$ is also a residue class modulo $\mathrm{P}(w_i\,dz)$. This class is determined independently of a decomposition of A into prime divisors, and furthermore, from the definition, this class is the residue class of the vector $(\zeta_1, \ldots, \zeta_g)$ modulo $\mathrm{P}(w_i\,dz)$, where

$$\zeta_i = \sum_{l=1}^{r} \int_{P_0}^{P_l} w_i\,dz - \sum_{m=1}^{s} \int_{P_0}^{Q_m} w_i\,dz. \tag{3.5}$$

As one varies the path from P_0 to P_l or Q_m, one gets all the vectors in $\bar{\mathfrak{v}}(A)$. We denote a vector in $\bar{\mathfrak{v}}(A)$, simply by

$$\mathfrak{v}(A) = (\zeta_1, \ldots, \zeta_g). \tag{3.6}$$

In particular, when the degree of A is 0, i.e., $r = s$, then (3.5) becomes simply

$$\zeta_i = \sum_{j=1}^{r} \int_{Q_j}^{P_j} w_i\,dz, \qquad i = 1, \ldots, g.$$

The path from Q_j to P_j in the above integral goes from Q_j to P_0 and then from P_0 to P_j. The collection of vectors, whose jth component ζ_j is obtained by varying paths from Q_j to P_j, makes up all the vectors in the residue class $\bar{\mathfrak{v}}(A)$. Therefore, if $n(A) = 0$, then $\mathfrak{v}(A)$ is defined independently of P_0 as well. From the definition, we clearly have

$$\bar{\mathfrak{v}}(AB) = \bar{\mathfrak{v}}(A) + \bar{\mathfrak{v}}(B)$$

for any divisors A and B. Hence, the map

$$A \mapsto \bar{\mathfrak{v}}(A) \tag{3.7}$$

is a homomorphism from the divisor group $\mathfrak{D}_0$ of degree 0 to $V_g(k)/\mathrm{P}(w_i\,dz)$.

The above $\bar{\mathfrak{v}}(A)$ and $\mathfrak{v}(A)$ depend upon the choice of the base $w_1\,dz, \dots, w_g\,dz$ for $\mathfrak{L}_0$ over k. For an arbitrary nonsingular complex matrix $\Lambda = (\lambda_{ij})$, change the base for $\mathfrak{L}_0$ by

$$w_j'\,dz = \sum_{i=1}^{g} \lambda_{ij} w_i\,dz, \qquad j = 1, \dots, g,$$

i.e.,

$$(w_1'\,dz, \dots, w_g'\,dz) = (w_1\,dz, \dots, w_g\,dz)\Lambda. \tag{3.8}$$

Then we have

$$\begin{aligned}\zeta_j' &= \sum_{l=1}^{r} \int_{P_0}^{P_l} w_j'\,dz - \sum_{m=1}^{s} \int_{P_0}^{Q_m} w_j'\,dz \\ &= \sum_{i=1}^{g} \lambda_{ij} \left(\sum_{l=1}^{r} \int_{P_0}^{P_l} w_i\,dz - \sum_{m=1}^{s} \int_{P_0}^{Q_m} w_i\,dz \right).\end{aligned}$$

Therefore,

$$\mathfrak{v}'(A) = (\zeta_1', \dots, \zeta_g') = (\zeta_1, \dots, \zeta_g)\Lambda = \mathfrak{v}(A)\Lambda. \tag{3.9}$$

Similarly, we obtain

$$\mathrm{P}(w_i'\,dz) = \mathrm{P}(w_i\,dz)\Lambda. \tag{3.10}$$

Namely, one obtains all the vectors in $\mathrm{P}(w_i'\,dz)$ by transforming all the vectors in $\mathrm{P}(w_i\,dz)$ by Λ.

From (3.10), all the vectors in a class in $V_g(k)/\mathrm{P}(w_i\,dz)$ are mapped by Λ to all the vectors in a class in $V_g(k)/\mathrm{P}(w_i'\,dz)$. Hence, Λ induces an isomorphism $\bar{\Lambda}$ from $V_g(k)/\mathrm{P}(w_i\,dz)$ onto $V_g(k)/\mathrm{P}(w_i'\,dz)$. Therefore, by (3.9) we obtain

$$\bar{\mathfrak{v}}'(A) = \bar{\mathfrak{v}}(A)\bar{\Lambda}. \tag{3.11}$$

That is, (3.11) describes the effect on $\bar{\mathfrak{v}}(A)$ of the base change of $\mathfrak{L}_0$. Then the kernel of the homomorphism in (3.7) coincides with the kernel of the homomorphism

$$A \mapsto \bar{\mathfrak{v}}'(A), \qquad A \in \mathfrak{D}_0. \tag{3.12}$$

The set of divisors A in $\mathfrak{D}_0$ satisfying $\bar{\mathfrak{v}}(A) = 0$ does not depend upon the choice of basis elements $w_1\,dz, \dots, w_g\,dz$.

We can carry out similar arguments replacing $V_g(k)$ by the $2g$-dimensional real vector space $V_{2k}(k_0)$. Let $u_1\,dz, \dots, u_{2g}\,dz$ be basis elements for $\mathfrak{L}_0$ over the field k_0 of real numbers. Let us denote the set of $2g$-dimensional real vectors

$$\left(\mathrm{Re}\left(\int_\gamma u_1\,dz\right), \dots, \mathrm{Re}\left(\int_\gamma u_{2g}\,dz\right)\right),$$

for all the closed analytic curves on $\mathfrak{R}$, by $\mathrm{P}_r(u_1\,dz, \dots, u_{2g}\,dz)$, or $\mathrm{P}_r(u_i\,dz)$. This set is called the real *periodic vector group*, or *real period group* for $u_i\,dz$. Then $\mathrm{P}_r(u_i\,dz)$ is a subgroup of $V_{2g}(k_0)$. For a base $\gamma_1, \dots, \gamma_{2g}$ for $B_1(\mathfrak{R})$, define

$$M = (\mu_{ij}), \quad \mu_{ij} = \mathrm{Re}\left(\int_{\gamma_j} u_i\,dz\right), \qquad i, j = 1, \dots, 2g.$$

Then $\mathrm{P}_r(u_i\,dz)$ is the totality of linear combinations of the column vectors of M with coefficients in the rational integers.

For an arbitrary degree-zero divisor $A = P_1 \cdots P_r/Q_1 \cdots Q_r$ of K, as above, let

$$\mathfrak{v}_r(A) = (\xi_1, \dots, \xi_{2g})$$

be the vector in $V_{2g}(k_0)$ whose ith component is given by

$$\xi_i = \mathrm{Re}\left(\sum_{j=1}^{r} \int_{Q_j}^{P_j} u_i\,dz\right), \qquad i = 1, \dots, 2g.$$

Then varying the path from Q_j to P_j, $\mathfrak{v}_r(A)$ is a residue class in $V_{2g}(k_0)/\mathrm{P}_r(u_i\,dz)$. We denote this class by

$$\bar{\mathfrak{v}}_r(A).$$

We have a homomorphism

$$A \mapsto \bar{\mathfrak{v}}_r(A) \tag{3.13}$$

from $\mathfrak{D}_0$ to $V_{2g}(k_0)$. Moreover, for another base $u'_1\,dz, \dots, u'_{2g}\,dz$ obtained from $u_1\,dz, \dots, u_{2g}\,dz$, by a nonsingular real matrix $\mathrm{P}_r(u_i\,dz), \mathfrak{v}_r(A)$, and $\bar{\mathfrak{v}}_r(A)$ are transformed similarly as in (3.9), (3.10), and (3.11), respectively. Therefore, the kernel of (3.13), i.e., the totality of elements A in $\mathfrak{D}_0$ satisfying $\bar{\mathfrak{v}}_r(A) = 0$ does not depend on the choice of basis elements $u_1\,dz, \dots, u_{2g}\,dz$ for $\mathfrak{L}_0$.

Let $w_1\,dz, \dots, w_g\,dz$ be basis elements for $\mathfrak{L}_0$ over k as before. Then

$$w_1\,dz, \dots, w_g\,dz, -\sqrt{-1}w_1\,dz, \dots, -\sqrt{-1}w_g\,dz$$

clearly form a base for $\mathfrak{L}_0$ over k_0. We rewrite them as $u'_1\,dz, \dots, u'_{2g}\,dz$. Define a map φ from $V_g(k)$ to $V_{2g}(k_0)$ by

$$\varphi((\zeta_1, \dots, \zeta_g)) = (\xi_1, \dots, \xi_g, \xi_{g+1}, \dots, \xi_{2g}),$$
$$\zeta_i = \xi_i + \sqrt{-1}\xi_{g+i}, \qquad i = 1, \dots, g.$$

Then as k_0-modules, φ is an isomorphism from $V_g(k)$ onto $V_{2g}(k_0)$, satisfying

$$\varphi(\mathrm{P}(w_i\,dz)) = \mathrm{P}_r(u'_i\,dz), \qquad \varphi(\mathfrak{v}(A)) = \mathfrak{v}'_r(A). \tag{3.14}$$

Hence, the kernel of the homomorphism in (3.7) coincides with the kernel of $A \mapsto \bar{\mathfrak{v}}'_r(A)$, and as a consequence, also with the kernel of (3.13). That is, in order to find the kernel of (3.7) one can compute the kernel of the homomorphism in (3.13) for a suitable set of basis elements $u_i\,dz$ for $\mathfrak{L}_0$ over k_0. For this purpose, choose subdivision curves α_i and β_i, $i = 1, \dots, g$, for a standard subdivision C of $\mathfrak{R}$. From Theorem 5.3, let

$$u_i\,dz = w_{\alpha_i}\,dz \qquad \text{and} \qquad u_{g+i}\,dz = w_{\beta_i}\,dz, \quad i = 1, \dots, g$$

be the corresponding basis elements for $\mathfrak{L}_0$ over k_0. Also take $\alpha_1, \dots, \alpha_g$, $\beta_1, \dots, \beta_g$ for a base of $B_1(\mathfrak{R})$. Then from (1.17) and (3.5) in Chapter 4, the real period matrix M with respect to $u_i\,dz$ is given by

$$\begin{array}{c} \alpha_1 \cdots\cdots \alpha_g \quad \beta_1 \cdots\cdots \beta_g \\ \left(\begin{array}{cccc|cccc} & & & & 1 & & & \\ & 0 & & & & 1 & & \\ & & & & & & \ddots & \\ & & & & & & & 1 \\ \hline -1 & & & & & & & \\ & -1 & & & & 0 & & \\ & & \ddots & & & & & \\ & & & -1 & & & & \end{array}\right) \end{array} \begin{array}{c} u_1\,dz \\ \vdots \\ u_g\,dz \\ u_{g+1}\,dz \\ \vdots \\ u_{2g}\,dz \end{array} \tag{3.15}$$

The real period group $\mathrm{P}_r(u_i\,dz)$ generated by the columns of M is the totality of $2g$-dimensional vectors of integer components. That is, representing $V_{2g}(k_0)$ as points in $2g$-dimensional Euclidean space, the above $\mathrm{P}_r(u_i\,dz)$ coincides with the set of the points on the rectangular lattice of unit intervals. Therefore, the condition $\overline{\mathfrak{v}}_r(A) = 0$ is equivalent to the assertion that

$$\mathrm{Re}\left(\sum_{j=1}^{r}\int_{Q_j}^{P_j} w_{\alpha_i}\,dz\right) \qquad \text{and} \qquad \mathrm{Re}\left(\sum_{j=1}^{r}\int_{Q_j}^{P_j} w_{\beta_i}\,dz\right), \quad i = 1, \dots, g$$

are rational integers. Whereas α_i and β_i, $i = 1, \dots, g$, form a base for $B_1(\mathfrak{R})$, the above assertion is equivalent to $\mathrm{Re}(\sum_{j=1}^{r}\int_{Q_j}^{P_j} w_\gamma\,dz)$ being a rational integer for all analytic curves γ, i.e.,

$$\exp\left(2\pi\sqrt{-1}\,\mathrm{Re}\left(\sum_{j=1}^{r}\int_{Q_j}^{P_j} w_\gamma\,dz\right)\right) = 1.$$

The left-hand side of the above is nothing but $(\overline{\gamma}, \overline{A})$ in Theorem 5.10. From this theorem, $B_1(\mathfrak{R})$ and $\overline{\mathfrak{D}_0}$ form a dual pair of topological Abelian groups for the character $(\overline{\gamma}, \overline{A})$. The condition $(\overline{\gamma}, \overline{A}) = 1$ for all γ means that $\overline{A}$ is the identity element in $\overline{\mathfrak{D}_0} = \mathfrak{D}_0/\mathfrak{D}_H$, i.e., A belongs to the principal divisor group $\mathfrak{D}_H$ of K. Consequently, $\overline{\mathfrak{v}_r}(A) = 0$ holds if and only if $A \in \mathfrak{D}_H$. Therefore, the kernel of (3.13), and hence the kernel of (3.7), are

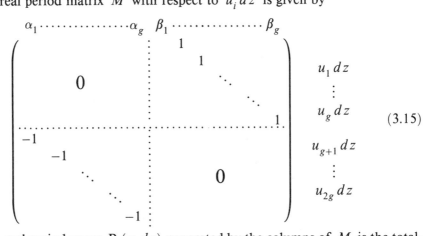

$\mathfrak{D}_H$. Moreover, Theorem 5.10 implies that all the characters of $B_1(\mathfrak{R})$ are obtained from $(\overline{\gamma}, \overline{A})$ by varying A. In particular, for basis elements α_i and β_i of $B_1(\mathfrak{R})$,

$$(\overline{\alpha}_i, \overline{A}) = \exp\left(2\pi\sqrt{-1}\,\mathrm{Re}\left(\sum_{j=1}^{r}\int_{Q_j}^{P_j} w_{\alpha_i}\,dz\right)\right),$$

$$(\overline{\beta}_i, \overline{A}) = \exp\left(2\pi\sqrt{-1}\,\mathrm{Re}\left(\sum_{j=1}^{r}\int_{Q_j}^{P_j} w_{\beta_i}\,dz\right)\right)$$

can take any values of absolute value 1. That is, given any vector in $V_{2g}(k_0)$ we can choose an $A = P_1 \cdots P_r/Q_1 \cdots Q_r$ and a path from Q_j to P_j, so that

$$\mathfrak{v}_r(A) = (\xi_1, \ldots, \xi_{2g}), \qquad \xi_i = \mathrm{Re}\left(\sum_{j=1}^{r}\int_{Q_j}^{P_j} u_i\,dz\right)$$

is the given vector in $V_{2g}(k_0)$. That is, the map in (3.13) is an onto map from $\mathfrak{D}_0$ to $V_{2g}(k_0)/\mathrm{P}(u_i\,dz)$. We have deduced the following theorem algebraically from Theorem 5.10.

THEOREM 5.12 (Abel's theorem). *Let $w_1\,dz, \ldots, w_g\,dz$ be basis elements for the set $\mathfrak{L}_0$ of differentials of the first kind in an algebraic function field K. Let $A = P_1 \cdots P_r/Q_1 \cdots Q_r$ be a divisor of degree 0 on K. Then the complex g-dimensional vector*

$$\mathfrak{v}(A) = (\zeta_1, \ldots, \zeta_g), \qquad \zeta_i = \sum_{j=1}^{r}\int_{Q_j}^{P_j} w_i\,dz \tag{3.16}$$

fulfills a residue class of $V_g(k)$ modulo the period group $\mathrm{P}(w_i\,dz)$ of $w_i\,dz$, varies A in terms of P_i and Q_i when the expansion of the paths from Q_j to P_j. Denote this residue class determined only by A as $\overline{\mathfrak{v}}(A)$. Then the map

$$A \mapsto \overline{\mathfrak{v}}(A)$$

is a homomorphism from the divisor group $\mathfrak{D}_0$ of degree 0 onto $V_g(k)/\mathrm{P}(w_i\,dz)$. The kernel is the principal divisor group $\mathfrak{D}_H$.

Thus,

$$\overline{\mathfrak{D}}_0 = \mathfrak{D}_0/\mathfrak{D}_H \cong V_g(k)/\mathrm{P}(w_i\,dz). \tag{3.17}$$

NOTE. The historically well-known theorem of Abel treats a more general case than Theorem 5.12. That is, Theorem 5.12 covers the special case of Abel's theorem concerned with Abelian integrals of the first kind. Abel proved only the necessary condition. The sufficient condition, i.e., the converse, was proved by Riemann and Clebsch. (See preface.)

As one can observe from the matrix in (3.15), $\mathrm{P}_r(u_i\,dz)$ contains $2g$ linearly independent vectors over the real field k_0. Hence, $\mathrm{P}_r(u_i'\,dz)$ and consequently $\mathrm{P}(w_i\,dz)$ contain $2g$ linearly independent vectors over k_0. With the usual topology, regard $V_g(k)$ as the $2g$-dimensional manifold S_{2g}. Consider the vector group $\mathrm{P}(w_i\,dz)$ as the topological transformation group of S_{2g}, i.e., the parallel transformation group by the vectors belonging to $\mathrm{P}(w_i\,dz)$. Then $\mathrm{P}(w_i\,dz)$ is a discontinuous transformation group of S_{2g}. The $2g$-dimensional manifold $S_{2g}(\mathrm{P}(w_i\,dz))$ obtained by identifying the equivalent points through $\mathrm{P}(w_i\,dz)$ is compact. Namely, it is the torus of dimension $2g$. A point on $S_{2g}(\mathrm{P}(w_i\,dz))$ corresponds to a residue class in $V_g(k)/\mathrm{P}(w_i\,dz)$ and also by Theorem 5.12, to an element of $\overline{\mathfrak{D}}_0$ in a one-to-one fashion. With the topology on $\overline{\mathfrak{D}}_0$ induced by this map from the topology on $S_{2g}(\mathrm{P}(w_i\,dz))$, $\overline{\mathfrak{D}}_0$ is also a $2g$-dimensional compact manifold. Let $\overline{A}$ be any element in $\overline{\mathfrak{D}}_0$, and let $\mathfrak{v}(A) = (\zeta_1, \dots, \zeta_g)$ be any vector in $\overline{\mathfrak{v}}(A)$. Since a point on $S_{2g} = V_g(k)$ locally corresponds homeomorphically to a point on $S_{2g}(\mathrm{P}(w_i\,dz))$, a sufficiently small neighborhood of $\overline{A}$ is mapped homeomorphically onto a neighborhood of $(\zeta_1, \dots, \zeta_g)$ by the map $\overline{A} \mapsto \mathfrak{v}(A)$. That is, we have a coordinate system in a neighborhood of $\overline{A}$.

In general, if a neighborhood U_P of each point P on a $2n$-dimensional manifold S is mapped homeomorphically by φ_P onto an open set in $V_n(k)$, and if on the intersection of U_P and $U_{P'}$, $\varphi_{P'}(Q)$ is a regular function of $\varphi_P(Q)$, and $\varphi_P(Q)$ is a regular function of $\varphi_{P'}(Q)$, then $\{\varphi_P, U_P\}$ is said to be an analytic function system for S. Then the $\varphi_P(Q) = (\eta_1, \dots, \eta_n)$ are called analytical coordinates at P. The above definition is a generalization of an R-function system in Definition 3.1 to a higher dimensional case. (The condition in Definition 3.1 that $\varphi_P(U_P)$ coincides with the interior of a disc is not essential.) An equivalence between analytic function systems on S may be defined as in Definition 3.2. A pair $\{S, \{\mathfrak{F}\}\}$ of an equivalence class $\{\mathfrak{F}\}$ of analytic function systems, is called an *analytic manifold* of complex dimension g and real dimension $2g$. An analytic manifold is a direct generalization of the concept of a Riemann surface. With this definition of an analytic manifold, $\overline{\mathfrak{D}}_0$ becomes an analytic manifold of complex dimension g. We take $\mathfrak{v}(A) = (\zeta_1, \dots, \zeta_g)$ as analytical coordinates at $\overline{A}$.

In our discussion we have fixed a base $w_1\,dz, \dots, w_g\,dz$ for $\mathfrak{L}_0$. By the base change in (3.8) from $w_i\,dz$ to $w_i'\,dz$, $\mathrm{P}(w_i\,dz)$, $\mathfrak{v}(A)$, $\overline{\mathfrak{v}}(\overline{A})$ are transformed by the linear transformation Λ, or $\overline{\Lambda}$, to $\mathrm{P}(w_i'\,dz)$, $\mathfrak{v}'(A)$, $\overline{\mathfrak{v}}'(\overline{A})$. Then a topology on $\overline{\mathfrak{D}}_0$ is induced from the compact manifold $S_{2g}(\mathrm{P}(w_i'\,dz))$ determined by $w_i'\,dz$, and analytical coordinates $\mathfrak{v}'(A) = (\zeta_1', \dots, \zeta_g')$ are induced on S. However, these two analytic function systems are equivalent. Namely, $\overline{\mathfrak{D}}_0$ is determined as an analytic manifold independently of the choice of a base for $\mathfrak{L}_0$. The manifold $J = \overline{\mathfrak{D}}_0$ is called the *Jacobian variety* of the algebraic function field K.

We next comment on the topology of the Jacobian variety J. For an arbitrary closed analytic curve γ on $\mathfrak{R}$, let

$$w_\gamma\, dz = \sum_{i=1}^{g} c_i w_i\, dz, \qquad c_i \in k.$$

Let $(\zeta_1, \ldots, \zeta_g)$ be a coordinate system by $w_i\, dz$ at $\overline{A}$ on $J = \overline{\mathfrak{D}}_0$. Then from the definition of ζ_i and (2.32) in Theorem 5.9, we have

$$(\overline{\sigma}, \overline{A}) = \exp\left(2\pi\sqrt{-1}\,\mathrm{Re}\left(\sum_{i=1}^{g} c_i\zeta_i\right)\right), \qquad \overline{\sigma} = \Psi^{-1}(\overline{\gamma}).$$

The above equation indicates that for a fixed $\overline{\sigma}$ the function of $\overline{A}(\overline{\sigma}, \overline{A})$ is continuous with respect to each coordinate ζ_i. Therefore, a neighborhood of the identity of $\overline{\mathfrak{D}}_0$ regarded as the character group of $\overline{\mathfrak{G}}$ is the set of $\overline{A}$ satisfying $|(\overline{\sigma}_i, \overline{A}) - 1| < \varepsilon$ for $\varepsilon > 0$ and finitely many elements $\overline{\sigma}_1, \ldots, \overline{\sigma}_n$ in $\overline{\mathfrak{G}}$. (See the book by Pontrjagin on topological groups cited earlier.) Hence, the identity map $\overline{A} \mapsto \overline{A}$ from $J = \overline{\mathfrak{D}}_0$ as a Jacobian variety to $X(\overline{\mathfrak{G}}) = \overline{\mathfrak{D}}_0$ as the character group of $\overline{\mathfrak{G}}$ is continuous. Since J and $X(\overline{\mathfrak{G}})$ are both compact, the inverse map is also continuous. In other words, the topology on the compact group $\overline{\mathfrak{D}}_0$ mentioned in Theorem 5.9, coincides with the topology on $J = \overline{\mathfrak{D}}_0$ as the Jacobian variety.

As a generalization of an analytic function on a Riemann surface, we will define an analytic function on an analytic manifold $S = S_{2n}$. When a complex-valued function f defined on S is expressed in a small neighborhood of each point P, as a quotient $f = f_1/f_2$, $f_2 \not\equiv 0$, of regular functions $f_1 = f_1(\zeta_1, \ldots, \zeta_n)$ and $f_2 = f_2(\zeta_1, \ldots, \zeta_n)$ of analytic coordinates $\zeta_1, \ldots, \zeta_n$, then f is called an analytic function on S. If an analytic function f is not identically zero, then $1/f$ is an analytic function on S. Hence the set $K(S)$ of all analytic functions on S is a field containing the field of complex numbers. In particular, when S is the Jacobian variety belonging to the algebraic function field K, $\mathfrak{K} = K(J)$ is called the *Abelian function field* belonging to K. An element of $\mathfrak{K}$, i.e., analytic function on J, is called an *Abelian function* belonging to K.

For a basis $\{w_i\, dz\}$ for $\mathfrak{L}_0$ of differentials of the first kind in K, the correspondence from a point $(\zeta_1, \ldots, \zeta_g)$ in $V_g(k)$ to the point $\overline{A}$ in $\overline{\mathfrak{D}}_0$ satisfying $\mathfrak{v}(A) = (\zeta_1, \ldots, \zeta_g)$, i.e.,

$$\overline{A} = \varphi((\zeta_1, \ldots, \zeta_g)),$$

is an analytic map from the g-dimensional analytic manifold $V_g(k)$ onto $J = \overline{\mathfrak{D}}_0$. For an arbitrary Abelian function $f(\overline{A})$ belonging to K, define

$$f^*(\zeta_1, \ldots, \zeta_g) = f(\varphi((\zeta_1, \ldots, \zeta_g))).$$

Then f^* is an analytic function on $V_g(k)$, satisfying

$$f^*(\zeta_1 + \omega_1, \ldots, \zeta_g + \omega_g) = f^*(\zeta_1, \ldots, \zeta_g) \tag{3.18}$$

for any vector $(\omega_1, \ldots, \omega_g)$ in $\mathrm{P}(w_i\,dz)$. That is, f^* is a periodic function having every element of $P(w_i\,dz)$ as its period. Conversely, if an analytic function f^* on $V_g(k)$ satisfies (3.18), then $f^*(\zeta_1, \ldots, \zeta_g) = f(\varphi(\zeta_1, \ldots, \zeta_g))$ is an analytic function on J, i.e., an Abelian function belonging to K. Therefore, the Abelian function field $\mathfrak{K}$ belonging to K is isomorphic to the set of analytic functions in $V_g(k)$ satisfying (3.18).

In order to study the Abelian function field K, it is important to study the so called elementary Abelian functions belonging to K. We can describe them as follows. Let $P_1^0, \ldots, P_g^0$ be g fixed points on the Riemann surface $\mathfrak{R} = \mathfrak{R}(K)$ of K. Then, from Theorem 2.23, each class $\overline{A}$ in $\overline{\mathfrak{D}}_0 = \mathfrak{D}_0/\mathfrak{D}_H$ contains a divisor of the form

$$\frac{P_1 \cdots P_g}{P_1^0 \cdots P_g^0}. \tag{3.19}$$

Hence, Theorem 5.12 implies that for an arbitrary point $(\zeta_1, \ldots, \zeta_g)$ in $V_g(k)$ there exist g points $P_1, \ldots, P_g$ such that

$$\zeta_i \equiv \sum_{j=1}^{g} \int_{P_j^0}^{P_j} w_i\,dz \mod \mathrm{P}(w_i\,dz), \qquad i = 1, \ldots, g. \tag{3.20}$$

The above $w_i\,dz$ in (3.20) are basis elements for $\mathfrak{L}_0$ over k and the congruence in (3.20) indicates that the difference of the vectors in $V_g(k)$ belongs to $\mathrm{P}(w_i\,dz)$.

We ask if there is another divisor of that form belonging to $\overline{A}$. In other words, we ask whether given $(\zeta_1, \ldots, \zeta_g)$ there is another set of $P_1, \ldots, P_g$ that satisfy (3.20). Suppose

$$\frac{Q_1 \cdots Q_g}{P_1^0 \cdots P_g^0}$$

is a divisor in $\overline{A}$ that differs from the one in (3.19). Then

$$P_1 \cdots P_g \equiv Q_1 \cdots Q_g \mod \mathfrak{D}_H. \tag{3.21}$$

Therefore, there exists an element z_1 in K such that

$$(z_1) = \frac{Q_1 \cdots Q_g}{P_1 \cdots P_g}. \tag{3.22}$$

By the assumption, $P_1 \cdots P_g \neq Q_1 \cdots Q_g$, z_1 is not constant. When $g = 1$, from (3.22) we have $(z_1) = P_1/Q_1$. By Theorem 2.7, $K = k(z_1)$ holds, which contradicts the assumption that $g = 1$. Hence, for the case $g = 1$, each class in $\overline{\mathfrak{D}}_0$ has precisely one divisor of the form P_1/P_1^0 (see the paragraph containing (7.19), §7, Chapter 2).

Next we will consider the case where $g \geq 2$. In this case z_1 is contained in $L(P_1 \cdots P_g)$. Hence, we have

$$l(P_1 \cdots P_g) \geq 2. \tag{3.23}$$

For a differential divisor W of K, the Riemann-Roch theorem (Theorem 2.13) implies

$$l(P_1 \cdots P_g) = n(P_1 \cdots P_g) - g + 1 + l\left(\frac{W}{P_1 \cdots P_g}\right).$$

Then from (3.23) we have

$$l\left(\frac{W}{P_1 \cdots P_g}\right) \geq 1. \tag{3.24}$$

By noting that $n(W) = 2g - 2$, we see that there exist an element z_2 in K and $g-2$ points $P_2', \dots, P_{g-2}'$ such that

$$(z_2) = \frac{P_1 \cdots P_g P_1' \cdots P_{g-2}'}{W}.$$

For another set of $g-2$ fixed points $P_{g+1}^0, \dots, P_{2g-2}^0$ besides $P^0, \dots, P_g^0$, the above equation can be rewritten as

$$\frac{W}{P_1^0 \cdots P_{2g-2}^0} \equiv \frac{P_1 \cdots P_g}{P_1^0 \cdots P_g^0} \frac{P_1' \cdots P_{g-2}'}{P_{g+1}^0 \cdots P_{2g-2}^0} \quad \bmod \mathfrak{D}_H. \tag{3.25}$$

Then let

$$\mathfrak{v}\left(\frac{W}{P_1^0 \cdots P_{2g-2}^0}\right) = (\lambda_1, \dots, \lambda_g). \tag{3.26}$$

We have from Theorem 5.12 that

$$\lambda_i \equiv \sum_{j=1}^{g} \int_{P_j^0}^{P_j} w_i\, dz + \sum_{l=1}^{g-2} \int_{P_{g+l}^0}^{P_l'} w_i\, dz \quad \bmod \mathrm{P}(w_i\, dz),$$

i.e.,

$$\zeta_i \equiv \lambda_i - \sum_{l=1}^{g-2} \int_{P_{g+l}^0}^{P_l'} w_i\, dz \quad \bmod \mathrm{P}(w_i\, dz), \qquad i = 1, \dots, g. \tag{3.27}$$

We can trace the above process backwards. Namely, if $\zeta_1, \dots, \zeta_g$ satisfy (3.27) for the constants λ_i and points $P_1', \dots, P_{g-2}'$ on $\mathfrak{R}$, then (3.25) holds from Theorem 5.12. Then (3.24) follows from (3.25), (3.23) follows from (3.24), and (3.22) follows from (3.23). That is, a class $\overline{A}$ in $\overline{\mathfrak{D}}_0$ contains more than one divisor of the form (3.19) if and only if its coordinates $(\zeta_1, \dots, \zeta_g)$ satisfy (3.27).

The totality $\mathfrak{X}$ of all the $\overline{A}$ on $J = \overline{\mathfrak{D}}_0$ containing more than one divisor of the form (3.19) is called the *singularity set* of J. In terms of coordinates, such a point is expressed as in (3.27). In order words, $\mathfrak{X}$ is the image of the map F' from the direct product $\mathfrak{R}^{(g-2)} = \mathfrak{R} \times \cdots \times \mathfrak{R}$ of $(g-2)$ closed Riemann surfaces $\mathfrak{R}$, defined by

$$F'(P'_1, \ldots, P'_{g-2}) = (\zeta_1, \ldots, \zeta_g),$$
$$\zeta_i = \lambda_i - \sum_{l=1}^{g-2} \int_{P^0_{g+l}}^{P'_l} w_i\, dz, \qquad i = 1, \ldots, g.$$

For a locally uniformizing variable t'_i for P'_i, the above ζ_i is clearly a regular function of $t'_1, \ldots, t'_{g-2}$. Therefore, the complex dimension of the connected closed subset $\mathfrak{X} = F'(\mathfrak{R}^{(g-2)})$ of J is at most $g-2$, and the real dimension of $\mathfrak{X}$ is at most $2g-4$. Hence, we have obtained the following theorem.

THEOREM 5.13. *For arbitrary fixed points $P^0_1, \ldots, P^0_g$ on an algebraic function field K of genus g, each class $\overline{A}$ of the divisor class group $\overline{\mathfrak{D}}_0$ of degree 0 of K, i.e., each point $\overline{A}$ on the Jacobian variety J of K, always contains a divisor of the form*

$$\frac{P_1 \cdots P_g}{P^0_1 \cdots P^0_g}.$$

Furthermore, except those points in the singularity set $\mathfrak{X}$, which is a connected subset of J of complex dimension at most $g-2$, each point $\overline{A}$ contains a unique divisor of the above form.

We may phrase the above theorem in analysis terms as follows.

THEOREM 5.14 (the theorem of Jacobi). *Let $P^0_1, \ldots, P^0_g$ be arbitrary points on a closed Riemann surface $\mathfrak{R}$, and let $\{w_1\, dz, \ldots, w_g\, dz\}$ be a base for the set $\mathfrak{L}_0$ of differentials of the first kind in $K = K(\mathfrak{R})$ over k. Then for given arbitrary complex numbers $\zeta_1, \ldots, \zeta_g$ there exist g points $P_1, \ldots, P_g$ satisfying*

$$\zeta_i \equiv \sum_{j=1}^{g} \int_{P^0_j}^{P_j} w_i\, dz \mod \mathrm{P}(w_i\, dz), \qquad i = 1, \ldots, g. \tag{3.28}$$

One can choose a path P^0_j to P_j in (3.28) so as to replace the congruence by an equality in (3.28). Except for the $(\zeta_1, \ldots, \zeta_g)$ that satisfy

$$\zeta_i \equiv \lambda_i - \sum_{l=1}^{g-2} \int_{P^0_{g+l}}^{P'_l} w_i\, dz \mod \mathrm{P}(w_i\, dz),$$

the above $P_1, \ldots, P_g$ are uniquely determined. Note that $\lambda_1, \ldots, \lambda_g$ are determined from P^0_i, and $P'_1, \ldots, P'_{g-2}$ are arbitrary points on $\mathfrak{R}$.

For $P^0_1, \ldots, P^0_g$ as above, Theorem 5.13 implies that for a point $\overline{A}$ of J that is not in the singularity set $\mathfrak{X}$, a unique set of $P_1, \ldots, P_g$ is determined. Then for an arbitrary function $z = z(P)$ in K, define the elementary symmetric polynomials of $z(P_i)$ by

$$s_1(\overline{A};\, z) = \sum_{i=1}^{g} z(P_i),$$
$$s_2(\overline{A};\, z) = \sum_{i<j} z(P_i)z(P_j),$$
$$\vdots \qquad \vdots$$
$$s_g(\overline{A};\, z) = \prod_{i=1}^{g} z(P_i).$$

We have g functions $s_1(\overline{A};\, z), \dots, s_g(\overline{A};\, z)$ defined on $J - \mathfrak{X}$. Then there exists a unique analytic function on J, i.e., an Abelian function of K, that coincides with $s_i(\overline{A};\, z)$ over $J - \mathfrak{X}$. We denote this analytic function again by $s_i(\overline{A},\, z)$. We call those analytic functions on J *elementary Abelian functions* belonging to K.

The fact that $s_i(\overline{A};\, z)$ can be extended to an analytic function on J can be proved by using Hartogs' theorem in the theory of functions of several complex variables. See, for example, S. Bochner and W. Martin, *Several complex variables*, Princeton University Press, Princeton, NJ, 1948. Riemann succeeded in explicitly expressing the function $s_i(\varphi((\zeta_1, \dots, \zeta_g));\, z)$ on $V_g(k)$ associated with $s_i(\overline{A};\, z)$ as a quotient of two integral functions of ζ_i. During the process, ϑ-functions are used. This achievement is considered as one of the most profound results in analysis. We can not afford the space to describe it in detail. However, we will describe the definition of a ϑ-function and show how to express an elementary Abelian function in terms of ϑ-functions.

NOTE. Even though Abelian functions and ϑ-functions are the climax of classical analysis, there are hardly any books on these topics written from the modern point of view. We list only the book, W. F. Osgood, *Lehrbuch der Functionentheorie*, II_2, Teubner, Berlin, 1924.

Let C be a standard subdivision of $\mathfrak{R}$, and let α_i and β_i $(i = 1, \dots, g)$ be the subdivision curves of C. Then there exists a base $\{w_1\, dz, \dots, w_g\, dz\}$ for $\mathfrak{L}_0$ satisfying

$$\int_{\alpha_j} w_i\, dz = \pi\sqrt{-1}\delta_{ij}, \qquad i, j = 1, \dots, g$$

from the isomorphism in (1.12) and (1.13). Let

$$\int_{\beta_j} w_i\, dz = \omega_{ij} = \mu_{ij} + \sqrt{-1}v_{ij}, \qquad \mu_{ij}, v_{ij} = \text{ real numbers},$$

be the periods for β_j. We have the following lemma.

LEMMA 5.4. *We have* $\omega_{ij} = \omega_{ji}$, *and for a suitable real number* $\mu > 0$, *we have*

$$\sum_{i,j=1}^{g} \mu_{ij} x_i x_j \leq -\mu(x_1^2 + \cdots + x_g^2) \tag{3.29}$$

for any real numbers $x_1, \ldots, x_g$.

PROOF. Let $w\,dz = w_i\,dz$ and $w'\,dz = w_j\,dz$ in Theorem 5.2. Then (1.5) implies $\omega_{ij} = \omega_{ji}$. Moreover, let $w\,dz = x_1 w_1\,dz + \cdots + x_2 w_g\,dz$ in the same theorem. Since

$$\xi_j = 0, \qquad \eta_j = \pi x_j,$$
$$\xi_{g+j} = \sum_{i=1}^{g} \mu_{ij} x_i, \qquad \eta_{g+j} = \sum_{i=1}^{g} v_{ij} x_i,$$

it follows from (1.6) in Theorem 5.2 that, unless $w\,dz = 0$, i.e., $x_1 = x_2 = \cdots = x_g = 0$, we obtain

$$\sum_{j=1}^{g} (\xi_j \eta_{g+j} - \xi_{g+j} \eta_j) = -\sum_{i,j=1}^{g} \mu_{ij} x_i x_j > 0.$$

That is, $\sum_{i,j=1}^{g} \mu_{ij} x_i x_j$ is a negative definite quadratic form. Since all the eigenvalues of the symmetric matrix (μ_{ij}) are negative, let $-\mu$ be the eigenvalue whose absolute value is the smallest. For this μ, (3.29) holds.

LEMMA 5.5. *When* $m_1, m_2, \ldots, m_g$ *range over all rational integers, for the above* ω_{ij} *define*

$$\vartheta(u_1, \ldots, u_g) = \sum_{m_i} \exp\left(\sum_{i,j=1}^{g} \omega_{ij} m_i m_j + 2 \sum_{i=1}^{g} m_i u_i \right). \tag{3.30}$$

Then the series on the right-hand side converges absolutely and uniformly for u_i *in a bounded region. Hence,* $\vartheta(u_1, \ldots, u_g)$ *is a regular function defined for all the complex values* $u_1, \ldots, u_g$.

PROOF. Let c be any positive real number, and let $|u_i| < c$. For $|m_i| \geq 4c/\mu$, we have

$$\left| \mathrm{Re}\left(2 \sum_{i=1}^{g} m_i u_i \right) \right| \leq 2 \sum_{i=1}^{g} |m_i|\,|u_i| \leq \frac{\mu}{2} \sum_{i=1}^{g} |m_i|^2.$$

On the other hand, (3.29) implies

$$\sum_{i,j=1}^{g} \mu_{ij} m_i m_j \leq -\mu \sum_{i=1}^{g} |m_i|^2.$$

Therefore, we obtain

$$\begin{aligned}\left|\exp\left(\sum_{i,j=1}^{g}\omega_{ij}m_im_j+2\sum_{i=1}^{g}m_iu_i\right)\right| &= \exp\left(\mathrm{Re}\left(\sum_{i,j=1}^{g}\omega_{ij}m_im_j+2\sum_{i=1}^{g}m_iu_i\right)\right)\\ &\le \exp\left(-\frac{\mu}{2}\sum_{i=1}^{g}|m_i|^2\right).\end{aligned}$$

Let $\sum'$ denote the sum over m_i satisfying $|m_i|\ge 4c/\mu$. Then we have

$$\begin{aligned}\sum{}'\left|\exp\left(\sum_{i,j=1}^{g}\omega_{ij}m_im_j+2\sum_{i=1}^{g}m_iu_i\right)\right| &\le \sum{}'\exp\left(-\frac{\mu}{2}\sum_{i=1}^{g}|m_i|^2\right)\\ \le \sum{}'\exp\left(-\frac{\mu}{2}\sum_{i=1}^{g}|m_i|\right) &\le 2^g\sum_{m_1,\dots,m_g=0}^{\infty}\exp\left(-\frac{\mu}{2}\sum_{i=1}^{g}m_i\right)\\ = 2^g\prod_{i=1}^{g}\left(\sum_{m_i=0}^{\infty}\exp\left(-\frac{\mu}{2}m_i\right)\right) &= 2\left(\frac{1}{1-\exp(-\mu/2)}\right)^g.\end{aligned}$$

Therefore, we conclude that $\sum'\exp(\sum_{i,j=1}^{g}\omega_{ij}m_im_j+2\sum_{i=1}^{g}m_iu_i)$ converges absolutely and uniformly for $|u_i|\le c$. Since the difference between $\sum$ and $\sum'$ is provided by finitely many terms, this completes the proof.

DEFINITION 5.5. The regular function $\vartheta(u_1,\dots,u_g)$ in the above is called the (Riemann) ϑ-function obtained from the matrix $(\omega_{i,j})$.

Let $z=z(P)$ be any analytic function on $\mathfrak{R}=\mathfrak{R}(K)$. Let $\overline{A}$ be a point on J, but not in $\mathfrak{X}$, and let $P_1\cdots P_g/P_1^0\cdots P_g^0$ be a divisor in $\overline{A}$. Then let $(\zeta_1,\dots,\zeta_g)$ be the coordinates for $\overline{A}$ given by (3.28) for the above $w_i\,dz$. We have the following important formula (see the book by Osgood):

$$s_g(\overline{A};z)=s_g(\varphi((\zeta_1,\dots,\zeta_g));z)=c\frac{\prod_{i=1}^{r}\vartheta(\dots,\zeta_l-c_l^{(i)},\dots)}{\prod_{i=1}^{r}\vartheta(\dots,\zeta_l-c_l'^{(i)},\dots)}. \tag{3.31}$$

In the above formula, r is the degree of the numerator and the denominator of the divisor (z), and c, $c_l^{(i)}$, $c_l'^{(i)}$ $(i=1,\dots,g)$ are constants determined by the zeros and the poles of z. Let $f^*(\zeta_1,\dots,\zeta_g)$ be the right-hand side of the above formula. From Lemma 5.5, the numerator and the denominator of f^* are integral functions of ζ_i. Hence, f^* is an analytic function on $V_g(k)$. Furthermore, as long as $\varphi((\zeta_1,\dots,\zeta_g))$ does not belong to the singularity set $\mathfrak{X}$, f^* coincides with $s_g(\varphi((\zeta_1,\dots,\zeta_g));z)$, then (3.18) holds. Since f^* is analytic, (3.18) holds for any $(\zeta_1,\dots,\zeta_g)$. Therefore, from f^* we obtain an analytic function on J, i.e., an Abelian function, as an extension of $s_g(\overline{A};\, z)$. For $s_1(\overline{A};\, z), \ldots, s_{g-1}(\overline{A};\, z)$, we proceed as follows. For an arbitrary complex number λ,

$$s_g(\overline{A};\, z+\lambda) = \prod_{i=1}^{g}(z(P_i)+\lambda) = \lambda^g + s_1(\overline{A};\, z)\lambda^{g-1} + \cdots + s_{g-1}(\overline{A};\, z)\lambda + s_g(\overline{A};\, z)$$

can be extended in terms of ϑ-functions to an analytic function on J. For g distinct complex numbers, λ_i, regard

$$\begin{aligned} s_g(\overline{A};\, z+\lambda_i) = \lambda_i^g + s_1(\overline{A};\, z)\lambda_i^{g-1} \\ + \cdots + s_{g-1}(\overline{A};\, z)\lambda_i + s_g(\overline{A};\, z), \qquad i = 1, \ldots, g \end{aligned}$$

as a system of linear equations for $s_1(\overline{A};\, z), s_2(\overline{A};\, z), \ldots, s_g(\overline{A};\, z)$. Solving the system, all the $s_i(\overline{A};\, z)$, $i = 1, \ldots, g$, are rational expressions of ϑ-functions. Hence, we obtain Abelian functions on J.

Next we will explain how the elementary Abelian functions $s_i(\overline{A};\, z)$ are related to the Abelian function field $\mathfrak{K}$. Since a thorough treatment requires too much preparation for our treatise, only the fundamental ideas will be given.

Let $s_i(\overline{A};\, z)$, $i = 1, \ldots, g$, be the Abelian functions of K obtained from a nonconstant analytic function z in K. Consider the number of points $\overline{A}$ on J satisfying

$$s_i(\overline{A};\, z) = c_i, \qquad i = 1, \ldots, g \tag{3.32}$$

for any given complex numbers $c_1, \ldots, c_g$. Excluding $\mathfrak{X}$, on which the value of $s_i(\overline{A};\, z)$ is undetermined, (3.32) can be solved for $\overline{A}$ as follows. From the definition of $s_i(\overline{A};\, z)$, we first find points $P_1, \ldots, P_g$ on $\mathfrak{R}$ which satisfy

$$\sum_{i=1}^{g} z(P_i) = c_1, \ \sum_{i<j} z(P_i)z(P_j) = c_2, \ldots, \ \prod_{i=1}^{g} z(P_i) = c_g. \tag{3.33}$$

Then $\overline{A}$ is the class of $A = P_1 \cdots P_g / P_1^0 \cdots P_g^0$. The values $z(P_1), \ldots, z(P_g)$ that satisfy (3.33) can be uniquely determined as the roots of

$$x^g - c_1 x^{g-1} + \cdots + (-1)^g c_g = 0.$$

If $n - [K : k(z)]$, then there are n P's that satisfy $z(P) = c$ for any complex number c (Theorem 4.7). Consequently, there are all together $m = n^g$ sets of $P_1, \ldots, P_g$ which satisfy (3.33). We omit the discussion whether all the $\overline{A}$ obtained from those m sets of $P_1, \ldots, P_g$ are indeed distinct points belonging to $J - \mathfrak{X}$. In general, one can prove that for given $c_1, \ldots, c_g$ there are m solutions for $\overline{A}$ to satisfy (3.32). Let us denote those points by

$$\overline{A}_1(c_1, \ldots, c_g), \ldots, \overline{A}_m(c_1, \ldots, c_g).$$

Then let $h = h(\overline{A})$ be an arbitrary Abelian function belonging to K, and let

$$\sigma_1(c_1, \dots, c_g), \dots, \sigma_m(c_1, \dots, c_g)$$

be the fundamental symmetric functions of

$$h(\overline{A}_1(c_1, \dots, c_g)), \dots, h(\overline{A}_m(c_1, \dots, c_g)).$$

One can prove that these functions $\sigma_1, \dots, \sigma_m$ are complex-valued functions defined for any $c_1, \dots, c_g$ except for a lower dimensional set, and are analytic with respect to c_i. From the theory of functions in several complex variables, each $\sigma_i(c_1, \dots, c_g)$ must be a rational function of $c_1, \dots, c_g$. In other words, $h(\overline{A})$ is a root of the polynomial

$$F(X) = X^m - \sigma_1(s_i(\overline{A}; z))X^{m-1} + \cdots + (-1)^m \sigma_m(s_i(\overline{A}; z)) \tag{3.34}$$

over the subfield $\mathfrak{K}_0 = k(s_1(\overline{A}; z), \dots, s_g(\overline{A}; z))$ of $\mathfrak{K}$, where $k(s_1(\overline{A}; z), \dots, s_g(\overline{A}; z))$ is the field obtained by adjoining $s_1(\overline{A}; z), \dots, s_g(\overline{A}; z)$ to the complex field. Since there are solutions to (3.32) for arbitrary $c_1, \dots, c_g$, we can conclude that $s_1(\overline{A}; z), \dots, s_g(\overline{A}; z)$ are algebraically independent over k. That is, $\mathfrak{K}_0$ is the rational function field of g variables with the coefficient field k, whereas an arbitrary element h in $\mathfrak{K}$ is a root of a polynomial of degree m over $\mathfrak{K}_0$. As was shown in the proof of Theorem 4.2, $\mathfrak{K}$ is an algebraic extension of $\mathfrak{K}_0$ of degree at most m.

For $z = z(P)$ in the above we will find $w = w(P)$ to satisfy $K = k(z, w)$, and we will study the elementary Abelian functions obtained from w:

$$s_i(\overline{A}; w), \qquad i = 1, \dots, g.$$

In general, a place P on K is determined by the values of $z(P)$ and $w(P)$. Hence, for the $m = n^g$ sets of $P_1, \dots, P_g$ satisfying (3.32), the values of

$$s_1(\overline{A}; w) = \sum_{i=1}^{g} w(P_i), \ s_2(\overline{A}; w) = \sum_{i<j} w(P_i)w(P_j), \dots,$$

i.e., the g complex numbers

$$s_1(\overline{A}_i(c_1, \dots, c_g); w), \dots, s_g(\overline{A}_i(c_1, \dots, c_g); w)$$

are distinct for i, $1 \le i \le m$. For a suitable linear combination of $s_i(\overline{A}; w)$

$$h(\overline{A}) = \sum_{i=1}^{g} \lambda_i s_i(\overline{A}; w), \qquad \lambda_i \in k,$$

and general $c_1, \dots, c_g$, $h(\overline{A}_i(c_1, \dots, c_g))$ takes different values for each distinct i, $1 \le i \le m$. That is, for given values of $s_i(\overline{A}; z)$ there are in general m corresponding values of $h(\overline{A})$. Hence, (3.34) is an irreducible polynomial. Therefore, we have $[\mathfrak{K} : \mathfrak{K}_0] \le m$. Consequently, we obtain

$$\begin{aligned}\mathfrak{K} = \mathfrak{K}_0(k) &= \mathfrak{K}_0(s_1(\overline{A}; w), \dots, s_g(\overline{A}; w)) \\ &= k(s_1(\overline{A}; z), \dots, s_g(\overline{A}; z), s_1(\overline{A}; w), \dots, s_g(\overline{A}; w)).\end{aligned}$$

We have outlined the proof of the following theorem.

THEOREM 5.15. *Let* $\mathfrak{K}$ *be the Abelian function field belonging to an algebraic function field* K. *For a nonconstant function* $z = z(P)$ *in* K, *let* $s_1(\overline{A}; z), \ldots, s_g(\overline{A}; z)$ *be the elementary Abelian functions obtained from* z. *Then*

$$\mathfrak{K}_0 = k(s_1(\overline{A}; z), \ldots, s_g(\overline{A}; z))$$

is the rational function field of g *variables* $x_i = s_i(\overline{A}; z)$ *over* k, *and* $\mathfrak{K}$ *is an algebraic extension field of degree* $m = n^g$ *over* $\mathfrak{K}_0$. *Note that* g *is the genus of* K, *and* $n = [K : k(z)]$. *Furthermore, for an arbitrary function* $w = w(P)$ *satisfying* $K = k(z, w)$, *we have*

$$\begin{aligned}\mathfrak{K} &= \mathfrak{K}_0(s_1(\overline{A}; w), \ldots, s_g(\overline{A}; w)) \\ &= k(s_1(\overline{A}; z), \ldots, s_g(\overline{A}; z), s_1(\overline{A}; w), \ldots, s_g(\overline{A}; w)).\end{aligned}$$

Therefore, any Abelian function belonging to K *can be expressed as a rational function of* ϑ*-functions.*

The above theorem leads to many results. Since $\mathfrak{K}$ is a finite extension of the rational function field $\mathfrak{K}_0$, $\mathfrak{K}$ is an algebraic function field of dimension g with the coefficient field k. For arbitrary complex numbers $\eta_1, \ldots, \eta_g$, the function

$$f_i(\zeta_1, \ldots, \zeta_g) = s_i(\varphi(\zeta_1 + \eta_1, \ldots, \zeta_g + \eta_g); z) \tag{3.35}$$

clearly satisfies (3.18). Hence, f_i is an Abelian function of K. Therefore, there exists a polynomial

$$H_i(Y; X_1, \ldots, X_g)$$

over k satisfying

$$H_i(s_i(\varphi(\zeta + \eta); z); s_1(\varphi(\zeta); z), \ldots, s_g(\varphi(\zeta); z)) = 0.$$

The coefficients of H_i are functions of $\eta_1, \ldots, \eta_g$. But the relations between ζ_i and η_i in (3.35) are symmetric. Therefore, for some polynomial over k:

$$H_i^*(Y; X_1, \ldots, X_g; X_1', \ldots, X_g'),$$

we expect that

$$H_i^*(s_i(\varphi(\zeta + \eta); z); s_1(\varphi(\zeta); z), \ldots, s_1(\varphi(\eta); z), \ldots) = 0 \tag{3.36}$$

should hold identically for ζ_i and η_i. In fact, one can rigorously prove (3.36), which is called the *algebraic addition formula* for the elementary Abelian functions $s_i(\overline{A}; z)$. In particular, if $g = 1$, then we get the well-known addition formula in the theory of elliptic functions. See §5 in this chapter.

NOTE. We define an algebraic function field K of dimension n with coefficient field k as follows:

(i) for n elements $x_1, \ldots, x_n$ algebraically independent over k, the field K is a finite extension of the rational function field $k(x_1, x_2, \ldots, x_n)$,

(ii) k is algebraically closed in K.

In the case where k is an algebraically closed field, we can simply say that K is an extension field of transcendence degree n over k. The above definition is a generalization of Definition 2.1. In the theory of algebraic function fields the importance of higher degree has been increasingly recognized with the recent development in algebraic geometry and in the theory of functions in several complex variables.

The problem of finding points $P_1, \dots, P_g$ on $\mathfrak{R}$ which satisfy (3.28) for given complex numbers $\zeta_1, \dots, \zeta_g$ is called Jacobi's *Umkehrproblem* (inverse problem). Theorem 5.14 answers the existence and the uniqueness of the solution of this problem. The starting point for the study of Abelian functions was to try to express $P_1, \dots, P_g$ as functions of $\zeta_1, \dots, \zeta_g$. As we showed in the sketch of a proof of Theorem 5.14, a point $\overline{A}$ in $J - \mathfrak{X}$ is uniquely determined by the values of $s_i(\overline{A}; z)$ and $s_i(\overline{A}; w)$. Then the divisor $A = P_1 \cdots P_g / P_1^0 \cdots P_g^0$ in $\overline{A}$ is uniquely determined. Hence, in order to express $P_1, \dots, P_g$ as functions of $\zeta_1, \dots, \zeta_g$, in a sense it is sufficient to express $s_i(\overline{A}; z)$ and $s_i(\overline{A}; w)$ as functions of $\zeta_1, \dots, \zeta_g$. It was Riemann's discovery that $s_i(\overline{A}; z)$ and $s_i(\overline{A}; w)$ are indeed rational functions of ϑ-functions of ζ_i. See the Preface.

§4. Extension fields

In Theorem 4.4 we briefly mentioned an extension field of an algebraic function field K over k and the corresponding closed Riemann surface. We will study them more closely here.

Let L be an arbitrary finite extension of K, i.e., L is also an algebraic function field over k. As we mentioned in the paragraph preceding Theorem 4.5, there are only finitely many branch points of L with respect to K. We will give an algebraic proof of this statement. Note, however, that we could have treated this topic following (5.23) in Chapter 2 and Theorem 2.22. See also the note at the end of §2, Chapter 4.

Let g be the genus of K and let $K_0 = k(z)$ for a nonconstant z in K. Then from (5.23) in Chapter 2

$$2g - 2 = \sum_P (e'_P - 1) - 2m, \tag{4.1}$$

where the summation on the right is over all the places on K, e'_P is the ramification index of P for K_0, and $m = [K : K_0]$. If the genus of L is g', then the same formula implies

$$2g' - 2 = \sum_{P'} (e'_{P'} - 1) - 2mn, \tag{4.2}$$

where P' is a place on L, $e'_{P'}$ is the ramification index of P' for K_0, and

$$n = [L : K] \quad \text{and} \quad mn = [L : K_0].$$

Then let $e_{P'}$ denote the ramification index of P' for K, and let P be the projection of P' on K. Then from Remark 2, §3, Chapter 2 we have

$$e'_{P'} = e_{P'}e_P \quad \text{and} \quad \sum{}' e_{P'} = n, \tag{4.3}$$

where $\sum'$ is over all the extensions P' on L of P. Hence (4.1) and (4.2) imply

$$\begin{aligned} 2g' - 2 - n(2g-2) &= \sum_{P'}(e_{P'}e_P - 1) - n\sum_P (e_P - 1) \\ &= \sum_{P'}(e_{P'}e_P - 1 - e_{P'}(e_P - 1)) \\ &= \sum_{P'}(e_{P'} - 1). \end{aligned}$$

That is,

$$2g' - 2ng = \sum_{P'}(e_{P'} - 1) - (2n-2), \tag{4.4}$$

which is called Hurwitz's Formula. From this formula, it is clear that there are only finitely many P' satisfying $e_{P'} > 1$, i.e., branch points. Hence,

$$\mathfrak{D}(L/K) = \prod_{P'} P'^{e_{P'}-1}$$

is a divisor of L, called the (relative) *different*. See Theorem 2.22.

Through the above algebraic method, (4.1), (4.2), and (4.3) can be used for any algebraic function field over an arbitrary algebraically closed field of characteristic 0. Hence, formula (4.4) and the finiteness of branch points of L/K are valid for such general algebraic function fields. Note also that the above is true for the characteristic $p \neq 0$ case. We will not give either a proof for the case or the essential (number theoretic) meaning of $\mathfrak{D}(L/K)$. See Hasse's paper mentioned at the beginning of §5 in Chapter 2.

When there exist no branch points for L/K, we call L/K an *unramified extension* of K. An unramified extension of an algebraic function field K over the field of complex numbers has the following meaning.

THEOREM 5.16. *Let L/K be an unramified extension, and let $P = f'(P')$ be the map from $\mathfrak{R}' = \mathfrak{R}(L)$ onto $\mathfrak{R} = \mathfrak{R}(K)$, induced by the projection of places from L to K. Then $\{\mathfrak{R}', f'; \mathfrak{R}\}$ is a covering Riemann surface for $\mathfrak{R}$.*

PROOF. From Theorem 4.4, f' is an analytic map from $\mathfrak{R}'$ to $\mathfrak{R}$. The ramification index (exponent) of P' for f' coincides with the algebraic ramification index of P' for K. Hence, by the assumption, a small neighborhood of P' is conformally mapped onto a neighborhood of $P = f'(P')$ in a one-to-one fashion. From (4.3) for a given P there are exactly $n = [L : K]$ points P' on $\mathfrak{R}'$ satisfying $P = f'(P')$. Therefore, for a sufficiently small connected neighborhood U of P, $f^{-1}(U)$ has n separated connected components corresponding to P' satisfying $f'(P') = P$. Each component is mapped by f' conformally and in a one-to-one fashion onto U. From the definition, f' is a covering map from $\mathfrak{R}'$ onto $\mathfrak{R}$. Q.E.D.

In general, let $\mathfrak{R} = \mathfrak{R}(K)$ be the closed Riemann surface of an algebraic function field K over k, and let $\{\mathfrak{R}^*, f^*; \mathfrak{R}\}$ be a simply connected covering Riemann surface of $\mathfrak{R}$. Then $\mathfrak{R}^*$ is either the Riemann sphere (complex t-sphere), a finite t-plane, or the interior $\{t; |t| < 1\}$ of the unit circle. Let G be the fundamental group of $\mathfrak{R}$ for $\mathfrak{R}^*$. Then from (5.8) in Chapter 3

$$\{\mathfrak{R}^*, f^*; \mathfrak{R}\} \cong \{\mathfrak{R}^*, f_G^*, \mathfrak{R}^*(G)\}.$$

Recall that $\mathfrak{R}^*(G)$ is the Riemann surface with the underlying space $S^*(G)$, and f_G^* is the analytic map. See §5, Chapter 3. Therefore, we have $\mathfrak{R} \cong \mathfrak{R}^*(G)$ and $K \cong K(\mathfrak{R}) \cong K(R^*(G))$. Hence, we may study $K = K(\mathfrak{R}^*(G))$, i.e., the analytic function field of the closed Riemann surface $\mathfrak{R}^*(G)$ to find the algebraic properties of K. For an arbitrary function $z = z(P)$ of K, let

$$z = z(t) = z(P_t), \quad P_t = f^*(t), \qquad t \in \mathfrak{R}^*. \tag{4.5}$$

As we mentioned at the beginning of §2, z can be considered a function $z(t)$ defined on $\mathfrak{R}^*$. Then K can be regarded as a subfield of the analytic function field $\widetilde{K} = K(\mathfrak{R}^*)$ of $\mathfrak{R}^*$.

As in §2, let $\mathfrak{G}$ be the Galois group of $\widetilde{K}/K$ corresponding to $G : G = \Phi(\mathfrak{G})$ and $\mathfrak{G} = \Phi^{-1}(G)$. For an arbitrary subgroup $\mathfrak{H}$ of $\mathfrak{G}$, let $H = \Phi(\mathfrak{H})$ be the corresponding subgroup of G. Then denote the Riemann surface belonging to the subgroup H by

$$\mathfrak{R}(\mathfrak{H}) = \mathfrak{R}^*(H),$$

where $\mathfrak{R}^*(H)$ is the Riemann surface having $S^*(H)$ as the underlying space and having f_H^* as its covering map. Then $\mathfrak{R}^*(H)$ is a covering Riemann surface of $\mathfrak{R}^*(G) = \mathfrak{R}(\mathfrak{G})$, where $f'_{G,H}$ is the covering map satisfying $f_G^* = f'_{G,H} f_H^*$. Let $u = u(P'')$, $P'' \ni \mathfrak{R}(\mathfrak{H})$, be an arbitrary function of

$$K(\mathfrak{H}) = K(\mathfrak{R}(\mathfrak{H})).$$

Then, as was the case for $K = K(\mathfrak{G})$, letting

$$u(t) = u(f_H^*(t)), \qquad t \in \mathfrak{R}^*,$$

we obtain a function $u(t)$ in $\widetilde{K}$. Identifying $u(t)$ with $u(P'')$, let

$$u = u(P'') = u(t) \qquad (P'' = f_H^*(t)). \tag{4.6}$$

That is, we consider an element of $K(\mathfrak{H})$ as an analytic function on $\mathfrak{R}(\mathfrak{H})$ and also as a function on $\mathfrak{R}^*$. We also identify $\mathfrak{R}(\mathfrak{H})$ with $\mathfrak{R}(K(\mathfrak{H}))$ as we described at the end of §2 in the previous chapter

$$K(\mathfrak{H}) = K(\mathfrak{R}(\mathfrak{H})), \qquad \mathfrak{R}(\mathfrak{H}) = \mathfrak{R}(K(\mathfrak{H})) \qquad (K = K(\mathfrak{G}),\ \mathfrak{R} = \mathfrak{R}(\mathfrak{G})).$$

As in the paragraph preceding Definition 5.1, the set of all functions in $\widetilde{K}$ that are invariant under all the automorphisms in $\mathfrak{H}$ is exactly $K(\mathfrak{H})$. Hence,

$$K(\mathfrak{H}) \supseteq K(\mathfrak{G}) = K .$$

More generally we have that

$$\mathfrak{H}_1 \subseteq \mathfrak{H}_2 \quad \text{implies} \quad K(\mathfrak{H}_1) \supseteq K(\mathfrak{H}_2) .$$

Let $z = z(P)$ be any element of K. Then $f_G^* = f'_{G,H} f_H^*$ and our conventions (4.5) and (4.6) imply that for an arbitrary point $P' = f_H^*(t)$ on $\mathfrak{R}(\mathfrak{H})$ and for the point $P = f'_{G,H}(P')$ on $\mathfrak{R} = \mathfrak{R}(\mathfrak{G})$ we have

$$z(P) = z(f'_{G,H}(P')) \tag{4.7}$$

from $z(f_G^*(t)) = z(t) = z(f_H^*(t))$. Namely, regarding $K = K(\mathfrak{G})$ and $K(\mathfrak{H})$ as subfields of $\widetilde{K}$, i.e., $K(\mathfrak{G}) \subseteq K(\mathfrak{H})$, (4.7) provides the relation between $z = z(P)$ as a function on $\mathfrak{R} = \mathfrak{R}(\mathfrak{G})$ and as an element of K, i.e., as an analytic function on $\mathfrak{R}(\mathfrak{H})$. When there is no cause for confusion, we denote the right-hand side of (4.7) simply by $z(P')$ when z is regarded as a function on $\mathfrak{R}(\mathfrak{H})$. Generally, for $K(\mathfrak{H}_1) \supseteq K(\mathfrak{H}_2)$ we use the same convention as above for an element in $K(\mathfrak{H}_2)$.

If the index $n = [\mathfrak{G} : \mathfrak{H}]$ of $\mathfrak{H}$ for $\mathfrak{G}$ is finite, then by Lemma 3.11, for any P on $\mathfrak{R}$ the number of points P' on $\mathfrak{R}(\mathfrak{H})$ satisfying $P = f'_{G,H}(P')$ is n. Then let $P_i = f'_{G,H}(P'_i)$ be the images of a sequence of mutually distinct points $P'_1, P'_2, \dots$ on $\mathfrak{R}(\mathfrak{H})$ under the map $f'_{G,H}$. Then there are infinitely many points among $P_1, P_2, \dots$ that are mutually distinct. Since $\mathfrak{R}$ is compact, there is an accumulation point P_0 in $\mathfrak{R}$. For a sufficiently small neighborhood of U, $\overset{-1}{f'_{G,H}}(U)$ is the union of neighborhoods $U^{(1)}, \dots, U^{(n)}$ of the inverse image $\{P_0^{(1)}, \dots, P_0^{(n)}\}$ of P_0 by $f'_{G,H}$ such that $U^{(i)}$ is homeomorphic to U for $i = 1, 2, \dots, n$. Hence, there is a point $P_0^{(i)}$ for some i that is an accumulation point $\{P'_i\}$ in $\mathfrak{R}(\mathfrak{H})$. That is, $\mathfrak{R}(\mathfrak{H})$ is compact. Then $K(\mathfrak{H})$ is also an algebraic function field over k and moreover, the map $f'_{G,H}$ from $\mathfrak{R}(\mathfrak{G})$ to $\mathfrak{R}(\mathfrak{H})$ is an unramified covering map. We conclude from Theorem 4.4 that $K(\mathfrak{H})/K$ is an unramified extension. Since $P = f'_{G,H}(P')$ is the projection on K of the place P' on $K(\mathfrak{H})$ (again by Theorem 4.4), and since there are exactly n points P' for a given P (for example, let $e_{P'} = 1$ in (4.3)), we obtain

$$n = [K(\mathfrak{H}) : K] .$$

Note that $\mathfrak{R}(\mathfrak{H})$ cannot be compact for $[\mathfrak{G} : \mathfrak{H}] = \infty$. This can be observed from the fact that the infinite set $\overset{-1}{f'_{G,H}}(P)$ does not have an accumulation point. Then $K(\mathfrak{H}) = K(\mathfrak{R}(\mathfrak{H}))$ is not an algebraic function field. Hence, $[K(\mathfrak{H}) : K] = \infty$ holds. Consequently, for any subgroup $\mathfrak{H}$ in $\mathfrak{G}$ we have

$$[\mathfrak{G} : \mathfrak{H}] = [K(\mathfrak{H}) : K] .$$

Next let L be an arbitrary unramified finite extension of K. Then for $\mathfrak{R}' = \mathfrak{R}(L)$, from Theorem 5.15 we have a covering Riemann surface $\{\mathfrak{R}', f'; \mathfrak{R}\}$ of $\mathfrak{R}$. From Lemma 3.11, for a subgroup $\mathfrak{H}$ of $\mathfrak{G}$ of index $[\mathfrak{G} : \mathfrak{H}] = [L : K]$,

$$\{\mathfrak{R}', f'; \mathfrak{R}\} \cong \{\mathfrak{R}(\mathfrak{H}), f'_{G,H}; \mathfrak{R}\}.$$

Then let φ' be an isomorphism from $\mathfrak{R}'$ to $\mathfrak{R}(\mathfrak{H})$ as in the above. Then φ' induces an isomorphism σ' between $K(\mathfrak{R}')$ and $K(\mathfrak{R}(\mathfrak{H})) = K(\mathfrak{H})$, defined by

$$u' = \sigma'(u), \quad \text{for } u' \in K(\mathfrak{R}'), \ u \in K(\mathfrak{H}),$$
$$u'(P') = u(\varphi'(P')) \quad \text{for } P' \in \mathfrak{R}'.$$

We denote the isomorphism induced by f' from $K = K(\mathfrak{G}) = K(\mathfrak{R})$ onto a subfield of $K(\mathfrak{R}')$ by

$$z' = \overline{\sigma}(z), \qquad z' \in \overline{\sigma}(K), \qquad z \in K,$$
$$z'(P') = z(f'(P')) \quad \text{for } P' \in \mathfrak{R}'.$$

Then from the definition of the isomorphism we have

$$f'_{G,H}\varphi' = f'.$$

Hence, $z'(P')$ can be written as

$$z(f'(P')) = z(f'_{G,H}(\varphi'(P'))).$$

Regarding K as a subfield of $K(\mathfrak{H})$ by (4.7), we see that the right-hand side above is $\sigma'(z)$ corresponding to $z = z(P')$. Namely, we have

$$\overline{\sigma}(z) = \sigma'(z), \qquad z \in K.$$

Consequently, we obtain

$$K(\mathfrak{H})/K \cong \sigma'(K(\mathfrak{H}))/\sigma'(K) \cong K(\mathfrak{R}')/\overline{\sigma}(K(\mathfrak{R})).$$

Since the far right side is, by Theorem 4.4, isomorphic to L/K, we obtain

$$K(\mathfrak{H})/K \cong L/K.$$

Therefore, we have obtained the following theorem.

THEOREM 5.17. *Let* $K, \mathfrak{R}, \widetilde{K}, \mathfrak{R}^*, \mathfrak{G}, \mathfrak{H}, K(\mathfrak{H}), K(\mathfrak{G})$ *be as in the above, and in particular, let* $\mathfrak{H}$ *be any subgroup of* $\mathfrak{G}$ *of finite index. Then* $K(\mathfrak{H})$ *is an unramified extension of* K *satisfying*

$$[K(\mathfrak{H}) : K] = [\mathfrak{G} : \mathfrak{H}]. \tag{4.8}$$

Conversely, for an arbitrary unramified finite extension L *over* K, *let* $\mathfrak{H}$ *be a subgroup of* $\mathfrak{G}$ *satisfying* (4.8). *Then we have*

$$L/K \cong K(\mathfrak{H})/K.$$

By this theorem, the algebraic study of the unramified extension L over K is reduced to the study of the extension $K(\mathfrak{H})/K$. In general, let L be an arbitrary intermediate field between K and $\widetilde{K}$. Denote the set of all automorphisms of $\widetilde{K}$ belonging to $\mathfrak{G}$ that do not change all the elements of L by $\mathfrak{G}(L)$. This is called the subgroup of $\mathfrak{G}$ corresponding to L. (Then $\mathfrak{G}(L)$ is clearly a subgroup of $\mathfrak{G}$.) For the extension $\widetilde{K}/K$ and the Galois group $\mathfrak{G}$, we have the following theorem paralleling the case of a finite Galois extension.

THEOREM 5.18. *Continuing with the same notations as above, we have the following.*

(i) *For a subgroup* $\mathfrak{H}$ *of* G *whose index* $n = [\mathfrak{G} : \mathfrak{H}]$ *is finite,* $K(\mathfrak{H})/K$ *is an extension of degree* n. *Moreover, we have*

$$\mathfrak{G}(K(\mathfrak{H})) = \mathfrak{H}. \tag{4.9}$$

(ii) *Let* L *be any intermediate field of* $\widetilde{K}/K$ *satisfying* $[L : K] = n < \infty$. *Then* $\mathfrak{G}(L)$ *is a subgroup of* $\mathfrak{G}$ *having index* n, *and we have*

$$K(\mathfrak{G}(L)) = L. \tag{4.10}$$

(iii) *From* (i) *and* (ii) *there is a one-to-one correspondence between all the subgroups in* $\mathfrak{G}$ *of finite index and all the finite extension fields over* K *in* $\widetilde{K}$. *The correspondence is given by* $L = K(\mathfrak{H})$ *and* $\mathfrak{H} = \mathfrak{G}(L)$. *Then* $\widetilde{K}$ *contains* $n = [L : K]$ *conjugate subfields of* L *over* K *which correspond to* n *conjugate subgroups of* $\mathfrak{H} = \mathfrak{G}(L)$ *in* $\mathfrak{G}$. *In particular,* L/K *is a Galois extension if and only if* $\mathfrak{H} = \mathfrak{G}(L)$ *is a normal subgroup of* $\mathfrak{G}$. *Furthermore, the Galois group* $\mathfrak{G}(L/K)$ *of the extension* L/K *is isomorphic to* $\mathfrak{G}/\mathfrak{G}(L)$:

$$\mathfrak{G}(L/K) \cong \mathfrak{G}/\mathfrak{G}(L).$$

PROOF. From the previous theorem, we only need to prove (4.9) in part (i). From the definition, we have $\mathfrak{G}(K(\mathfrak{H})) \supseteq \mathfrak{H}$. Then let σ be an arbitrary element in $\mathfrak{G}$ but not in $\mathfrak{H}$. Let $H = \Phi(\mathfrak{H})$ be the subgroup of the fundamental group G for $\{R^*, f_G^*; \mathfrak{R}\}$ corresponding to $\mathfrak{H}$, and let $\varphi = \Phi(\sigma)$. Since $\varphi \notin H$, $f_H^* \neq f_H^*\varphi$ holds. Hence, there is a point t on $\mathfrak{R}$ such that $f_H^*(t) \neq f_H^*(\varphi(t))$. For the distinct points $f_H^*(t)$ and $f_H^*(\varphi(t))$ on the Riemann surface $\mathfrak{R}(\mathfrak{H})$, by Theorem 3.11 there exists a function $u(P')$ in $K(\mathfrak{H}) = K(\mathfrak{R}(\mathfrak{H}))$ such that

$$u(f_H^*(t)) \neq u(f_H^*(\varphi(t))). \tag{4.11}$$

From (4.11) we have $u(t) \neq u(\varphi(t))$, where u is regarded as a function in $\widetilde{K}$. Hence, we have

$$\sigma(u) \neq u \quad \text{for } \sigma \notin \mathfrak{G}(K(\mathfrak{H})).$$

That is, we showed $\sigma \notin \mathfrak{G}(K(\mathfrak{H}))$ for $\sigma \notin \mathfrak{H}$, which completes the proof of (i).

We will prove (ii). By the definition we have $K(\mathfrak{G}(L)) \supseteq L$. From (4.8) (even if $[\mathfrak{G} : \mathfrak{G}(L)] = \infty$), we have

$$[\mathfrak{G} : \mathfrak{G}(L)] = [K(\mathfrak{G}(L)) : K] \geq [L : K]. \tag{4.12}$$

From our assumption $n = [L : K] < \infty$, L/K is a simple extension

$$L = K(u). \tag{4.13}$$

Let $F(X)$ be an irreducible polynomial in $K[X]$ having u as a root. Then an element σ in $\mathfrak{G}$ belongs to $\mathfrak{G}(L)$ if and only if $\sigma(u) = u$. Hence, when σ runs through all the elements in $\mathfrak{G}$, $\sigma(u)$ takes $[\mathfrak{G} : \mathfrak{G}(L)]$ distinct elements in $\widetilde{K}$. Since we have

$$F(\sigma(u)) = \sigma(F(u)) = 0,$$

all the $\sigma(u)$ are roots of $F(X) = 0$. Therefore, $[\mathfrak{G} : \mathfrak{G}(L)]$ cannot be greater than the degree n of $F(X)$, i.e.,

$$[\mathfrak{G} : \mathfrak{G}(L)] \leq n = [L : K].$$

Lastly we will prove (iii). Decompose $\mathfrak{G}$ as follows

$$\mathfrak{G} = \sigma_1\mathfrak{G}(L) + \cdots + \sigma_n\mathfrak{G}(L).$$

Then by the above, all the roots of $F(X) = 0$ are given by $\sigma_1(u), \dots, \sigma_n(u)$. Therefore, $\widetilde{K}$ contains n conjugate fields $K(\sigma_i(u)) = \sigma_i(L)$ of $L = K(u)$ over K, $i = 1, \dots, n$. Then the corresponding subgroups are the conjugate subgroups $\sigma_i\mathfrak{G}(L)\sigma_i^{-1}$ of $\mathfrak{G}(L)$. Hence, in particular, L/K is a Galois extension if and only if $\mathfrak{G}(L)$ is a normal subgroup of $\mathfrak{G}$. In this case, since we have $\sigma(L) = L$ for any element σ in $\mathfrak{G}$, for the induced automorphism $\overline{\sigma}$ of L/K, the map

$$\sigma \mapsto \overline{\sigma}$$

is clearly a homomorphism from $\mathfrak{G}$ to the Galois group $\mathfrak{G}(L/K)$. From the definition, the set of σ corresponding to the identity map in $\mathfrak{G}(L/K)$ is nothing but $\mathfrak{G}(L)$. The isomorphism theorem implies that $\mathfrak{G}/\mathfrak{G}(L)$ is isomorphic to a subgroup of $\mathfrak{G}(L/K)$. However, by (4.8) the order of $\mathfrak{G}/\mathfrak{G}(L)$ equals the degree of L/K, i.e., the order of $\mathfrak{G}(L/K)$. We have

$$\mathfrak{G}/\mathfrak{G}(L) \cong \mathfrak{G}(L/K) \quad \text{Q.E.D.}$$

Notice that from the proof of (i), for an arbitrary subgroup $\mathfrak{H}$ of $\mathfrak{G}$ (even of nonfinite index), part (i) of the above theorem holds. However, part (ii) is not valid for an arbitrary intermediate field between K and $\widetilde{K}$. The one-to-one correspondence exists only between subgroups of $\mathfrak{G}$ and finite extension fields of K. Next we will give an example of $K(\mathfrak{G}(L)) \neq L$. Let $\xi_1, \dots, \xi_{2g}$ be $2g$ linearly independent real numbers over the field of rational numbers. Let $w\,dz$ be the differential of the first kind whose real parts of periods with respect to α_i and β_i $(i = 1, \dots, g)$ are given by ξ_i and ξ_{g+i}, i.e.,

$$\xi_i = \operatorname{Re}\left(\int_{\alpha_i} w\,dz\right) \quad \text{and} \quad \xi_{g+i} = \operatorname{Re}\left(\int_{\beta_i} w\,dz\right), \qquad i = 1, \dots, g.$$

Then for an arbitrary closed analytic curve

$$\gamma \sim \sum_{i=1}^{g} n_i \alpha_i + \sum_{i=1}^{g} n_{g+i} \beta_i, \qquad n_i = \text{ rational integer},$$

we have

$$\operatorname{Re}\left(\int_\gamma w\, dz\right) = \sum_{i=1}^{2g} n_i \xi_i .$$

Hence, $\int_\gamma w\, dz = 0$ holds only if $\gamma \sim 0$. Then let u be the additive function in $\widetilde{K}$ obtained from $w\, dz$. Theorem 5.6 implies

$$\sigma(u) = u + \omega_\sigma, \qquad \omega_\sigma = -\int_\gamma w\, dz, \quad \overline{\gamma} = \Psi(\overline{\sigma}). \tag{4.14}$$

Therefore, $\sigma(u) = u$ holds, i.e., $\omega_\sigma = 0$, only when σ belongs to the commutator subgroup $\mathfrak{G}'$ of $\mathfrak{G}$. Hence, we have

$$\mathfrak{G}' = \mathfrak{G}(L),$$

where $L = K(u)$. From (4.14) u cannot be a root of an algebraic equation of finite degree. Hence, $K(u)$ is a simple transcendental extension of K. On the other hand, $\mathfrak{G}/\mathfrak{G}'$ is a free Abelian group generated by $2g$ elements. Therefore, there exists a subgroup $\mathfrak{H}$ such that

$$\mathfrak{G}' \subseteq \mathfrak{H} \subseteq \mathfrak{G} \quad \text{and} \quad 1 < [\mathfrak{G} : \mathfrak{H}] < \infty .$$

Then $K(\mathfrak{H})$ is contained in $K(\mathfrak{G}')$ and $K(\mathfrak{H})/K$ is a finite algebraic extension. Since any intermediate field $L' \neq K$ of the simple transcendental extension $K(u)/K$ must contain a transcendental element for K, $K(\mathfrak{G}')$ must differ from $K(u)$. That is,

$$K(\mathfrak{G}(L)) \neq L .$$

The intermediate field $\overline{K} = K(\mathfrak{G}')$ corresponding to the commutator subgroup $\mathfrak{G}'$ is an important field. The field $\overline{K}$ contains all the additive and multiplicative functions in $\widetilde{K}$, as one can observe from (2.7) and (2.12). Even though the index of $\mathfrak{G}'$ for $\mathfrak{G}$ is infinite, as the proof of part (iii) in Theorem 5.17 shows, $\overline{\mathfrak{G}} = \mathfrak{G}/\mathfrak{G}'$ may be considered as the automorphism group of $\overline{K}/K$. That is, let σ be an arbitrary element in $\mathfrak{G}$. Then for a function u in $\overline{K}$, $\sigma(u)$ depends only upon the residue class $\overline{\sigma}$ of σ in $\mathfrak{G}/\mathfrak{G}'$. Defining $\overline{\sigma}(u) = \sigma(u)$, $\overline{\mathfrak{G}} = \mathfrak{G}/\mathfrak{G}'$ becomes the automorphism group of $\overline{K}/K$. In the above sense, we call $\overline{\mathfrak{G}}$ the Galois group of $\overline{K}/K$.

Let L be any finite extension of K contained in $\overline{K}$, and let $\mathfrak{H} = \mathfrak{G}(L)$. Since $L \subseteq \overline{K}$, we have $\mathfrak{H} = \mathfrak{G}(L) \supseteq \mathfrak{G}(\overline{K}) = \mathfrak{G}(K(\mathfrak{G}')) = \mathfrak{G}'$. Since $\mathfrak{G}/\mathfrak{G}'$ is an Abelian group, $\mathfrak{H}$ containing $\mathfrak{G}'$ is a normal subgroup of $\mathfrak{G}$, and $\mathfrak{G}/\mathfrak{H}$ is also an Abelian group. Hence, from Theorem 5.17, L/K is a Galois extension, and its Galois group $\mathfrak{G}(L/K) \cong \mathfrak{G}/\mathfrak{H}$ is the Abelian group. Namely, L/K is an Abelian extension of K. Conversely, if L/K is a finite

Abelian extension, then $\mathfrak{G}/\mathfrak{H}$ is Abelian. Therefore, we obtain $\mathfrak{H} \supseteq \mathfrak{G}'$ and $L = K(\mathfrak{H}) \subseteq K(\mathfrak{G}') = \overline{K}$. Consequently, all the finite extensions of K contained in $\overline{K}$ are Abelian extensions of finite degree over K. Furthermore, all those finite extensions correspond to the subgroups $\mathfrak{H}$ in $\mathfrak{G}$ of finite indices containing $\mathfrak{G}'$ in a one-to-one fashion and also correspond to the subgroups $\overline{\mathfrak{H}} = \mathfrak{H}/\mathfrak{G}'$ in $\overline{\mathfrak{G}} = \mathfrak{G}/\mathfrak{G}'$ of finite index. According to Theorem 5.9, the topological Abelian group $\overline{\mathfrak{G}}$ forms a dual pair with the divisor class group $\overline{\mathfrak{D}}_0$ of degree 0 of K with respect to the integral character (2.32). Therefore, from the theory of characters of topological Abelian groups, subgroups $\overline{\mathfrak{H}}$ of $\overline{\mathfrak{G}}$ and subgroups $\overline{\mathfrak{B}}$ of $\overline{\mathfrak{D}}_0$ are in a one-to-one correspondence through the orthogonality relation. (See the book by Pontrjagin on topological groups cited earlier for the theory of characters of topological Abelian groups.) We say that $\overline{\mathfrak{H}}$ and $\overline{\mathfrak{B}}$ form an orthogonal pair if the set of elements $\overline{A}$ in $\overline{\mathfrak{D}}_0$ satisfying

$$(\overline{\sigma}, \overline{A}) = 1 \tag{4.15}$$

for every $\overline{\sigma}$ in $\overline{\mathfrak{H}}$ is $\overline{\mathfrak{B}}$, and conversely the totality of $\overline{\sigma}$ satisfying (4.15) for every $\overline{A}$ in $\overline{\mathfrak{B}}$ is $\overline{\mathfrak{H}}$. Then $\overline{\mathfrak{B}}$ is isomorphic to the character group of $\overline{\mathfrak{G}}/\overline{\mathfrak{H}}$, and $\overline{\mathfrak{H}}$ is isomorphic to the character group of $\overline{\mathfrak{D}}_0/\overline{\mathfrak{B}}$. In particular, let $\overline{\mathfrak{H}}$ be a subgroup of finite index in $\overline{\mathfrak{G}}$ corresponding to the Abelian extension L/K. Then

$$\overline{\mathfrak{G}}/\overline{\mathfrak{H}} \cong \mathfrak{G}/\mathfrak{H} \cong \mathfrak{G}(L/K)$$

is a finite group, and its character group is isomorphic to $\overline{\mathfrak{G}}/\overline{\mathfrak{H}}$ itself. Hence

$$\overline{\mathfrak{B}} \cong \mathfrak{G}(L/K) \tag{4.16}$$

is a finite subgroup of $\overline{\mathfrak{D}}_0$. Conversely, an $\overline{\mathfrak{H}}$ that is orthogonal to a finite subgroup $\overline{\mathfrak{B}}$ of $\overline{\mathfrak{D}}_0$ has a finite index in $\overline{\mathfrak{G}}$ since $\overline{\mathfrak{G}}/\overline{\mathfrak{H}} \cong \overline{\mathfrak{B}}$. Thus, there is a one-to-one correspondence between finite Abelian extensions L/K and finite subgroups $\overline{\mathfrak{B}}$ in $\overline{\mathfrak{D}}_0$ such that the corresponding groups are isomorphic.

The corresponding pair L and $\overline{\mathfrak{B}}$ has a more direct relation as follows. Let $\mathfrak{M}_L$ be the set of all multiplicative functions in the strict sense belonging to L. A multiplicative function v belongs to $\mathfrak{M}_L$ if and only if for every element σ in $\mathfrak{H} = \mathfrak{G}(L)$

$$\sigma(v) = v\,.$$

If we let $(v) = A$, then from (2.28) the above is equivalent to

$$(\overline{\sigma}, \overline{A}) = 1\,, \qquad \overline{\sigma} \in \overline{\mathfrak{H}} = \mathfrak{H}/\mathfrak{G}'\,,$$

i.e., $\overline{A}$ is contained in $\overline{\mathfrak{B}}$. Thus, $\overline{\mathfrak{B}}$ is orthogonal to $\overline{\mathfrak{H}}$. Let $\overline{\mathfrak{B}} = \mathfrak{B}/\mathfrak{D}_H$, where $\mathfrak{B} \subseteq \mathfrak{D}_0$. Since $\mathfrak{M}_L$ clearly contains all the nonzero elements of k, the set of divisors (v) of v belonging to $\mathfrak{M}_L$ coincides with $\mathfrak{B}$. We have established the relation between L and $\overline{\mathfrak{B}}$. Moreover, we can show that L is obtained from K by adjoining the elements of $\mathfrak{M}_L$. Let L' be the field obtained from K by adjoining all the elements of $\mathfrak{M}_L$. Since $L' \subseteq L$ clearly holds, the totality of multiplicative functions in the strict sense belonging to

L' is contained in $\mathfrak{M}_L$. On the other hand, since $\mathfrak{M}_L$ is contained in L', we have $\mathfrak{M}_L \subseteq \mathfrak{M}_{L'}$. Hence $\mathfrak{M}_L = \mathfrak{M}_{L'}$ holds. From what we proved above, the subgroups $\overline{\mathfrak{B}}$ and $\overline{\mathfrak{B}}'$ corresponding to L and L', respectively, are the same. Since the correspondence between L and $\overline{\mathfrak{B}}$ is one-to-one, $L = L'$. We have proved the following theorem.

THEOREM 5.19. *Let L be an arbitrary finite Abelian extension of K contained in $\overline{K}$, and let $\mathfrak{B}$ be the totality of the divisors of multiplicative functions in the strict sense in L. Then $\mathfrak{B}$ clearly contains the principal divisor group $\mathfrak{D}_H$ of K, and the subgroup $\overline{\mathfrak{B}} = \mathfrak{B}/\mathfrak{D}_H$ of the divisor class group $\overline{\mathfrak{D}}_0$ of degree 0 for K is a finite group isomorphic to the Galois group $\mathfrak{G}(L/K)$ of L/K. Furthermore, there is a one-to-one correspondence between all the finite Abelian extensions L of K contained in $\widetilde{K}$ and all the finite subgroups $\overline{\mathfrak{B}}$ of $\overline{\mathfrak{D}}_0$*

$$L \leftrightarrow \overline{\mathfrak{B}}.$$

In particular, L is obtained by adjoining all the multiplicative functions in L to K.

(See M. Moriya, *Algebraische Funktionenköper und Riemannsche Flächen*, Jour. Fac. Sci. Hokkaido Univ. Ser. I, **9** (1941) for the general theory of extensions mentioned in this section.)

In the above theorem, we considered only the finite Abelian extensions of K in $\widetilde{K}$. By Theorem 5.17 and part (iii) of Theorem 5.18, any finite unramified Abelian extension of K is isomorphic to a unique L/K in $\widetilde{K}$ over K. Hence, all the finite unramified Abelian extensions in $\widetilde{K}$ correspond in a one-to-one fashion to all the finite groups in $\overline{\mathfrak{D}}_0$. In other words, finite unramified Abelian extensions over K are captured by the divisor class group $\overline{\mathfrak{D}}_0$ through the multiplicative functions. For example, let n be any natural number. Then the number of finite unramified Abelian extensions of degree n over K (up to isomorphism) is the number of subgroups in $\overline{\mathfrak{D}}_0$ of order n. On the other hand, from Theorem 5.9, $\overline{\mathfrak{D}}_0$ is isomorphic to the direct sum of $2g$ copies of $\mathbb{R}^+/\mathbb{Z}^+$. Therefore, the above number can be found easily by a group theoretical consideration. For example, when $n = p$ is a prime number, the number is given by $(p^{2g} - 1)/(p - 1)$. (The reader is encouraged to prove this.) As this example shows, the structure of $\overline{\mathfrak{D}}_0$ is determined by the genus g of K. Hence, the number of finite unramified Abelian extensions over K depends upon only the genus of K. The same thing can be said for any unramified extension if Theorem 5.17 and Theorem 5.18 are considered. Recall that there are infinite nonisomorphic algebraic function fields of the same genus (See the paragraph following Theorem 4.9.). Contrasted with that, the above result is remarkable. Note also that our results on finite unramified Abelian extensions correspond to one of the fundamental results in class field theory. Namely, an arbitrary finite Abelian extension (in particular, a finite unramified Abelian extension) over a given algebraic number field is governed by the ideal-class group (in particular, absolute ideal-class group) of the ground field.

The latter part of Theorem 5.19 can be generalized to the case of a Galois extension using representation functions.

THEOREM 5.20. *Let L/K be an arbitrary finite Galois extension in $\widetilde{K}$ over K. Then L is obtained by adjoining to K all the representation functions belonging to L.*

PROOF. Let $\mathfrak{H} = \mathfrak{G}(L)$. Then $\mathfrak{G}/\mathfrak{H}$ is a finite group isomorphic to the Galois group of L/K. There exists a representation $\{M_\sigma\}$ of $\mathfrak{G}$ such that the kernel of $\{M_\sigma\}$ is $\mathfrak{H}$, i.e., such that the set of all σ satisfying

$$M_\sigma = E, \quad \text{where } E = \text{ identity matrix},$$

coincides with $\mathfrak{H}$. For the $\{M_\sigma\}$, define $Z^* = Z^*(t)$ using Theorem 5.11 so as to satisfy

$$\sigma(Z^*) = Z^* M_\sigma, \qquad \sigma \in \mathfrak{G}.$$

Since $|Z^*| \neq 0$, $\sigma(Z^*) = Z^*$ holds if and only if $M_\sigma = E$, i.e., $\sigma \in \mathfrak{H}$. Then let L' be the field obtained by adjoining all the entries of Z^* to K. We obtain

$$\mathfrak{H} = \mathfrak{G}(L').$$

Therefore, from Theorem 5.17, we obtain

$$L' = K(\mathfrak{G}(L')) = K(\mathfrak{H}) = K(\mathfrak{G}(L)) = L.$$

Since each row of Z^* gives representation functions belonging to $\{M_\sigma\}$, L is clearly the field obtained by adjoining all the representative functions in L to K. Q.E.D.

Let us call the kernel of the representation obtained from representative functions $u_1, \dots, u_n$ of $\widetilde{K}$ for $\mathfrak{G}$ the kernel of u_i. Then, from the above proof, the intersection of the kernels of all the representative functions contained in L is precisely $\mathfrak{H} = \mathfrak{G}(L)$. Hence, a finite Galois extension in $\widetilde{K}$ over K, and in particular, any finite unramified Galois extension, is determined by the representation functions. Since the representation functions themselves are not functions in K, the above result is not as strong as our earlier one: by Theorem 5.19, a finite unramified Abelian extension of K is determined by the group $\overline{\mathfrak{D}}_0$ in K. In the case of an Abelian extension, we have not only the relationship between L and the multiplicative functions in the strict sense, but from Theorem 5.8 we also have the correspondence between the multiplicative functions in the strict sense and the divisors of degree 0 of K. This fact made it possible to relate L to the group $\overline{\mathfrak{D}}_0$ in K. Based on the above observation, A. Weil extended the notion of divisors of K, related them to representation functions and generalized the correspondence in Theorem 5.8 between divisor classes and the representations of $\mathfrak{G}$. Weil succeeded in capturing all the unramified Galois extensions of K by the set of generalized divisor classes in K. See A. Weil, *Généralisation des fonctions abéliennes*, Jour. Math. Pure et Appl., **17** (1938). Weil also noticed that from Theorem 5.9, as a set the Jacobian variety $J = \overline{\mathfrak{D}}_0$ coincides with the set of all characters of $\overline{\mathfrak{G}}$, and that the Abelian function field belonging to K is nothing but the analytic function field defined on the set of those characters. Using general representations, as a generalization of characters of $\mathfrak{G}$, Weil defined the *hyperabelian function field* as the analytic functions on the space of representation classes as points. There are yet many problems to be solved in this area. See H. Tōyama, *On a non-Abelian theory of algebraic functions*, Kōdai Math. Sem. Report, **2** (1949).

So far, we have concentrated on unramified extension fields. Our discussion has been based on Theorem 5.17 where the unramified extension is considered as an intermediate field of $\widetilde{K} = K(\mathfrak{R}^*)$ and K. We can extend our results to the case where an extension of K has ramified points. Let $P_1, \dots, P_l$ be arbitrary places on K, and let $n_1, \dots, n_l$ be arbitrary natural numbers. Assume that every extension Q_{ij} of P_i to an extension field L of K has ramification index divisible by n_i, and that all places of L other than Q_{ij} are unramified. Then such an L corresponds, in the sense of Theorems 5.17, 5.18, to a subgroup $\mathfrak{H}$ of the Galois group $\mathfrak{G}$ of $\widetilde{K} = K(\mathfrak{R}^*)$ for K defined by a group G of motions (including transformations with fixed points) of $\mathfrak{R}_1^* = \{t; |t| < 1\}$ determined by P_i and n_i. (We assume that at least one n_i is not 1. If $n_1 = n_2 = \cdots = n_l = 1$, then it is the unramified extension case in the above.) In other words, even if there are branch points, by specifying the ramifiedness, such an extension can be considered as an intermediate field between K and a certain analytic function field $\widetilde{K}$. In fact, while Weil generalized the notion of divisors, he also extended the results from the unramified case to the ramified case. See Weil's paper cited earlier.

§5. Elliptic function fields

We do not intend to give a thorough description of the grand classical theory of elliptic functions. We attempt to clarify the general theory through the simpler and explicit case of genus 1, and to explain the peculiarity of its being of genus one. Hence, we will not cover some of the important results for elliptic functions. For example, we will not treat Jacobi's elliptic functions.

Let K be an elliptic function field over the field of complex numbers, i.e., an algebraic function field of genus 1 over k. A nonconstant element in K, considered as an analytic function, is called an *elliptic function*. The set $\mathfrak{L}_0$ of differentials of the first kind is one dimensional over k. Let $w\,dz$ ($\neq 0$) be an arbitrary differential. Then from Theorem 5.14, for an arbitrary complex number ζ there exists a unique P on $\mathfrak{R} = \mathfrak{R}(K)$ satisfying

$$\zeta \equiv \int_{P_0}^{P} w\,dz \mod \mathrm{P}(w\,dz). \tag{5.1}$$

See the last paragraph of §3 for the uniqueness. Note that P_0 is a fixed point on $\mathfrak{R}$ and $\mathrm{P}(w\,dz)$ is the period group of $w\,dz$. For basis elements α_1 and β_1 for the first homology group $B_1(\mathfrak{R})$ of $\mathfrak{R}$, let

$$\omega_1 = \int_{\alpha_1} w\,dz \quad \text{and} \quad \omega_2 = \int_{\beta_1} w\,dz.$$

Then we have

$$\mathrm{P}(w\,dz) = \{m_1\omega_1 + m_2\omega_2 ;\, m_1, m_2 = 0, \pm 1, \pm 2, \dots\}.$$

As in §2, let $\mathfrak{R}^*$ be a simply connected covering Riemann surface of $\mathfrak{R}$ and let $u = u(t)$ be an integrating function of $w\,dz$. Then we can rephrase the above as follows. For an arbitrary complex number ζ there exists a unique t on $\mathfrak{R}^*$ such that

$$u(t) = \zeta. \tag{5.2}$$

Since $u(t)$ is a regular function on $\mathfrak{R}^*$, by (5.2) $\mathfrak{R}^*$ is conformally mapped onto the finite u-plane $\mathfrak{R}^*_\infty = V_1(k)$ in a one-to-one fashion. Since the Jacobian variety J of K is isomorphic to the analytic manifold obtained from $V_1(k)$ by identifying the equivalent points for $\mathrm{P}(w\,dz)$, J is, in this case, isomorphic to $\mathfrak{R}^*_\infty(\omega_1, \omega_2)$ in Theorem 3.19. Let f^* be the covering map from $\mathfrak{R}^*$ to $\mathfrak{R}$ and let f^*_∞ be the covering map from $\mathfrak{R}^*_\infty$ to $\mathfrak{R}^*_\infty(\omega_1, \omega_2)$. Denote by φ the one-to-one correspondence in (5.1) between a point P on $\mathfrak{R}$ and a point on $\mathfrak{R}^*_\infty(\omega_1, \omega_2)$ with the coordinate ζ. Then we have

$$\varphi(f^*(t)) = f^*_\infty(u(t)).$$

Therefore, by Theorem 3.16 φ is an isomorphism from $\mathfrak{R}$ to $\mathfrak{R}^*_\infty(\omega_1, \omega_2)$ and we have

$$\{\mathfrak{R}^*, f^*; \mathfrak{R}\} \cong \{\mathfrak{R}^*_\infty, f^*_\infty; \mathfrak{R}^*_\infty(\omega_1, \omega_2)\}. \tag{5.3}$$

Conversely, if $\mathfrak{R}$ is isomorphic to the canonical type $\mathfrak{R}^*_\infty(\omega_1, \omega_2)$, then it is clearly a closed Riemann surface of genus one. Hence, we have obtained the following theorem.

THEOREM 5.21. *The closed Riemann surface $\mathfrak{R}$ of an elliptic function field K is isomorphic to a canonical type $\mathfrak{R}^*_\infty(\omega_1, \omega_2)$ in Theorem* 3.19. *Conversely, if $\mathfrak{R} \cong \mathfrak{R}^*_\infty(\omega_1, \omega_2)$ holds, then $K = K(\mathfrak{R})$ is an elliptic function field. Moreover, an Abelian function field $\mathfrak{K}$ belonging to K is an algebraic function field that is isomorphic to K.*

By this theorem, the Abelian function field belonging to an elliptic function field does not provide anything new, but it coincides with the original elliptic function field. This is one of the major differences between the case of genus one and the case of genus $g \geq 2$.

Namely, for the study a general elliptic function field K, choose any complex numbers ω_1 and ω_2 satisfying $\operatorname{Im}(\omega_2/\omega_1) > 0$ and let

$$\mathfrak{R} = \mathfrak{R}^*_\infty(\omega_1, \omega_2).$$

Then, by Theorem 5.21, we can study the analytic function field $K(\mathfrak{R})$. For a simply connected covering Riemann surface of $\mathfrak{R}$, we can choose the finite u-plane $\mathfrak{R}^*_\infty$. As in §2, for a function $z = z(P)$ in $K = K(\mathfrak{R})$, let

$$z(u) = z(P_u), \quad \text{where } P_u = f^*(u), \ u \in \mathfrak{R}^*_\infty.$$

Then we can regard z as a function in $\widetilde{K} = K(\mathfrak{R}^*)$. Then K is the set of all functions in $\widetilde{K}$ that are invariant under the Galois group $\widetilde{\mathfrak{G}}$ of $\widetilde{K}/K$. (See the paragraph following (4.16).) In this case, the fundamental group for $\mathfrak{R}^*$ of $\mathfrak{R}$ is

$$G(\omega_1, \omega_2) = \{\varphi(u) = u + m_1\omega_1 + m_2\omega_2 ; m_1, m_2 = 0, \pm 1, \pm 2, \dots\}.$$

Hence, K can be characterized as those analytic functions on the finite u-plane that satisfy

$$z(u+\omega) = z(u), \quad \omega = m_1\omega_1 + m_2\omega_2, \quad m_1, m_2 = 0, \pm 1, \pm 2, \dots. \tag{5.4}$$

Next we will explicitly determine $z(u)$ satisfying (5.4).

LEMMA 5.6. *Let μ be any real number greater than 2. Then we have*

$$\sum{}' \frac{1}{|\omega|^\mu} < \infty, \tag{5.5}$$

where $\sum'$ is the sum for all ω, excluding 0, in

$$\mathrm{P}(\omega_1, \omega_2) = \{\omega = m_1\omega_1 + m_2\omega_2 ; m_1, m_2 = 0, \pm 1, \pm 2, \dots\}.$$

PROOF. Since ω_1 and ω_2 satisfy $\operatorname{Im}(\omega_1/\omega_2) > 0$, $P(\omega_1, \omega_2)$ forms a lattice in the complex plane as shown in Figure 5.4. Let $L_1, L_2, \dots$ be similar parallelograms indicated in bold lines in Figure 5.4 with the center the origin so that L_m contains $8m$ elements of $\mathrm{P}(\omega_1, \omega_2)$. Let r be the shortest distance from 0 to L_1. Then the shortest distance from 0 to L_m is mr. Therefore, the sum of $1/|\omega|^\mu$ over ω on L_m does not exceed

$$\frac{8m}{(mr)^\mu} = \frac{8}{r^\mu} \cdot \frac{1}{m^{\mu-1}}.$$

Hence, for $\mu - 1 > 1$, we have

$$\sum{}' \frac{1}{|\omega|^\mu} \le \sum_{m=1}^{\infty} \frac{8}{r^\mu} \cdot \frac{1}{m^{\mu-1}} < \infty.$$

Next consider the series

$$\sum_{\omega} \frac{1}{(u-\omega)^l}, \qquad \omega \in \mathrm{P}(\omega_1, \omega_2) \tag{5.6}$$

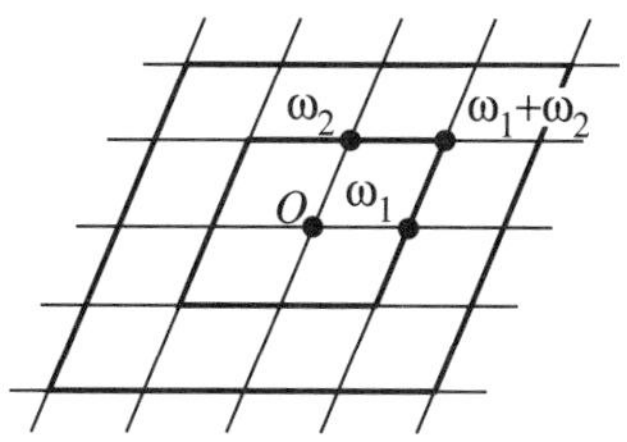

FIGURE 5.4

for a natural number $l \geq 3$. Let c be any positive real number and let $|u| \leq c$. Then for ω satisfying $|\omega| \geq 2c$ we have

$$\left|\frac{u-\omega}{\omega}\right| \geq 1 - \left|\frac{u}{\omega}\right| \geq \frac{1}{2},$$

i.e.,

$$\left|\frac{1}{u-\omega}\right| \leq \frac{2}{|\omega|}. \tag{5.7}$$

From (5.5)

$$\sum_{|\omega|\geq 2c} \frac{1}{|u-\omega|^l} \leq 2^l \sum_{|\omega|\geq 2c} \frac{1}{|\omega|^l} < \infty.$$

Hence, $\sum_{|\pi|\geq 2c} 1/(u-\omega)^l$ converges absolutely and uniformly for $|u| \leq c$, defining a regular function of u for $|u| \leq c$. There are only finitely many $\omega \in \mathrm{P}(\omega_1, \omega_2)$ with $|\omega| < 2c$. Since we took c to be arbitrary, series (5.6) has a pole of order l at $u = \omega$ and defines a function in $\widetilde{K}$ which is regular everywhere else. Such a function clearly satisfies (5.4). Thus, we can obtain many functions belonging to K. In particular,

$$\mathfrak{p}'(u) = -2\sum_{\omega} \frac{1}{(u-\omega)^3}$$

is a function in K. Next integrate the right-hand side term by term from 0 to u. However, we integrate the term for $\omega = 0$ from ∞ to u. Namely,

$$\begin{aligned} \mathfrak{p}(u) &= \int_\infty^u \frac{-2}{u^3}\,du + \sum_{\omega}{}' \int_0^u \frac{-2}{(u-\omega)^3}\,du \\ &= \frac{1}{u^2} + \sum_{\omega}{}' \left(\frac{1}{(u-\omega)^2} - \frac{1}{\omega^2}\right). \end{aligned} \tag{5.8}$$

By the theorem about the term by term integration of a convergent series, $\mathfrak{p}(u)$ is also an analytic function on $\mathfrak{R}^*$. As the notation suggests, $\mathfrak{p}'(u)$ is equal to the derivative $d\mathfrak{p}/du$ of $\mathfrak{p}(u)$. Furthermore,

$$\begin{aligned} \mathfrak{p}(-u) &= \frac{1}{(-u)^2} + \sum_{\omega}{}' \left(\frac{1}{(-u-\omega)^2} - \frac{1}{\omega^2}\right) \\ &= \frac{1}{u^2} + \sum_{-\omega}{}' \left(\frac{1}{(u-(-\omega))^2} - \frac{1}{(-\omega)^2}\right). \end{aligned}$$

Since $-\omega$, $\omega \neq 0$, runs over all the nonzero elements in $\mathrm{P}(\omega_1, \omega_2)$ without repeating, the right-hand side equals $\wp(u)$, i.e.,

$$\wp(-u) = \wp(u). \tag{5.9}$$

On the other hand, we have

$$\frac{d}{du}(\wp(u+\omega_1) - \wp(u)) = \wp'(u+\omega_1) - \wp'(u) = 0.$$

Hence, $c = \wp(u+\omega_1) - \wp(u)$ is a constant. Let $u = -\omega_1/2$. Since $u = -\omega_1/2$ is not a pole of $\wp(u)$, (5.9) implies

$$c = \wp\left(\frac{\omega_1}{2}\right) - \wp\left(\frac{\omega_1}{2}\right) = 0,$$

i.e.,

$$\wp(u+\omega_1) = \wp(u).$$

Similarly, we obtain $\wp(u+\omega_2) = \wp(u)$. Hence, generally

$$\wp(u+\omega) = \wp(u), \quad \omega = m_1\omega_1 + m_2\omega_2, \quad m_1, m_2 = 0, \pm 1, \pm 2, \ldots.$$

That is, $\wp(u)$ also belongs to K. The function $\wp(u)$ is called the *Weierstrass* $\wp$-*function.*

From (5.8), $\wp(u)$ has only poles of order 2 at $u = \omega$. Hence, as a function on $\mathfrak{R}$, $\wp(u)$ is a function having only a pole of order 2 at the point $P_0 = f^*(\omega) = f^*(0)$. Similarly, $\wp'(u)$ has only a pole of order 3 at P_0. Therefore, $[K : k(\wp)] = 2$ and $[K : k(\wp')] = 3$ imply

$$K = k(\wp, \wp'),$$

since $[K : k(\wp, \wp')]$ divides $[K : k(\wp)]$ and $[K : k(\wp')]$, so that $[K : k(\wp, \wp')] = 1$. (See the paragraphs following Definition 2.17.)

Next we will find an irreducible equation over k which relates $\wp$ and $\wp'$. First, expand $\wp(u)$ and $\wp'(u)$ in a neighborhood of $u = 0$ using (5.8) to find

$$\begin{aligned} \wp(u) &= \frac{1}{u^2} + c_2u^2 + c_3u^4 + \cdots + c_nu^{2n-2} + \cdots, \\ \wp'(u) &= -\frac{2}{u^3} + 2c_2u + 4c_3u^3 + \cdots + (2n-2)c_nu^{2n-3} + \cdots, \end{aligned} \tag{5.10}$$

where

$$c_n = (2n-1)\sum_{\omega}{}' \frac{1}{\omega^{2n}}, \qquad n \geq 2.$$

Then

$$\wp'(u)^2 - 4\wp(u)^3 + 20c_2\wp(u) = -28c_3 + \cdots,$$

where the right-hand side is a power series in u beginning $-28c_3$ and is regular at $u = 0$. Hence, the left-hand side, as a function in K, is regular everywhere in $\mathfrak{R}$ and is consequently constant by Theorem 2.7. We obtain

$$\wp'(u)^2 = 4\wp(u)^3 - 20c_2\wp(u) - 28c_3.$$

Following Weierstrass' notation, let

$$\begin{aligned}\gamma_2 &= 20c_2 = 60\sum{}' \frac{1}{\omega^4} \\ \gamma_3 &= 28c_3 = 140\sum{}' \frac{1}{\omega^6}.\end{aligned} \tag{5.11}$$

Then we have

$$\mathfrak{p}'^2 = 4\mathfrak{p}^3 - \gamma_2\mathfrak{p} - \gamma_3.$$

This is precisely the Weierstrass canonical form found in Chapter 2. Then substituting (5.11) into the above, we find that the invariant δ in Theorem 2.30

$$\delta = \frac{\gamma_2^3}{\gamma_2^3 - 27\gamma_3^2}$$

is a homogeneous function of (ω_1, ω_2) of degree 0. That is,

$$\delta(\lambda\omega_1, \lambda\omega_2) = \delta(\omega_1, \omega_2), \qquad \lambda \in k, \ \lambda \neq 0.$$

Consequently, δ is a function of $\tau = \omega_1/\omega_2$, where $\operatorname{Im}(\tau) > 0$. Let

$$j(\tau) = \delta(1, \omega_2/\omega_1) = \delta(\omega_1, \omega_2).$$

Then $j(\tau)$ is a function relating the invariant δ, determining the elliptic function field K, with ω_1 and ω_2, determining the closed Riemann surface $\mathfrak{R}$ of K. Note that ω_1 and ω_2 are arbitrary complex numbers satisfying $\operatorname{Im}(\omega_2/\omega_1) > 0$. Hence, $j(\tau)$ is defined in the upper half plane $\operatorname{Im}(\tau) > 0$ of the τ-plane. As we showed in §7, Chapter 2, for any complex number δ there exists an elliptic function field having δ as its invariant. The function $j(\tau)$ takes on all complex values for $\operatorname{Im}(\tau) > 0$. On the other hand, from Theorem 3.19, $\mathfrak{R}_\infty^*(\omega_1, \omega_2)$ and $\mathfrak{R}_\infty^*(\omega_1', \omega_2')$ are isomorphic if and only if, for $\tau = \omega_2/\omega_1$ and $\tau' = \omega_2'/\omega_1'$,

$$\tau' = \frac{c + d\tau}{a + b\tau}, \quad ad - bc = 1, \qquad a, b, c, d = \text{ rational integers}. \tag{5.12}$$

Whereas from Theorem 4.5, we have $\mathfrak{R}_\infty^*(\omega_1, \omega_2) \cong \mathfrak{R}_\infty^*(\omega_1', \omega_2')$ if and only if their elliptic function fields K and K' are isomorphic, and by Theorem 2.30, K and K' are isomorphic if and only if their invariants coincide. As a consequence, for complex numbers τ and τ' satisfying $\operatorname{Im}(\tau) > 0$ and $\operatorname{Im}(\tau') > 0$,

$$j(\tau) = j(\tau')$$

holds if and only if (5.12) holds for τ and τ'.

The totality G^* of linear transformations of τ defined in (5.12) for rational integers a, b, c, d satisfying $ad - bc = 1$ forms a group, called the *modular group*. A transformation such as in (5.12) is called a *modular transformation*. Since any modular transformation maps the upper half plane to itself, G^* may be considered as a topological transformation group of the

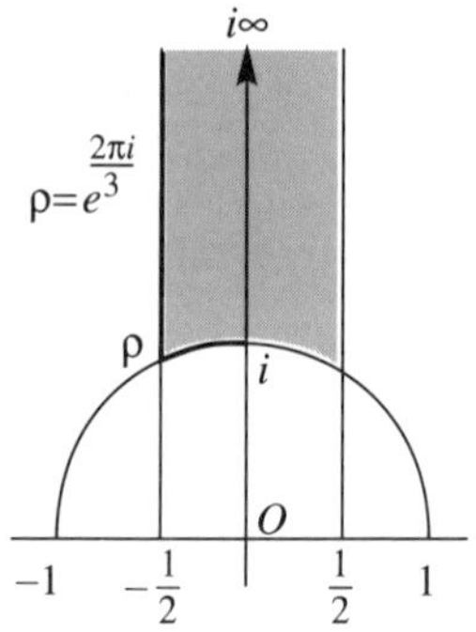

FIGURE 5.5

upper half plane. Then G^* is a discontinuous transformation group (with fixed points) in the sense described in the paragraph preceding Lemma 3.9 in §5, Chapter 3. One can prove that the shaded part of Figure 5.5 is a fundamental region. Hence, by the above results, for τ in the fundamental region and on the bold lined boundary in Figure 5.5, $j(\tau)$ takes on all complex values without repeating any values. One can also prove that $j(\tau)$ is a regular function of τ for $\mathrm{Im}(\tau) > 0$. In general, an analytic function $f(\tau)$ which is invariant under all the modular transformations satisfying a certain condition in a neighborhood of ∞ is called a modular function. The above $j(\tau)$ is a modular function, and one can prove that an arbitrary modular function can be written as a rational expression of $j(\tau)$. A modular function is essentially an analytic function on the closed Riemann surface of genus 0 obtained by identifying some boundaries of the above fundamental region including ∞. Then this Riemann surface is mapped isomorphically by $j = j(\tau)$ onto the complex j-plane (the Riemann sphere). We will not go into detail on this topic.

Let ω_1 and ω_2 satisfy $\mathrm{Im}(\omega_2/\omega_1) > 0$ as before. We will study the elliptic function field $K = K(\mathfrak{R})$ of $\mathfrak{R} = \mathfrak{R}^*_\infty(\omega_1, \omega_2)$. From the above, we have

$$K = k(\mathfrak{p}, \mathfrak{p}') \quad \text{and} \quad [K : k(\mathfrak{p})] = 2.$$

Therefore, any element $z = z(P)$ in K can be written as

$$z = r_1(\mathfrak{p}) + r_2(\mathfrak{p})\mathfrak{p}', \tag{5.13}$$

where r_1 and r_2 are rational expressions with complex coefficients. By differentiating (5.9) we obtain

$$\mathfrak{p}'(-u) = -\mathfrak{p}'(u). \tag{5.14}$$

Replacing u with $-u$ in $z(u)$ in (5.13), we have

$$z(-u) = r_1(\mathfrak{p}(u)) - r_2(\mathfrak{p}(u))\mathfrak{p}'(u).$$

Hence,

$$r_1(\mathfrak{p}(u)) = \frac{1}{2}(z(u) + z(-u)) \quad \text{and} \quad r_2(\mathfrak{p}(u)) = \frac{1}{2\mathfrak{p}'(u)}(z(u) - z(-u)).$$

That is, r_1 and r_2 in (5.13) are determined by $z(u)$. In particular, if $z = z(u)$ is an even function, i.e., $z(-u) = z(u)$, then $z = r_1(\mathfrak{p})$. If $z = z(u)$ is an odd function, i.e., $z(-u) = -z(u)$, then $z = r_2(\mathfrak{p})\mathfrak{p}'$. For example, denoting the nth derivative of $\mathfrak{p}(u)$ with respect to u by $\mathfrak{p}^{(n)}(u)$, $z = \mathfrak{p}^{(n)}(z)$ is an even function for n even, and $z = \mathfrak{p}^{(n)}(z)$ is an odd function for n odd.

For an arbitrary complex number a, $z(u) = \mathfrak{p}(u + a)$ clearly satisfies (5.4). Namely, $z(u) = \mathfrak{p}(u + a)$ belongs to K. Hence, $z(u) = \mathfrak{p}(u + a)$ can be expressed in terms of $\mathfrak{p}(u)$ and $\mathfrak{p}'(u)$. To find the explicit form we consider the function

$$\varphi(u, a) = \frac{1}{2}\frac{\mathfrak{p}'(u) - \mathfrak{p}'(a)}{\mathfrak{p}(u) - \mathfrak{p}(a)}. \tag{5.15}$$

We assume $P_a \neq P_0$, i.e., $a \neq 0 \mod \mathrm{P}(\omega_1, \omega_2)$. In a neighborhood of $u = 0$ we have

$$\varphi(u, a) = \frac{1}{2}\frac{-2/u^3 - \mathfrak{p}'(a) + 2c_2u + \cdots}{1/u^2 - \mathfrak{p}(a) + c_2u^2 + \cdots} = -\frac{1}{u} - \mathfrak{p}(a)u + \cdots. \tag{5.16}$$

Then $\varphi(u, a)$ has a pole of order 1 at P_0. Another possible pole of $\varphi(u, a)$ can take place only at $P = P_u$, where

$$\mathfrak{p}(u) - \mathfrak{p}(a) = 0. \tag{5.17}$$

It is clear that $u = a$ and $u = -a$ are solutions of (5.17) (see (5.9)). Theorem 4.7 implies that there are $2 = [K : k(\mathfrak{p})]$ solutions on $\mathfrak{R}$ for (5.17). Hence, if $P_a \neq P_{-a}$, then P_a and P_{-a} are all the solutions of (5.17), while P_a and P_{-a} are zeros of $\mathfrak{p}(u) - \mathfrak{p}(a)$ of order 1. Since $u = a$ satisfies $\mathfrak{p}'(u) - \mathfrak{p}(a) = 0$, $P = P_a$ is a point where $\varphi(u, a)$ is regular. On the other hand, the expansion at $u = -a$ is given by

$$\begin{aligned}\frac{1}{2}\frac{\mathfrak{p}'(u) - \mathfrak{p}'(a)}{\mathfrak{p}(u) - \mathfrak{p}(a)} &= \frac{1}{2}\frac{(\mathfrak{p}'(-a) - \mathfrak{p}'(a)) + \mathfrak{p}''(-a)(u - a) + \cdots}{(\mathfrak{p}(-a) - \mathfrak{p}(a)) + \mathfrak{p}'(-a)(u + a) + \cdots} \\ &= \frac{1}{2}\frac{2\mathfrak{p}'(-a) + \mathfrak{p}''(-a)(u + a) + \cdots}{\mathfrak{p}'(-a)(u + a) + \mathfrak{p}''(-a)(u + a)^2/2 + \cdots} \\ &= \frac{1}{u + a} + b_1(u + a) + \cdots.\end{aligned} \tag{5.18}$$

Thus, $\varphi(u, a)$ has poles of order 1 only at P_0 and P_{-a}.

In the above, we assumed $P_a \neq P_{-a}$, but $P_a = P_{-a}$ can happen only when (excluding P_0)

$$a \equiv \frac{\omega_1}{2}, \frac{\omega_2}{2}, \frac{\omega_1 + \omega_2}{2} \mod \mathrm{P}(\omega_1, \omega_2).$$

Since $\wp'(\omega_1/2) = \wp'(\omega_1/2 - \omega_1) = \wp'(-\omega_1/2) = -\wp'(\omega_1/2)$ and etc. hold, we have

$$\wp'\left(\frac{\omega_1}{2}\right) = \wp'\left(\frac{\omega_2}{2}\right) = \wp'\left(\frac{\omega_1+\omega_2}{2}\right) = 0. \tag{5.19}$$

Hence, P_a is a zero of $\wp(u)-\wp(a)$ of order 2. Therefore, in this case also, P_a and P_{-a} are the only solutions for (5.18). Since $\wp'''(u)$ is an odd function, as before we have $\wp'''(\omega_1/2) = 0$. Therefore, one can verify that the expansion of $\varphi(u, a)$ at $u = -a$ is given as in (5.18).

The function

$$f(u) = \wp(u+a) + \wp(u) - \varphi(u, a)^2$$

belongs to K, and possibly has poles at P_0 and P_{-a}. For instance, from (5.10) and (5.16) at $u = 0$, we have

$$\begin{aligned} f(u) &= \wp(a) + \wp'(a)u + \cdots + \frac{1}{u^2} + c_2u^2 + \cdots - \frac{1}{u^2} - 2\wp(a) + \cdots \\ &= -\wp(a) + \cdots . \end{aligned} \tag{5.20}$$

Hence, $f(u)$ is regular at $u = 0$. Similarly, one can show from (5.18) that $f(u)$ is regular at $u = -a$. Consequently, $f(u)$ in K is regular for all the points on $\mathfrak{R}$, so that $f(u)$ is a constant. By (5.20), this constant is clearly $-\wp(a)$. Therefore, we have

$$f(u) = \wp(u+a) + \wp(u) - \varphi(u, a)^2 = -\wp(a).$$

Replacing a by v, we can rewrite the above as

$$\wp(u+v) = -\wp(u) - \wp(v) + \frac{1}{4}\left(\frac{\wp'(u) - \wp'(v)}{\wp(u) - \wp(v)}\right)^2 . \tag{5.21}$$

Even though we assumed, in the middle of the above proof, that $P_a \neq P_0$, (5.21) is a symmetric equation in u and v. Hence, we can eliminate this assumption. However, if $u = 0$ or $v = 0$, we need to consider the expansions of the functions in (5.21) in a neighborhood of $u = 0$ or $v = 0$.

Equation (5.21) is called the *addition theorem* of the $\wp$-function. As in the earlier remark, in the case of an elliptic function field, the Abelian function field of K coincides essentially with K. (5.21) above is a special case of the addition theorem of Abelian functions in (3.36), i.e., a property of the elliptic function field as an Abelian function field.

Letting v converge to u in (5.21), we get

$$\wp(2u) = -2\wp(u) + \frac{1}{4}\left(\frac{\wp''(u)}{\wp'(u)}\right)^2 . \tag{5.22}$$

Since the right-hand side of (5.22) equals

$$-2\wp(u) + \frac{1}{4}\left(\frac{d}{du}\log \wp'(u)\right)^2 ,$$

using $\mathfrak{p}'^2 = 4\mathfrak{p}^3 - \gamma_2\mathfrak{p} - \gamma_3$, we obtain

$$\mathfrak{p}(2u) = -2\mathfrak{p}(u) + \left(\frac{12\mathfrak{p}(u)^2 - \gamma_2}{4\mathfrak{p}(u)^3 - \gamma_2\mathfrak{p}(u) - \gamma_3}\right)^2. \tag{5.23}$$

We will investigate differentials in K. As we described in §7, Chapter 2, the set $\mathfrak{L}_0$ of all differentials of the first kind in K is given by

$$a\frac{d\mathfrak{p}}{\mathfrak{p}'}, \qquad a \in k.$$

Since $\mathfrak{p}' = d\mathfrak{p}/du$, the above becomes

$$a\,du, \qquad a \in k.$$

Therefore, the set of integrating functions of the differentials of the first kind on $\mathfrak{R}^* = \mathfrak{R}^*_\infty$ is given as

$$au + b, \qquad a, b \in k.$$

In particular, u itself is an integrating function. Then let

$$z = \mathfrak{p}(u), \qquad w = \frac{1}{\mathfrak{p}'(u)} = \frac{1}{\sqrt{4z^3 - \gamma_2 z - \gamma_3}}.$$

Then we have

$$u = \int_{P_0}^{P} w\,dz = \int_{P_0}^{P} \frac{dz}{\sqrt{4z^3 - \gamma_2 z - \gamma_3}}, \qquad P = P_u = f^*(u). \tag{5.24}$$

The upper and lower limits of the integral on the right-hand side may be replaced by $z = \mathfrak{p}(u)$ and $\infty = \mathfrak{p}(0)$, respectively. With this replacement, (5.24) is called an *elliptic integral of the first kind*. The problem: for a given u find $z = \mathfrak{p}(u)$ satisfying (5.24), i.e., to find the inverse function of the elliptic integral is Jacobi's Umkehrproblem (see the last paragraph of §3). Jacobi's theorem (Theorem 5.14) confirms the existence of a solution for (5.24) for any u. But in this case, this is obvious from $P = f^*(u)$. Note that we began with $\mathfrak{p}(u)$ to derive an elliptic integral so that the Umkehrproblem is clearly solved. During the time of Abel and Jacobi, the theory began with the elliptic integral, regarding an elliptic function as the inverse function of the elliptic integral. This problem must have been a hard problem to solve. See Preface. Also from Theorem 5.14, for the period group $\mathrm{P}(w\,dz)$ of $w\,dz = du$, there is a one-to-one correspondence through (5.24) between residue classes $\overline{u} \bmod \mathrm{P}(w\,dz)$ containing u and points P on $\mathfrak{R}$. Since f^* is the covering map from $\mathfrak{R}^*_\infty$ to $\mathfrak{R} = \mathfrak{R}^*_\infty(\omega_1, \omega_2)$, $f^*(u) = f^*(u')$ holds if and only if

$$u \equiv u' \mod \mathrm{P}(\omega_1, \omega_2).$$

Hence,

$$\mathrm{P}(w\,dz) = \mathrm{P}(\omega_1, \omega_2)$$

must hold. Using the above, we can rephrase Abel's theorem for K as follows. Namely, a divisor of K of degree 0

$$A = \frac{P_{a_1} \cdots P_{a_r}}{P_{b_1} \cdots P_{b_r}}$$

is a principal divisor of K if and only if

$$a_1 + \cdots + a_r \equiv b_1 + \cdots + b_r \mod \mathrm{P}(\omega_1, \omega_2). \tag{5.25}$$

For example, since $\mathfrak{p}'(u)$ has a pole at P_0 of order 3, and has zeros of order 1 at $P_{\omega_1/2}$, $P_{\omega_2/2}$, $P_{(\omega_1+\omega_2)/2}$ (see (5.19)), we have

$$\frac{\omega_1}{2} + \frac{\omega_2}{2} + \frac{\omega_1 + \omega_2}{2} \equiv 0 + 0 + 0 \mod \mathrm{P}(\omega_1, \omega_2).$$

Next we will consider differentials of the second kind in K. Since $\mathfrak{p}(u)$ and $\mathfrak{p}'(u)$ have only poles of order 2 and 3, respectively, at $P = P_0$, one can construct a function, as a product of $\mathfrak{p}(u)$ and $\mathfrak{p}'(u)$, having a pole only at P_0 of arbitrary order (≥ 2). Then we get the following differentials of the second kind having the only pole at P_0:

$$\mathfrak{p}\, du, \mathfrak{p}\, du, \mathfrak{p}^2\, du, \dots,$$

where $du = d\mathfrak{p}/\mathfrak{p}'$. It is clear that we can obtain differentials of the second kind, having a pole at a general point P_a, by replacing $\mathfrak{p}(u)$ and $\mathfrak{p}'(u)$ with $\mathfrak{p}(u-a)$ and $\mathfrak{p}'(u-a)$. By Theorem 5.6, one can obtain additive functions in $\widetilde{K}$ from the integrating functions of the above differentials. The integrating function of $-\mathfrak{p}\, du$ is particularly important. With a proper constant added, we define an integrating function $\zeta(u)$ of $-\mathfrak{p}\, du$ by

$$\zeta(u) = \frac{1}{u} - \int_0^u \sum{}' \left(\frac{1}{(u-\omega)^2} - \frac{1}{\omega^2} \right) du = \frac{1}{u} + \sum{}' \left(\frac{1}{u-\omega} + \frac{1}{\omega} + \frac{u}{\omega^2} \right).$$

Notice $d\zeta/du = -\mathfrak{p}(u)$. Since $\zeta(u)$ is an additive function, let

$$\zeta(u + \omega_1) = \zeta(u) + \eta_1 \quad \text{and} \quad \zeta(u + \omega_2) = \zeta(u) + \eta_2, \qquad \eta_1, \eta_2 \in k.$$

Let $\overline{V}_4$ be the parallelogram with vertices $(\pm\omega_1 \pm \omega_2)/2$ in the u-plane. Then the image of $\overline{V}_4$ under f^* provides a standard subdivision of $\mathfrak{R}$. Let α_1, β_1, α_1^{-1}, β_1^{-1} be the images of sides I, II, III, IV in Figure 5.6 under f^*. Then the periods of $du = d\mathfrak{p}/\mathfrak{p}'$ for α_1 and β_1 are clearly ω_1 and ω_2, respectively. Furthermore, the periods of $-\mathfrak{p}\, du$ for α_1 and β_1 are the periods η_1 and η_2 of $\zeta(u)$, respectively. Then apply Theorem 5.5 to du and $\mathfrak{p}\, du$. Note that $-\mathfrak{p}\, du$ has a pole only at P_0, and in a neighborhood of P_0 we have the expansion

$$-\mathfrak{p}\, du = -\left(\frac{1}{u^2} + \cdots \right) du.$$

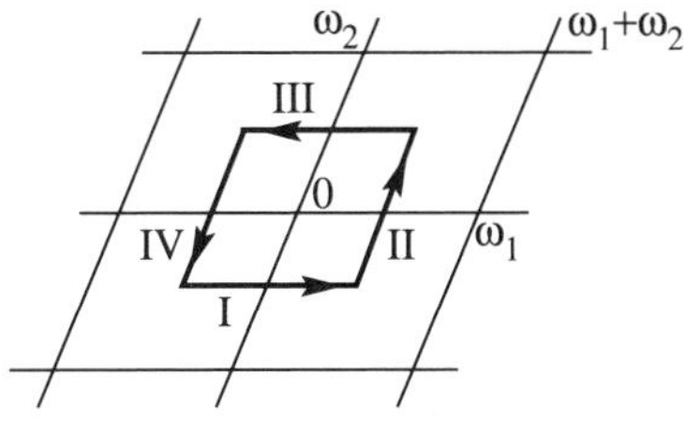

FIGURE 5.6

From (1.21) we have
$$\eta_1\omega_2 - \eta_2\omega_1 = 2\pi\sqrt{-1}. \tag{5.26}$$
We can obtain (5.26) as well by integrating $\zeta(u)\,du$ along the boundary of $\overline{V}_4$. Relation (5.26) is called the *Legendre relation*. Weierstrass extended this relation to the case of a general algebraic function field.

Let $h(u)$ be an arbitrary additive function in $\widetilde{K}$ and let
$$h(u+\omega_1) = h(u) + \lambda_1 \quad \text{and} \quad h(u+\omega_2) = h(u) + \lambda_2, \qquad \lambda_1, \lambda_2 \in k.$$
Since from (5.26) there exist a and b satisfying
$$a\omega_1 + b\eta_1 = \lambda_1 \quad \text{and} \quad a\omega_2 b + b\eta_2 = \lambda_2,$$
define
$$w_1(u) = h(u) - au - b\zeta(u). \tag{5.27}$$
Then
$$w_1(u+\omega_1) = w_1(u) \quad \text{and} \quad w_1(u+\omega_2) = w_1(u).$$
Therefore, in general, we have
$$w_1(u+\omega) = w_1(u) \quad \text{for } \omega \in \mathrm{P}(\omega_1, \omega_2).$$
That is, $w_1 = w_1(u)$ is a function in K. From (5.27) we have
$$h(u) = au + b\zeta(u) + w_1(u), \qquad a, b \in k \quad \text{and} \quad w_1(u) \in K.$$
Conversely, a function $h(u)$ such as above is clearly an additive function in $\widetilde{K}$. Thus, all the additive functions in $\widetilde{K}$ can be obtained as in the above, and one sees that the most significant additive functions are u and $\zeta(u)$.

As for differentials of the third kind, we can easily obtain them from $\varphi(u, a)$ in (5.15). The function $\varphi(u, a)$ has poles of order 1 only at P_0 and P_{-a}, and the expansions at $u = 0$ and $u = -a$ are given by (5.16) and (5.18), respectively. Hence,
$$(\varphi(u, -a) - \varphi(u, -b))\,du \tag{5.28}$$
is a differential of the third kind in K whose residues at P_a and P_b are 1 and -1, respectively.

As we described in the above, differentials of the first kind, the second kind, and the third kind can be expressed explicitly by $\wp(u)$ and $\wp'(u)$. Therefore, from Theorem 2.26, any differential can be found explicitly. Moreover, from

Theorem 5.7, taking exp of the integral of the sum of $a\,du$ and a differential as in (5.28), one gets a multiplicative function in $\widetilde{K}$. However, it is better to begin with another function to describe such a function explicitly. Instead of taking the exponential of the Abelian integral of the third kind, we will take the exp of the integral of $\zeta(u)$:

$$\begin{aligned}\int_\infty^u \frac{du}{u} + \int_0^u \sum_\omega{}' \left(\frac{1}{u-\omega} + \frac{1}{\omega} + \frac{u}{\omega^2}\right) du \\ = \log u + \sum_\omega{}' \left(\log\left(1 - \frac{u}{\omega}\right) + \frac{u}{\omega} + \frac{u^2}{2\omega^2}\right).\end{aligned} \tag{5.29}$$

As a consequence, we obtain

$$\sigma(u) = u \prod_\omega{}' \left\{\left(1 - \frac{u}{\omega}\right) e^{u/\omega + u^2/2\omega^2}\right\}. \tag{5.30}$$

Since $\zeta(u)$ is a convergent series, the right-hand side of (5.29) is term by term integrable and convergent as before. The right-hand side of (5.30) is a convergent infinite product. Hence, $\sigma(u)$ is regular everywhere in $\mathfrak{R}^*$ and has poles of order 1 at $u = \omega$ for $\omega \in \mathrm{P}(\omega_1, \omega_2)$. Even though the function in (5.29) is multivalued, $\sigma(u)$ is a single-valued regular function belonging to $\widetilde{K}$. This is because all the residues of $\zeta(u)\,du$ are rational integers, as was shown in the proof of Theorem 5.7.

We will explain the effect on $\sigma(u)$ of the transformations in the fundamental group $G = G(\omega_1, \omega_2)$. By the definition,

$$\frac{d \log \sigma(u)}{du} = \zeta(u).$$

Hence,

$$\frac{d \log \sigma(u + \omega_i)}{du} = \frac{d \log \sigma(u)}{du} + \eta_i, \qquad i = 1, 2.$$

That is,

$$\frac{d}{du} \log \frac{\sigma(u+\omega_i)}{\sigma(u)} = \eta_i, \qquad i = 1, 2.$$

Therefore, we have

$$\sigma(u + \omega_i) = \sigma(u) \exp(\eta_i u + a_i), \qquad i = 1, 2, \tag{5.31}$$

where a_i is a constant. In order to determine a_i, replace u by $-u$ in (5.30) to obtain

$$\begin{aligned}\sigma(-u) &= -u \prod_\omega{}' \left\{\left(1 + \frac{u}{\omega}\right) e^{-u/\omega + u^2/2\omega^2}\right\} \\ &= -u \prod_\omega{}' \left\{\left(1 - \frac{u}{(-\omega)}\right) e^{u/(-\omega) + u^2/2(-\omega)^2}\right\} \\ &= -\sigma(u).\end{aligned}$$

That is, $\sigma(u)$ is an odd function. Substitute $u = -\omega_i/2$ in (5.31) to obtain $\sigma(-\omega_i/2) = -\sigma(\omega_i/2) \neq 0$. Hence, $\exp(-\eta_i\omega_i/2 + a_i) = -1$. Therefore, we obtain

$$\sigma(u+\omega_i) = -\sigma(u)\exp\left(\eta_i u + \frac{\eta_i\omega_i}{2}\right), \qquad i = 1, 2. \tag{5.32}$$

Let $v = v(u)$ be an arbitrary multiplicative function for the Galois group $\mathfrak{G}$ of $\widetilde{K}/K$, and set

$$v(u+\omega_i) = \chi_{\omega_i} v(u), \quad \chi_{\omega_i} = e^{c_i}, \qquad i = 1, 2.$$

Note that in §2 we wrote $\sigma(v) = \chi_\sigma v$ for σ in $\mathfrak{G}$. We need to distinguish an element σ of $\mathfrak{G}$ from a function $\sigma(u)$ to use the notation χ_{ω_i} with respect to $G(\omega_1, \omega_2) = \Phi(\mathfrak{G})$. Since $G(\omega_1, \omega_2)$ is the free Abelian group generated by $\varphi_1(u) = u + \omega_1$ and $\varphi_2(u) = u + \omega_2$, the multiplier of $v(u)$ is uniquely determined by χ_{ω_1} and χ_{ω_2}. By (5.26), there are uniquely determined complex numbers a and b that satisfy

$$a\omega_1 + b\eta_1 = c_1 \quad \text{and} \quad a\omega_2 + b\eta_2 = c_2. \tag{5.33}$$

Then let

$$v_1(u) = e^{au}\frac{\sigma(a+b)}{\sigma(u)}.$$

From (5.32) and (5.33), we have

$$v_1(u+\omega_i) = e^{c_i}v_1(u), \qquad i = 1, 2.$$

Then the function $w_2(u) = v(u)/v_1(u)$ satisfies

$$w_2(u+\omega_i) = w_2(u), \qquad i = 1, 2.$$

Hence, ω_2 belongs to K. That is, an arbitrary multiplicative function in $\widetilde{K}$ can be written in the form

$$v(u) = e^{au}\frac{\sigma(u+b)}{\sigma(u)}w_2(u), \qquad a, b \in k, \; w_2(u) \in K. \tag{5.34}$$

One can verify that conversely, for $v(u)$ defined as in (5.34) in terms of arbitrary a, b, and $w_2(u)$ $(w_2(u) \neq 0)$, $v(u)$ is a multiplicative function in $\widetilde{K}$ having

$$\chi_{\omega_i} = e^{c_i}, \qquad i = 1, 2$$

as its multiplier obtained from (5.33). Considering the zeros of $\sigma(u)$, we see that the divisor of $v(u)$ is given by

$$(v) = \frac{P_{-b}}{P_0}(w_2). \tag{5.35}$$

For pure imaginary numbers $c_1 = 2\pi\sqrt{-1}\xi_1$ and $c_2 = 2\pi\sqrt{-1}\xi_2$, we can solve (5.33) using (5.26) to obtain

$$a = \xi_2\eta_1 - \xi_1\eta_2 \quad \text{and} \quad b = \xi_1\omega_1 - \xi_2\omega_2.$$

By substituting those into (5.34), we obtain all the multiplicative functions in the strict sense having

$$\chi_{\omega_i} = \exp(2\pi\sqrt{-1}\xi_i), \qquad i = 1, 2$$

as their characters. The divisor class group $\overline{\mathfrak{D}}_0$ of degree 0 of K contains exactly one divisor of the form P_{-b}/P_0, $b \in k$, in each class. Then Theorem 5.8 asserting the one-to-one correspondence between the classes in $\overline{\mathfrak{D}}_0$ and the classes of multiplicative functions in the strict sense is trivial from (5.35).

Let $z = z(P)$ be any function in K and denote the divisor of z by

$$(z) = \frac{P_1 \cdots P_r}{Q_1 \cdots Q_r}.$$

For $P_i = P_{a_i}$ and $Q_i = Q_{b_i}$, from (5.25) in Abel's theorem, we have

$$a_1 + \cdots + a_r \equiv b_1 + \cdots + b_r \mod \mathrm{P}(\omega_1, \omega_2).$$

Since a_i and b_i are determined only as $\bmod \mathrm{P}(\omega_1, \omega_2)$, we can assume

$$a_1 + \cdots + a_r = b_1 + \cdots + b_r.$$

With that choice of a_i and b_i, the function

$$z_1(u) = \frac{\sigma(u - a_1) \cdots \sigma(u - a_r)}{\sigma(u - b_1) \cdots \sigma(u - b_r)} \tag{5.36}$$

is a function in $\widetilde{K}$, and moreover from (5.32) we obtain

$$z_1(u + \omega_i) = z_1(u) \exp\left(\eta_i\left(\sum_{j=1}^{r} a_j - \sum_{j=1}^{r} b_j\right)\right) = z_1(u), \qquad i = 1, 2.$$

Hence, $z_1(u)$ is a function in K. Since the divisor of z_1 is clearly equal to the divisor (z) of z by (5.36), we obtain

$$z(u) = cz_1(u) = c\frac{\sigma(u - a_1) \cdots \sigma(u - a_r)}{\sigma(u - b_1) \cdots \sigma(u - b_r)}, \qquad c \in k, \ c \neq 0. \tag{5.37}$$

Namely, any function of K can be expressed in terms of $\sigma(u)$ as above. By noting

$$\exp(\eta_i u) = c_i' \frac{\sigma(u + \omega_i)}{\sigma(u)}, \qquad c_i' \in k, \ i = 1, 2,$$

we see, from (5.34) and the above, that an arbitrary multiplicative function in $\widetilde{K}$ can be written as a product of $\sigma(u - a)$.

Since the periods of the differential of the first kind $du = d\wp/\wp'$ for α_1 and β_1 are ω_1 and ω_2 as we showed earlier, we see that for

$$u' = \frac{\pi\sqrt{-1}}{\omega_1} u, \qquad du' = \frac{\pi\sqrt{-1}}{\omega_1} du$$

the periods of du' for α_1 and β_1 are given by

$$\pi\sqrt{-1} \quad \text{and} \quad \pi\sqrt{-1}\frac{\omega_2}{\omega_1} = \omega_{11}. \tag{5.38}$$

Then from (3.30), a ϑ-function can be defined by

$$\vartheta(u) = \sum_{m=-\infty}^{\infty} \exp(\omega_{11}m^2 + 2mu).$$

One can prove the following relation between $\vartheta(u)$ and the above $\sigma(u)$:

$$\sigma(u) = \exp\left(c' - \frac{\pi\sqrt{-1}}{\omega_1}u + \frac{\eta_1}{2\omega_1}u^2\right)\vartheta\left(\frac{\pi\sqrt{-1}}{\omega_1}\left(u - \frac{\omega_1 + \omega_2}{2}\right)\right), \qquad c' \in k.$$

Substitute this relation into the right-hand side of (5.37) to obtain

$$z(u) = c_1 \frac{\prod_{i=1}^{r} \vartheta\left(\frac{\pi\sqrt{-1}}{\omega_1}\left(u - \frac{\omega_1+\omega_2}{2} - a_i\right)\right)}{\prod_{i=1}^{r} \vartheta\left(\frac{\pi\sqrt{-1}}{\omega_1}\left(u - \frac{\omega_1+\omega_2}{2} - b_i\right)\right)}, \qquad c_1 \in k.$$

As we mentioned earlier, for an elliptic function field, an Abelian function belonging to the elliptic function field coincides essentially with the original elliptic function. In other words, (3.31) is the generalization of the above formula to the case of genus g. The right-hand side of the above formula contains $u' = \pi\sqrt{-1}u/\omega_1$ in the ϑ-function, and it may appear to be different from the expression in (3.31). However, notice that formula (3.31) was obtained through the normalized du' having periods as in (5.38). Instead of $\vartheta(u)$, in the elliptic function theory, the function

$$\vartheta_1(v) = c'' \exp\left(-\frac{\eta_1}{2\omega_1}u_2\right)\sigma(u), \qquad v = \frac{u}{\omega_1}, \quad c'' \in k$$

is often used. Moreover, functions $\vartheta_2(v)$, $\vartheta_3(v)$, $\vartheta_0(v)$ similar to $\vartheta_1(v)$ are defined. They are called Jacobi's ϑ-functions. See textbooks on elliptic functions for those functions and for the connections between these functions and Jacobi's elliptic functions $\operatorname{sn}(u)$, $\operatorname{cn}(u)$, $\operatorname{dn}(u)$, etc.

We have described how to express an additive function and a multiplicative function in $\widetilde{K}$ in terms of $\zeta(u)$ and $\sigma(u)$. Next we will consider representation functions in $\widetilde{K}$. We will start with the proof of Theorem 5.11 for the case $g = 1$ which we skipped earlier. We can state the theorem for the elliptic function field as follows, by using the fundamental group $G(\omega_1, \omega_2)$ for $\mathfrak{R}^*$ of $\mathfrak{R}$, or by the isomorphic period group $\mathrm{P}(\omega_1, \omega_2)$. Let

$$\omega \mapsto M_\omega, \qquad \omega \in \mathrm{P}(\omega_1, \omega_2)$$

be an arbitrary representation of $\mathrm{P}(\omega_1, \omega_2)$. We need to find a nonsingular matrix with entries in $\widetilde{K}$

$$Z^*(u) = (z^*_{ij}(u)), \qquad i, j = 1, \dots, n$$

such that

$$Z^*(u+\omega) = (z^*_{ij}(u+\omega)) = Z^*(u)M_\omega, \qquad \omega \in \mathrm{P}(\omega_1, \omega_2). \tag{5.39}$$

Since $\mathrm{P}(\omega_1, \omega_2)$ is a free Abelian group generated by ω_1 and ω_2, it is sufficient to confirm (5.39) only for ω_1, ω_2, and for

$$M_1 = M_{\omega_1} \quad \text{and} \quad M_2 = M_{\omega_2}.$$

Let C be any nonsingular matrix with entries in k, and let

$$Z_1^*(u) = Z(u)C \quad \text{and} \quad M'_\omega = C^{-1}M_\omega C.$$

Then (5.39) becomes

$$Z_1^*(u+\omega) = Z_1^*(u)M'_\omega.$$

Hence, we can replace M_ω by any equivalent representation $C^{-1}M_\omega C$ to solve (5.39). Since $M_1M_2 = M_2M_1$ holds, therefore, we may assume that M_1 and M_2 can be decomposed as

$$M_1 = \begin{pmatrix} M_1^{(1)} & & 0 \\ & \ddots & \\ 0 & & M_1^{(r)} \end{pmatrix} \quad \text{and} \quad M_2 = \begin{pmatrix} M_2^{(1)} & & 0 \\ & \ddots & \\ 0 & & M_2^{(r)} \end{pmatrix},$$

where $M_1^{(i)}$ and $M_2^{(i)}$ are upper triangle matrices

$$M_1^{(i)} = \begin{pmatrix} \varepsilon_1^{(i)} & * & * \\ & \ddots & * \\ 0 & & \varepsilon_1^{(i)} \end{pmatrix}, \qquad M_2^{(i)} = \begin{pmatrix} \varepsilon_2^{(i)} & * & * \\ & \ddots & * \\ 0 & & \varepsilon_2^{(i)} \end{pmatrix}.$$

If one can find a nonsingular $Z^{(i)}(u)$ for $M_1^{(i)}$ and $M_2^{(i)}$ such that

$$Z^{(i)}(u+\omega_1) = Z^{(i)}(u)M_1^{(i)}, \qquad Z^{(i)}(u+\omega_2) = Z^{(i)}(u)M_2^{(i)}, \tag{5.40}$$

then

$$Z^*(u) = \begin{pmatrix} Z^{(1)}(u) & & 0 \\ & \ddots & \\ 0 & & Z^{(r)}(u) \end{pmatrix}$$

satisfies (5.39). Hence, it is sufficient to solve (5.40). Furthermore, let $v(u)$ be a multiplicative function in $\widetilde{K}$ having

$$\chi_{\omega_1} = \varepsilon_1^{(i)} \quad \text{and} \quad \chi_{\omega_2} = \varepsilon_2^{(i)}$$

as the multipliers, and let $\overline{Z}^{(i)}(u)$ be a matrix satisfying (5.39) for the presentation determined by $(1/\varepsilon^{(i)})M_1^{(i)}$ and $(1/\varepsilon_2^{(i)})M_2^{(i)}$. Then

$$Z^{(i)}(u) = v(u)\overline{Z}^{(i)}(u)$$

clearly satisfies (5.40). Therefore, the problem is reduced to the following. For both matrices M_1 and M_2 of the type

$$\begin{pmatrix} 1 & * & * & * \\ & 1 & * & * \\ & & \ddots & * \\ 0 & & & 1 \end{pmatrix},$$

we need to solve (5.39). Then let

$$M_1 = E + N_1 \quad \text{and} \quad M_2 = E + N_2 \qquad (E = \text{ identity matrix}),$$

where both N_1 and N_2 have the form

$$\begin{pmatrix} 0 & * & * & * \\ & 0 & * & * \\ & & \ddots & * \\ 0 & & & 0 \end{pmatrix}.$$

Then we have

$$\begin{gathered} N_1^n = N_2^n = 0, \\ M_i^{\pm m} = E \pm \binom{m}{1} N_i \pm \cdots \pm \binom{m}{n-1} N_i^{n-1}, \\ \|M_i^{\pm m}\| \leq \|E\| + \binom{m}{1}\|N_i\| + \cdots + \binom{m}{n-1}\|N_i\|^{n-1}, \\ i = 1, 2, \qquad m = 1, 2, \ldots . \end{gathered}$$

Therefore, for some $c_1 > 0$ we have

$$\|M_i^{\pm m}\| \leq c_1 m^{n-1}, \qquad i = 1, 2, \ m = 1, 2, \ldots, \tag{5.41}$$

where $\|M\|$ is the norm of the matrix M as defined in (2.34).

For a sufficiently large natural number l, let

$$Z^*(u) = \sum_{\omega} \frac{M_\omega}{(u - \omega)^l}, \tag{5.42}$$

where $\sum_\omega$ indicates the sum over all elements ω in $\mathrm{P}(\omega_1, \omega_2)$. For an arbitrary positive real number c, if $|u| \leq c$ and $|\omega| \leq 2c$, then from (5.7) we have

$$\frac{1}{|u - \omega|} \leq \frac{2}{|\omega|}.$$

Let $L_1, L_2, \ldots$ be as in Figure 5.4, and let the point ω on L_m be expressed as

$$\omega = m_1\omega_1 + m_2\omega_2.$$

Then we clearly have

$$|m_1| \leq m \quad \text{and} \quad |m_2| \leq m.$$

Hence, we obtain the following by $M_\omega = M_1^{m_1} M_2^{m_2}$ and (5.41):

$$\|M_\omega\| \le \|M_1^{m_1}\| \, \|M_2^{m_2}\| \le c_1^2 |m_1|^{n-1} |m_2|^{n-1} \le c_1^2 m^{2n-2}.$$

Therefore, for a sufficiently large m and $|\omega| \ge 2c$, we have

$$\frac{\|M_\omega\|}{|u-\omega|^l} \le \frac{2^l c_1^2 m^{2n-2}}{(mr)^l} = \left(\frac{2}{r}\right)^l c_1^2 \frac{1}{m^{l-2n+2}},$$

where r is the shortest distance from the origin to L_1.

Since there are $8m$ ω on L_m, the sum of $\|M_\omega\|/|u-\omega|^l$ for these $8m$ ω's does not exceed

$$8\left(\frac{2}{r}\right)^l c_1^2 \frac{1}{m^{l-2n+1}}.$$

Therefore, for $l - 2n + 1 > 1$, e.g., $l = 2n + 1$, the sum

$$\sum{}' \frac{M_\omega}{(u-\omega)^l},$$

excluding finitely many ω on L_m, converges absolutely and uniformly for $|u| \le c$. Since c is an arbitrary positive real number, all the entries of $Z^*(u)$ in (5.42) belong to $\widetilde{K}$. Furthermore, from the above proof,

$$\sum_{\omega \ne 0}{}' \frac{M_\omega}{(u-\omega)^l}$$

is a matrix whose entries are regular functions in a neighborhood of $u = 0$. Hence, from

$$Z^*(u) = \frac{E}{u^l} + \sum_{\omega \ne 0}{}' \frac{M_\omega}{(u-\omega)^l},$$

$|Z^*(u)|$ has a pole of order nl at $u = 0$. That is, $|Z^*(u)| \ne 0$. Since we have

$$Z^*(u+\omega') = \sum_{\omega} \frac{M_\omega}{(u+\omega'-\omega)^l} = \sum_{\omega-\omega'} \frac{M_{\omega-\omega'} M_{\omega'}}{(u-(\omega-\omega'))^l} = Z^*(u) M_{\omega'},$$

the matrix $Z^*(u)$ satisfies (5.39), completing the proof of Theorem 5.11.

Lastly, we will consider an extension field, in particular an unramified extension, of an elliptic function field K. Let $K = K(\mathfrak{R})$, $\mathfrak{R} = \mathfrak{R}_\infty^*(\omega_1, \omega_2)$, and let

$$G(\omega_1, \omega_2) = \{\varphi(u) = u + m_1\omega_1 + m_2\omega_2;\ m_1, m_2 = 0, \pm 1, \pm 2, \dots\}$$

be the fundamental group of $\mathfrak{R}$ for the finite u-plane $\mathfrak{R}_\infty^*$. Also let $\mathfrak{G}$ be the Galois group of $\widetilde{K} = K(\mathfrak{R}_\infty^*)$ over K, and let

$$\mathrm{P}(\omega_1, \omega_2) = \{\omega = m_1\omega_1 + m_2\omega_2;\ m_1, m_2 = 0, \pm 1, \pm 2, \dots\}$$

be the period group of the differential du of the first kind. Then we obviously have

$$\mathfrak{G} \cong G(\omega_1, \omega_2) \cong \mathrm{P}(\omega_1, \omega_2).$$

From Theorems 5.17 and 5.18, the correspondence between a finite unramified extension L of K and a subgroup P' of a finite index in $\mathrm{P}(\omega_1, \omega_2)$ is one-to-one. Given such a subgroup P', there exists a basis ω_1, ω_2 of the period group and suitable natural number n_1 and n_2 such that P' is generated by

$$\omega_1' = n_1\omega_1 \quad \text{and} \quad \omega_2' = n_2\omega_2,$$

i.e.,

$$\mathrm{P}' = \mathrm{P}(\omega_1', \omega_2') = \{\omega' = m_1\omega_1' + m_2\omega_2';\ m_1, m_2 = 0, \pm 1, \pm 2, \dots\}.$$

Then we have

$$[\mathrm{P}(\omega_1, \omega_2) : \mathrm{P}(\omega_1', \omega_2')] = n_1 n_2.$$

The correspondence between L and P' implies that L is precisely the analytic function field on $\mathfrak{R}' = \mathfrak{R}_\infty^*(\omega_1', \omega_2')$. This is because L is the set of functions in $\widetilde{K}$ for which each element of P' is a period. Thus,

$$L = K(\mathfrak{R}') \quad \text{and} \quad [L : K] = n_1 n_2. \tag{5.43}$$

Therefore, any finite unramified extension L of an elliptic function field K is an elliptic function field, and furthermore L/K is an Abelian extension. Hurwitz's formula (4.4) also shows that an unramified extension of an elliptic function field is again an elliptic function field: since $\sum_{P'}(e_{P'} - 1) = 0$ and $g = 1$ in (4.4), we obtain $g' = 1$.

Since the Galois group $\mathfrak{G}(L/K)$ of L/K is, by Theorem 5.18, isomorphic to $\mathrm{P}(\omega_1, \omega_2)/\mathrm{P}(\omega_1', \omega_2')$, $\mathfrak{G}(L/K)$ is isomorphic to the product of cyclic groups of order n_1 and n_2. The differential du in K is also a differential in $\widetilde{K}$. When du is regarded as a differential of the first kind in L, its period group is $\mathrm{P}' = \mathrm{P}(\omega_1', \omega_2')$.

Let $z = \wp(u|\omega_1, \omega_2)$ be the $\wp$-function in (5.8) constructed from the period group $\mathrm{P}(\omega_1, \omega_2)$, and let $z_1 = \wp(u|\omega_1', \omega_2')$ be the $\wp$-function given by $\mathrm{P}(\omega_1', \omega_2')$. Then z_1 is certainly a function in L, and we have

$$L = K(z_1). \tag{5.44}$$

The proof goes as follows. Let $w = \wp'(u|\omega_1, \omega_2)$ and let $w_1 = \wp'(\omega_1', \omega_2')$. Then w_1/w is an even function of u belonging to L. Hence, from the paragraph containing (5.13), w_1/w has a rational function in z_1. Namely, $w_1 \in K(z_1)$. On the other hand, we have $L = k(z_1, w_1)$. Hence, (5.44) holds.

Since z is an even function of u, z is a rational function of z_1 as well, that is,

$$z = r(z_1).$$

Multiplying both sides by the denominator of the right-hand side, we get an equation

$$F_1(z, z_1) = 0.$$

Note that the polynomial $F_1(X, Y)$ has degree 1 in X and, by (5.43) and (5.44), degree n_1n_2 in Y.

In particular, for $n_1 = n_2 = n$ we have

$$z_1 = \mathfrak{p}(u|n\omega_1, n\omega_2') = \frac{1}{n^2}\mathfrak{p}\left(\frac{u}{n}\,\middle|\,\omega_1, \omega_2\right).$$

Hence, we obtain $L = K(z_1) = K(\mathfrak{p}(u/n|\omega_1, \omega_2))$. Then let

$$F(X, Y) = F_1\left(X, \frac{1}{n^2}Y\right).$$

We have an irreducible polynomial $F(z, Y)$ over K having $\mathfrak{p}(u/n|\omega_1, \omega_2)$ as a root. Then the equation

$$F(z, Y) = 0$$

for Y is called the general n-division equation for K. Since L/K is an Abelian extension, $F(z, Y) = 0$ is an Abelian equation, so that $F(z, Y) = 0$ is solvable by radicals for K. The roots are given by the conjugates of $\mathfrak{p}(u/n|\omega_1, \omega_2)$ over K

$$\mathfrak{p}\left(\frac{u + m_1\omega_1 + m_2\omega_2}{n}\,\middle|\,\omega_1, \omega_2\right), \qquad m_1, m_2 = 0, 1, \dots, n-1.$$

Then $\mathfrak{p}(u/n|\omega_1, \omega_2)$ is called n-division value of $\mathfrak{p}(u|\omega_1, \omega_2)$. Clearly the function $\mathfrak{p}(nu|\omega_1, \omega_2)$, which satisfies (5.4), belongs to K. Next we determine whether, for a given complex number a, $\mathfrak{p}(au|\omega_1, \omega_2)$ belongs to K or not. From (5.4), we have $\mathfrak{p}(au|\omega_1, \omega_2) \in K$ if and only if we have

$$\mathfrak{p}(a(u + \omega)|\omega_1, \omega_2) = \mathfrak{p}(au|\omega_1, \omega_2), \qquad \omega \in \mathrm{P}(\omega_1, \omega_2). \tag{5.45}$$

If $u = 0$, then the right-hand side is ∞. Hence, $a\omega$ must be a pole of $\mathfrak{p}(u|\omega_1, \omega_2)$,

$$a\omega \in \mathrm{P}(\omega_1, \omega_2). \tag{5.46}$$

Conversely, (5.46) clearly implies (5.45). Note also (5.46) is equivalent to

$$a\omega_1 = p\omega_1 + qw_2 \quad\text{and}\quad a\omega_2 = r\omega_1 + s\omega_2, \tag{5.47}$$

where p, q, r, s are rational integers. Eliminating ω_1 and ω_2 from the above equations, we have

$$\begin{vmatrix} p-a & q \\ r & s-a \end{vmatrix} = 0.$$

Namely, a is a root of

$$X^2 - (p+s)X + ps - qr = 0.$$

Hence, if a is not a rational integer, then a must be an integer of a quadratic field. Then since $q \neq 0$ must hold, we have from $a = p + q\tau$ $(\tau = \omega_2/\omega_1)$

$$R_0(a) = R_0(\tau), \tag{5.48}$$

for the field R_0 of rational numbers. Since our assumption is $\mathrm{Im}(\tau) > 0$, (5.48) is an imaginary quadratic field. Consequently, a is also imaginary. The transformation

$$u \mapsto au$$

induced by such an a is called a *complex multiplication* of K.

Conversely, if $\tau = \omega_2/\omega_1$ determining $\mathfrak{R} = \mathfrak{R}(K)$ is an imaginary quadratic number, then the elliptic function field K has complex multiplication. Moreover, the set of a that satisfy (5.45) forms a subring of the algebraic integers of the imaginary quadratic field $R_0(\tau)$. Thus, in general, there is a profound relationship between imaginary quadratic fields and elliptic functions. The theory of complex multiplication clarifies the relationship. See H. Weber, *Lehrbuch der Algebra*, III, Braunschweig, F. Vieweg, 1908. For a generalization to elliptic function fields over a general coefficient field, see Hasse's papers, *Zur Theorie der abstrakten elliptischen Funktionenkörper*, I, II, III, Crelle's Jour. **175** (1936).

Appendix

As we mentioned in the Preface, we explain here a theory concerning a general differential and its residue based on Tate's idea. See J. Tate, *Residues of differentials on curves*, Ann. École Norm. Sup. (1968), 149–159. This method seems to have applications beyond algebraic function theory.

§1. Finitepotent map and its trace

1.1. Let k be an arbitrary field, let V be a vector space over k, and let $\mathrm{End}(V)$ be the ring of endomorphisms of V. As we can define the trace $T(f)$ of a linear map $f\colon V \to V$, i.e., $f \in \mathrm{End}(V)$, for a finite dimensional vector space V, we can define $T(f)$ for a certain class of f even if V is an infinite dimensional space.

DEFINITION. When a linear map $f\colon V \to V$ satisfies

$$\dim f^n(V) < +\infty$$

for a sufficiently large natural number n, f is called a *finitepotent map.*

LEMMA A.1. *A linear map f is finitepotent if and only if there exists a subspace W of V such that*

(i) $\dim W < +\infty$,
(ii) $f(W) \subseteq W$,
(iii) *the map* $f_{V/W}\colon V/W \to V/W$ *induced by f is nilpotent.*

The proof is trivial.

Such a subspace W as above is called a core space of the finitepotent map f. If W_1 and W_2 are core spaces of f, then $W = W_1 + W_2$ is also a core space. Moreover, the induced map $f_{W/W_1}\colon W/W_1 \to W/W_1$ is nilpotent. Hence,

$$T(f_{W_1}) = T(f_W),$$

where $T(f_{W_1})$ and $T(f_W)$ are the traces of the maps f_{W_1} and f_W induced by f on the finite dimensional spaces W_1 and W, respectively. Similarly, we obtain

$$T(f_{W_2}) = T(f_W).$$

That is, for an arbitrary core space W for f, the value $T(f_W)$ depends only upon f and is independent of the choice of a particular core space W. Therefore, we can uniquely define the trace $T(f)$ of a finitepotent map f by

$$T(f) = T(f_W).$$

We will discuss several properties of the above trace.

LEMMA A.2.

(i) *If* $\dim V < +\infty$, *then every linear map* $f\colon V \to V$ *is finitepotent, and the above* $T(f)$ *coincides with the usual one.*

(ii) *If* f *is nilpotent, then* f *is a finitepotent map and*

$$T(f) = 0.$$

(iii) *Let* f *be finitepotent, and let* U *be an invariant subspace of* $f (f(U) \subseteq U)$. *Then the induced maps*

$$f_{V/U}\colon V/U \to V/U \quad \text{and} \quad f_U\colon U \to U$$

are both finitepotent, satisfying

$$T(f) = T(f_{V/U}) + T(f_U).$$

PROOF. Since (i) and (ii) are clear, we will prove (iii). Let W be a core space for f, and let $W' = (W + U)/U$ and $W'' = W \cap U$. Then W' and W'' are core spaces for $f_{V/U}$ and f_U, respectively. Hence, $f_{U/V}$ and f_U are finitepotent. Since we have an isomorphism $W' \cong W/W''$, $T(f_{W'}) = T(f_{W/W''})$ holds. Hence, the above formula is obtained from $T(f_W) = T(f_{W/W''}) + T(f_{W''})$.

REMARK. We sometimes write $T_{U/V}(f)$ and $T_U(f)$ for $T(f_{V/U})$ and $T(f_U)$, respectively. Then the above becomes

$$T_V(f) = T_{V/U}(f) + T_U(f).$$

Notice that properties (i), (ii), and (iii) for a finitepotent map f uniquely characterize the function $T(f)$.

LEMMA A.3. *Let* f *and* g *be elements of* $\mathrm{End}(V)$. *If* fg *is finitepotent, then* gf *is also finitepotent, and*

$$T(fg) = T(gf).$$

PROOF. The finitepotentness of fg implies that $W = (fg^n)(V)$ stabilizes for all sufficiently large n. Then W is a core space for fg. On the other hand, $(fg)^{n+1}(V) = g(fg)^n f(V) \subseteq g(W)$ holds. Therefore, fg is also finitepotent. Hence, for a sufficiently large n, $W' = (gf)^n(V)$ is a core space for fg. Then $W' \subseteq g(W)$, and $\dim W' \leq \dim g(W) \leq \dim W$. Similarly, we get $W \subseteq f(W')$ and $\dim W \leq \dim W'$. We conclude $\dim W = \dim W'$, and $g: W \to W'$ is an isomorphism. Since we clearly have the following commutative diagram

$$\begin{array}{ccc} W & \xrightarrow{fg} & W \\ \downarrow g & & \downarrow g \\ W' & \xrightarrow{gf} & W' \end{array}$$

we obtain $T((fg)_W) = T((gf)_{W'})$, i.e., $T(fg) = T(gf)$.

DEFINITION. Let F be a k-subspace of $\mathrm{End}(V)$. If there exists a natural number n such that for arbitrary maps $f_1, f_2, \ldots, f_n$ in F satisfying

$$\dim f_1 \cdots f_n(V) < +\infty,$$

then F is called a finitepotent subspace of $\mathrm{End}(V)$.

Note that any f belonging to a finitepotent subspace F is finitepotent. Then $T(f)$ is defined.

LEMMA A.4. *For a finitepotent subspace F of* $\mathrm{End}(V)$,

$$\begin{array}{cccc} T: & F & \longrightarrow & k \\ & \cup\!\!\!\!| & & \cup\!\!\!\!| \\ & f & \mapsto & T(f) \end{array}$$

is a k-linear map.

PROOF. It is sufficient to prove the case where F is spanned by a finite number of elements (e.g., by two elements) over k. Let n be a natural number such as in the above definition. Let F^n be the subspace of $\mathrm{End}(V)$ spanned by all the products $f_1 \cdots f_n$ of n elements in F. Let $g_1, \ldots, g_m$ be basis elements for F over k. Then F^n is spanned by finitely many products $g_{i_1} \cdots g_{i_n}$, $1 \le i_1, \ldots, i_n \le m$. Hence, by the definition of finitepotentness, we have

$$\dim F^n(V) < +\infty.$$

Therefore, $W = F^n(V)$ is a core space for all the f in F. Since $T(f) = T(f_W)$, the lemma is proved.

1.2. We will apply what we obtained in the above to a vector space with a linear topology, i.e., a linear topological space. We will briefly describe properties of such spaces. For detailed proofs, the reader is referred to S. Lefschetz, *Algebraic topology*, Amer. Math. Soc. Colloq. Pub. 27, 1942, Chapter II, §6.

When a vector space V is given a Hausdorff topology satisfying the following conditions (i) and (ii), then V is said to be a linear topological space, and the topology is called a linear topology.

(i) As an additive group, V is a topological group with the topology.

(ii) There exists a neighborhood system of 0 consisting of subspaces $\{W_\alpha\}$.

For example, the discrete topology on V is a linear topology.

For a linear topological space V, if an arbitrary family of closed linear submanifolds $\{F_\alpha\}$ of finite intersection property has a nonempty intersection, then V is said to be a linearly compact linear topological space. (A submanifold is a residue class of V modulo a subspace.) We abbreviate "linearly compact" as "l.c." Then we have the following important properties. (a) An arbitrary product of l.c. spaces is an l.c. space. (b) A linear topological space V is l.c. and discrete if and only if $\dim V < +\infty$ holds.

When a linear topological space V contains an l.c. open subspace, V is said to be locally linearly compact, abbreviated as l.l.c. For an l.l.c. space, one can define the dual space to establish a duality theorem similar to the duality theorem for locally compact Abelian groups. For a linear topological space V and a closed subspace W of V, V is l.l.c. (or l.c.) if and only if V/W and W are l.l.c. (or l.c.).

1.3. Let V be an l.l.c. linear topological space, and let E be the totality of all the continuous endomorphisms of V. Then E is a subalgebra of $\mathrm{End}(V)$ over k. Let E_1 be the set of elements f in E such that the closure of $\mathrm{Im}\, f = f(V)$ is l.c., and also let E_2 be the totality of elements f in E such that $\mathrm{Ker}\, f$ is an open subspace of V. Let $E_0 = E_1 \cap E_2$. Then it is easy to show that E_0, E_1, and E_2 are two-sided ideals of E. Let U be an arbitrary l.c. open subspace of V, and let π be a projection from V onto U. Then, we have

$$\pi \in E_1 \quad \text{and} \quad 1 - \pi \in E_2,$$

which yields

$$E = E_1 + E_2.$$

Lemma A.5. *For $f \in E_1$ and $g \in E_2$, we have*

$$\dim gf(V) < +\infty.$$

Proof. Since the closure W of $\mathrm{Im}\, f$ is l.c., and $U = \ker g$ is an open subspace in V, $W/(W \cap U)$ is l.c. and discrete. Hence, $\dim W/(W \cap U) < +\infty$ holds. Since we have $f(V) \subseteq W$ and $g(W \cap U) = 0$, $\dim gf(V) \leq \dim W/(W \cap U) < +\infty$.

Since we have $E_0 = E_1 \cap E_2$, the above lemma implies that E_0 is a finitepotent subspace of $\mathrm{End}(V)$. From Lemma A.4, we can define the linear map

$$\begin{array}{cccc} T\colon & E_0 & \longrightarrow & k \\ & \cup\!\!\!\shortmid & & \cup\!\!\!\shortmid \\ & f & \mapsto & T(f). \end{array}$$

Lemma A.6. *If $f \in E_1$ and $g \in E_2$, then fg and gf belong to E_0 such that*

$$T(fg) = T(gf).$$

Proof. This is clear from Lemma A.3.

§2. The differential and its residue

2.1. We will begin with a definition of a general differential. Let R be a commutative ring with an identity containing a field k as a subfield, i.e., $1 \in k \subseteq R$. Since R is a k-module, we can define the tensor product $R \otimes_k R$. Then define an R-module structure on $R \otimes_k R$ by

$$x(y \otimes z) = xy \otimes z \quad \text{for } x, y, z \in R.$$

Let A be the submodule of $R \otimes_k R$ generated by

$$x \otimes yz - xy \otimes z - xz \otimes y.$$

Then A is an R-subgroup of $R \otimes_k R$, and as an R-module, is generated by

$$1 \otimes yz - y \otimes z - z \otimes y, \qquad y, z \in R.$$

Define

$$\begin{array}{ccc} \Omega_{R/k} & = & R \otimes_k R/A \\ \cup\!\!\!\!\!| & & \cup\!\!\!\!\!| \\ dx & = & 1 \otimes x \mod A, \qquad x \in R. \end{array}$$

Then $\Omega_{R/k}$ is an R-module generated by all the elements dx, where $x \in R$. Since we have

$$x\,dy = x \otimes y \mod A, \qquad x, y \in R,$$

$\Omega_{R/k}$ is, as an additive group, generated by the elements $x\,dy$.

DEFINITION. An element of $\Omega_{R/k}$ is called a *differential* of R/k. In particular, dx is called the differential of x.

From the definition of dx, we have

$$d(x+y) = dx + dy \quad \text{and} \quad d(xy) = x\,dy + y\,dx, \qquad x, y \in R.$$

For $a \in k$,

$$\begin{aligned} 1 \otimes a = a \otimes 1 &= -(a \otimes 1 \cdot 1 - a \otimes 1 - a \otimes 1) \\ &\equiv 0 \mod A. \end{aligned}$$

From the above, we obtain

$$da = 0 \quad \text{and} \quad d(ax) = a\,dx, \qquad a \in k,\ x \in R.$$

We will explain one important property of $\Omega_{R/k}$. Let D be a map from R to an R-module V $(1 \cdot v = v,\ v \in V)$, $D\colon R \to V$ such that

$$D(x+y) = D(x) + D(y), \qquad D(xy) = xD(y) + yD(x), \qquad x, y \in R,$$
$$D(a) = 0, \qquad a \in k.$$

Then D is said to be a k-derivation of R with values in V. We denote the set of such k-derivations by $\mathfrak{D}(R/k, V)$. With the operation

$$(xD + yD')z = xD(z) + yD'(z), \qquad x, y, z \in R,$$

$\mathfrak{D}(R/k, V)$ is an R-module.

Let

$$f\colon \Omega_{R/k} \to V$$

be an arbitrary R-homomorphism, then

$$\begin{array}{cccc} f \circ d\colon & R & \to & V \\ & \Psi & & \Psi \\ & x & \mapsto & f(dx) \end{array}$$

defines a k-derivation. Conversely, for any $D \in \mathfrak{D}(R/k, V)$ there exists a unique $f\colon \Omega_{R/k} \to V$ such that $D = f \circ d$ holds. Therefore, we obtain

$$\mathrm{Hom}_R(\Omega_{R/k}, V) \overset{\approx}{\to} \mathfrak{D}(R/k, V).$$

From the above isomorphism, one can learn the properties of differentials through derivations.

2.2. Let R/k be as above, and let V be an R-module satisfying the following two conditions:

(i) V is an l.l.c. linear topological space over k.

(ii) For any $x \in R$, the linear map

$$\begin{array}{cccc} \varphi_x\colon & V & \to & V \\ & \Psi & & \Psi \\ & v & \mapsto & xv \end{array}$$

is continuous. We call such an R-module V as above a continuous R-module.

Let V be as in the above, and as in 1.3, let E be the set of continuous endomorphisms of V. Then for any element $x, y \in R$, from (ii), we have $\varphi_x, \varphi_y \in E$.

Since we also have $E = E_1 + E_2$, there are f and g in E_1 such that

$$f \equiv \varphi_x \quad \text{and} \quad g \equiv \varphi_y \quad \mod E_2.$$

LEMMA A.7. *The commutators* $[f, g](= fg - gf)$, $[f, \varphi_y], [\varphi_x, g]$ *belong to* E_0, *and furthermore,*

$$T([f, g]) = T([f, \varphi_y]) = T([\varphi_x, g]).$$

PROOF. We obviously have $[f, g] \in E_1$. Since R is commutative, we have $xy = yx$ and $\varphi_x\varphi_y = \varphi_y\varphi_\alpha$. Hence, we get

$$[f, g] \equiv [\varphi_x, \varphi_y] \equiv \mod E_2.$$

Therefore, we have $[f, g] \in E_0 = E_1 \cap E_2$. Similarly, $[f, \varphi_y]$ and $[\varphi_x, g]$ belong to E_0. Moreover, $f \in E_1$ and $g - \varphi_y \in E_2$. Then Lemma A.6 implies $T([f, g - \varphi_y]) = 0$. Namely, $T([f, g]) = T([f, \varphi_y])$. Similarly, we can show $T([f, g]) = T([\varphi_x, g])$.

From the above lemma, the common value of these traces depends only upon x and y and is independent of the choice of f and g. We denote this common value by $\{x, y\}$:

$$\{x, y\} = T([f, g]) = T[(f, \varphi_y]) = T([\varphi_x, g]).$$

Clearly, $\{x, y\}$ is bilinear in x, y. Hence, we have the following linear map

$$\begin{array}{ccc} R \otimes_k R & \longrightarrow & k \\ \Psi & & \Psi \\ x \otimes y & \longmapsto & \{x, y\}. \end{array}$$

Let x, y, f, g be as above, and for $z \in R$ let

$$h \in E_1 \quad \text{and} \quad h \equiv \varphi_z \mod E_2.$$

Then we have

$$gh \in E_1 \quad \text{and} \quad gh \equiv \varphi_y \varphi_z \equiv \varphi_{yz} \mod E_2.$$

Therefore, we obtain

$$\begin{aligned} \{x, yz\} &= T([f, gh]) = T(fgh - ghf), \\ \{xy, z\} &= T([fg, h]) = T(fgh - hfg), \\ \{zx, y\} &= T([hf, g]) = T(hfg - ghf). \end{aligned}$$

Hence, $zx = xz$ implies

$$\{x, yz\} - \{xy, z\} - \{xz, y\} = 0.$$

That is, the above map $R \otimes_k R \to k$ takes

$$x \otimes yz - xy \otimes z - xz \otimes y \mapsto 0.$$

Hence, the map $R \otimes_k R \to k$ induces a linear map

$$\begin{array}{rcl} \mathrm{Res}_V : \Omega_{R/k} = R \otimes_k R/A & \longrightarrow & k \\ \Psi & & \Psi \\ \omega & \longmapsto & \mathrm{Res}_V(\omega). \end{array}$$

Definition. $\mathrm{Res}_V(\omega)$ is called the *residue* of ω for V.

For $\omega = x\,dy = x \otimes y \mod A$, where x, $y \in R$, we have

$$\begin{aligned} \mathrm{Res}_V(x\,dy) &= \{x, y\} = T([f, g]) \\ &= T([f, \varphi_y]) = T([\varphi_x, g]). \end{aligned}$$

Since $[f, g] + [g, f] = 0$, we have

$$\mathrm{Res}_V(x\,dy) + \mathrm{Res}_V(y\,dy) = 0, \qquad x, y \in R.$$

In particular, for $y = 1$ we have $dy = 0$. Hence,

$$\mathrm{Res}_V(dx) = 0, \qquad x \in R.$$

Remark. Tate defined a residue without the linear topology on the R-module V. Thus, his results are more general than ours. However, his argument is more complex than ours. For an application to an algebraic function field, it seems better to assume that V is an l.l.c. linear space.

2.3. We will prove some of the main properties of a residue.

THEOREM A.1. *Let V be a continuous R-module, and let W be a closed R-subgroup of V. Then V/W and W are continuous R-modules such that*

$$\mathrm{Res}_V(\omega) = \mathrm{Res}_{V/W}(\omega) + \mathrm{Res}_W(\omega), \qquad \omega \in \Omega_{R/k}.$$

PROOF. It is enough to prove the theorem for $\omega = x\,dy$, $x, y \in R$. Let f and g be elements of E_1 such that $f \equiv \varphi_x$ and $g \equiv \varphi_y \mod E_2$. Then we have

$$f_{V/W}, g_{V/W} \in E_1(V/W),$$
$$f_{V/W} \equiv \varphi_{x,V/W}, \qquad g_{V/W} \equiv \varphi_{y,V/W} \quad \mod E_2(V/W),$$
$$f_W, g_W \in E_1(W), \quad f_W \equiv \varphi_{x,W}, \quad g_W \equiv \varphi_{y,W} \qquad \mod E_2(W),$$

where $E_1(V/W)$ is the E_1-space for V/W, etc. Since we have $[f_{V/W}, g_{V/W}] = [f, g]_{V/W}$ and $[f_W, g_W] = [f, g]_W$, the theorem follows from Lemma A.2, iii).

COROLLARY. *When V is the direct sum of closed R-subgroups V_1 and V_2, we have*

$$\mathrm{Res}_V(\omega) = \mathrm{Res}_{V_1}(\omega) + \mathrm{Res}_{V_2}(\omega), \qquad \omega \in \Omega_{R/k}.$$

THEOREM A.2. *If V is either l.c., or discrete, then*

$$\mathrm{Res}_V(\Omega_{R/k}) = 0.$$

PROOF. If V is l.c., then $E_1 = E$. We may take $f = \varphi_x$ and $g = \varphi_y$. Since $[\varphi_x, \varphi_y] = 0$ holds, we have

$$\mathrm{Res}_V(x\,dy) = T([\varphi_x, \varphi_y]) = 0.$$

If V is discrete, then $E_2 = E$ holds. We may let $f = g = 0$. Hence,

$$\mathrm{Res}_V(x\,dy) = T([f, g]) = 0.$$

LEMMA A.8. *For x and y in R, if there exists an l.c. open subspace U in V such that either $x(U + yU + y^2U) \subseteq U$ or $y(U + xU + x^2U) \subseteq U$ holds, then we have*

$$\mathrm{Res}_V(x\,dy) = 0.$$

In particular, when there exists U such that $xU \subseteq U$ and $yU \subseteq U$, we have

$$\mathrm{Res}_V(x\,dy) = 0.$$

PROOF. Since $\mathrm{Res}_V(x\,dy) = \mathrm{Res}_V(y\,dx) = 0$ holds, it is enough to prove the case where $x(U + yU + y^2U) \subseteq U$. Let π be a projection from V to U, then $\pi\varphi_x \in E_1$ and $\pi\varphi_x \equiv \varphi_x \mod E_2$. Therefore, we obtain

$$\mathrm{Res}_V(x\,dy) = T(\psi), \qquad \psi = [\pi\varphi_x, \varphi_y] = \pi\varphi_x\varphi_y - \varphi_y\pi\varphi_x.$$

Let $W = U + yU$, then $\psi_{V/W} = 0$ and $\psi_W = 0$. Therefore,

$$T(\psi) = T_V(\psi) = T_{V/W}(\psi) + T_W(\psi) = 0.$$

LEMMA A.9. *For $x \in R$, we always have*

$$\mathrm{Res}_V(x^n\, dx) = 0, \qquad n = 0, 1, \dots.$$

In particular, for a regular element x (an invertible element) in R, we have

$$\mathrm{Res}_V(x^n\, dx) = 0 \quad \textit{for all } n, \ \textit{but } n = -1.$$

PROOF. We have already treated $\mathrm{Res}_V(dx) = 0$. Let $n \geq 1$, and let $f_1 \in E_1$ such that $f \equiv \varphi_x \mod E_2$. Then we have $f^n \in E_1$ and $f^n \equiv \varphi_x^n \equiv \varphi_{x^n} \mod E_2$. Therefore, we have

$$\mathrm{Res}_V(x^n\, dx) = T([f^n, f]) = 0.$$

If x is a regular element, then we have

$$x\, d(x^{-1}) + x^{-1}\, dx = d(1) = 0,$$
$$dx = -x^2\, d(x^{-1}), \quad \text{and} \quad x^n\, dx = -(x^{-1})^{-2-n}\, d(x^{-1}).$$

From these we obtain $\mathrm{Res}_V(x^n\, dx) = 0$ for all $n \leq -2$.

LEMMA A.10. *For a regular element x in R, let U be an l.c. open subspace of V such that $xU \subseteq U$ holds. Then,*

$$\mathrm{Res}_V(x^{-1}\, dx) = \dim(U/xU) 1_k.$$

PROOF. Let π be a projection from V onto U. Then we have

$$\mathrm{Res}_V(x^{-1}\, dx) = T(\psi), \qquad \psi = [\pi\varphi_x^{-1}, \varphi_x] = \pi - \varphi_x \pi \varphi_x^{-1}.$$

From $xU \subseteq U$ we obtain

$$\psi_{V/U} = 0, \quad \psi_{U/xU} = 1, \quad \psi_{xU} = 0.$$

The formula in Lemma A.10 follows from $T_V = T_{V/U} + T_{U/xU} + T_{xU}$. Notice that $\dim U/xU < +\infty$ since U/xU is l.c. and discrete.

2.4. Next we will consider the case where the above commutative ring R is contained in a commutative ring S. We will examine the relationship between $\Omega_{R/k}$ and $\Omega_{S/k}$ and their connections to the residues.

Let B be the S-submodule of $S \otimes_k S$ defining $\Omega_{S/k}$. The induced map $R \otimes_k R \to S \otimes_k S$ from the injection $R \to S$ induces

$$\begin{array}{ccc} \Omega_{R/k} & \longrightarrow & \Omega_{S/k} \\ \cup\!\!\!\!\!\!\;\text{ }& & \cup\!\!\!\!\!\!\;\text{ } \\ x\, dy = x \otimes y \mod A & \mapsto & x\, dy = x \otimes y \mod B. \end{array}$$

The above map may not be injective.

For an arbitrary continuous S-module X, X is also a continuous R-module. Hence, we can define

$$\mathrm{Res}_X : \Omega_{R/k} \to k \quad \text{and} \quad \mathrm{Res}_X : \Omega_{S/k} \to k.$$

For x and y in R, by definition $\mathrm{Res}_X(x\,dy)$ depends only on the linear maps on X induced by x and y. It makes no difference in the value of $\mathrm{Res}_X(x\,dy)$ whether $x\,dy$ is regarded as a differential in $\Omega_{R/k}$ or $\Omega_{s/k}$. Therefore, in general for

$$\begin{array}{ccc} \Omega_{R/k} & \to & \Omega_{S/k} \\ \Psi & & \Psi \\ \omega & \mapsto & \omega', \end{array}$$

we have

$$\mathrm{Res}_X(\omega) = \mathrm{Res}_X(\omega').$$

Since R is a subring of S, S is an R-module. We will assume that S is a free R-module of finite rank n. For an arbitrary continuous R-module V, let

$$X = S \otimes_R V.$$

Then X is an S-module. For a fixed free base, $\xi_1, \dots, \xi_n$ for S/R, we have

$$\begin{array}{ccc} V^n & \xrightarrow{\approx} & X \\ \Psi & & \Psi \\ (v_1, \dots, v_n) & \mapsto & \xi_1 \otimes v_1 + \cdots + \xi_n \otimes v_n. \end{array}$$

By the assumption, V is an l.l.c. linear topological space over k. With the topology on X induced by the above isomorphism, X also becomes an l.l.c. linear topological space. It is easy to show that the linear topology on X does not depend upon the particular choice of the base, $\xi_1, \dots, \xi_n$. In general, for an arbitrary $f \in \mathrm{End}(X)$, there exists a unique set of n^2 f_{ij} in $\mathrm{End}(V)$, $i, j = 1, 2, \dots, n$, such that

$$f\left(\sum_{i=1}^{n} \xi_i \otimes v_i\right) = \sum_{i,j=1}^{n} \xi_i \otimes f_{ij}(v_j)$$

for all $v_1, \dots, v_n$ in V. Therefore, we obtain a k-algebra isomorphism

$$\begin{array}{ccc} \mathrm{End}(X) & \xrightarrow{\approx} & M_n(\mathrm{End}(V)) \\ \Psi & & \Psi \\ f & \longmapsto & (f_{ij}), \end{array}$$

where M_n is the ring of $n \times n$ matrices. This isomorphism induces

$$E(X) \xrightarrow{\approx} M_n(E(V)).$$

For an arbitrary element ξ in S, let

$$\xi\xi_j = \sum_{i=1}^{n} x_{ij}\xi_i, \qquad j = 1, \dots, n,$$

and let

$$\begin{array}{cccc} \psi_\xi\colon & X & \to & X \\ & \text{Ⓦ} & & \text{Ⓦ} \\ & w & \mapsto & \xi\cdot w. \end{array}$$

Then for $\psi_\xi \in E(X)$, we have

$$\psi_\xi \mapsto (\varphi_{x_{ij}}) \in M_n(E(V)),$$

where for $x \in R$, as before,

$$\begin{array}{cccc} \varphi_x\colon & V & \to & V \\ & \text{Ⓦ} & & \text{Ⓦ} \\ & v & \mapsto & x\cdot v. \end{array}$$

Since $\varphi_x \in E(V)$ from the assumption, the isomorphism $E(X) \cong M_n(E(V))$ implies $\psi_\xi \in E(X)$. That is, ψ_ξ is continuous. Hence, $X = S \otimes_R V$ is a continuous S-module.

Let E' be the inverse image of $M_n(E_0(V))$ under $E(X) \overset{\approx}{\to} M_n(E(V))$. Then E' is a subspace of $E(X)$ such that $E' \subseteq E_0(X)$. Hence, we can define the trace

$$T_X\colon E' \to k.$$

For an arbitrary element f in E', let $f \mapsto (f_{ij})$, $f_{ij} \in E_0(V)$, $i, j = 1, \dots, n$. Write the matrix (f_{ij}) as the sum of a diagonal matrix f_0, a strictly upper triangular matrix f_1, and a strictly lower triangular matrix f_2.

$$f = f_0 + f_1 + f_2.$$

Then, f_0, f_1, f_2 belong to E', and f_1 and f_2 are nilpotent. Therefore, we obtain the following:

$$T_X(f_1) = T_X(f_2) = 0 \quad \text{and} \quad T_X(f) = T_X(f_0).$$

On the other hand, by the definition, we have

$$T_X(f_0) = \sum_{i=1}^{n} T_V(f_{ii}).$$

Hence, we obtain the formula

$$T_X(f) = \sum_{i=1}^{n} T_V(f_{ii}).$$

Since X is a continuous S-module,

$$\operatorname{Res}_X\colon \Omega_{S/k} \to k$$

is defined.

THEOREM A.3. *For $\xi \in S$ and $y \in R$, we have*

$$\mathrm{Res}_X(\xi\, dy) = \mathrm{Res}_V(T_{S/R}(\xi)\, dy),$$

where $T_{S/R}$ is the trace of the R-algebra S.

PROOF. Let U be an l.c. open subspace of V, and let

$$W = \xi_1 \otimes U + \cdots + \xi_n \otimes U.$$

Then W is an l.c. open subspace of X. For a projection π from V onto U, let the isomorphism $\mathrm{End}(X) \overset{\approx}{\to} M_n(\mathrm{End}(V))$ be given by

$$\pi' \mapsto \begin{pmatrix} \pi & & & 0 \\ & \pi & & \\ & & \ddots & \\ 0 & & & \pi \end{pmatrix}.$$

Then π' is a projection from X onto W. On the other hand,

$$\psi_\xi \mapsto (\varphi_{x_{ij}}) \quad \text{and} \quad \psi_y \mapsto \begin{pmatrix} \varphi_y & & & 0 \\ & \ddots & & \\ & & \ddots & \\ 0 & & & \varphi_y \end{pmatrix},$$

hence we obtain

$$[\pi' \psi_\xi, \psi_y] \mapsto ([\pi\varphi_{x_{ij}}, \varphi_y]).$$

Therefore we get the following from the formula preceding Theorem A.3:

$$\begin{aligned} \mathrm{Res}_X(\xi\, dy) &= T_X([\pi'\psi_\xi, \psi_y]) = \sum_{i=1}^{n} T_V([\pi\varphi_{x_{ii}}, \varphi_y]) \\ &= \sum_{i=1}^{n} \mathrm{Res}_V(x_{ii}\, dy) = \mathrm{Res}_V\left(\left(\sum_{i=1}^{n} x_{ii}\right) dy\right) \\ &= \mathrm{Res}_V(T_{S/R}(\xi)\, dy). \end{aligned}$$

§3. Differentials and residues in an algebraic function field

3.1. We will apply the above general theory of differentials and residues to an algebraic function field.

Let K be an algebraic function field over the ground field k. For a place P on K, let K_P be the completion of K, let $\mathfrak{o}_P$ be the valuation ring of K_P, and let $\mathfrak{p}_p$ be the prime ideal of P. Since $k \subseteq K \subseteq K_P$, K_P is a vector space over k. A system of fundamental neighborhoods of O for the P-adic topology is provided by the powers of $\mathfrak{p}_P^m$, $m > 0$. Since $\mathfrak{p}_P^m$ is a subspace of K_P over k, the topology on K_P is a linear topology. On the other hand, by the expansion theorem (Theorem 1.6), $\mathfrak{o}_P$ is a direct product of 1-dimensional l.c. vector spaces k. Hence, $\mathfrak{o}_P$ is an l.c. open subspace of K_P. Namely, K_P is an l.l.c. vector space.

Since the multiplication in the topological field K_P is continuous, K_P is a continuous K_P-module in the sense of 2.2, and in particular, K_P is a continuous K-module. Hence, the residue $\mathrm{Res}_{K_P}(\omega)$ of ω in $\Omega_{K_P/k}$ or in $\Omega_{K/k}$ can be defined. We write this residue as

$$\mathrm{Res}_P(\omega),$$

which is called the residue of ω at the place P.

LEMMA A.11. *For ξ and η in K_P, if*

$$\min(\nu_P(\xi), \nu_P(\xi)+\nu_P(\eta), \nu_P(\xi)+2\nu_P(\eta)) \geq 0$$

or

$$\min(\nu_P(\eta), \nu_P(\eta)+\nu_P(\xi), \nu_P(\eta)+2\nu_P(\xi)) \geq 0,$$

then

$$\mathrm{Res}_P(\xi\, d\eta) = 0.$$

PROOF. Our assumption implies

$$\xi(\mathfrak{o}_P + \eta\mathfrak{o}_P + \eta^2\mathfrak{o}_P) \subseteq \mathfrak{o}_P \quad \text{or} \quad \eta(\mathfrak{o}_P + \xi\mathfrak{o}_P + \xi^2\mathfrak{o}_P) \subseteq \mathfrak{o}_P.$$

Since $\mathfrak{o}_P$ is an l.c. open subspace of K_P, Lemma A.8 immediately implies the conclusion of this lemma.

COROLLARY. *The map*

$$\begin{array}{ccc} K_P \times K_P & \longrightarrow & k \\ \Psi & & \Psi \\ (\xi, \eta) & \longmapsto & \mathrm{Res}_P(\xi, d\eta) \end{array}$$

is continuous.

Let the degree of P be 1, i.e., $n(P)=1$, and let π be a prime element of K_P. For any elements ξ and η in K_P, by the expansion theorem we have

$$\xi = \sum_{-\infty \ll n < \infty} a_n \pi^n,$$

$$\eta = \sum_{-\infty \ll n < \infty} b_n \pi^n, \qquad a_n, b_n \in k.$$

For the formal derivative of η

$$\frac{d\eta}{d\pi} = \sum_{-\infty \ll n < \infty} n b_n \pi^{n-1},$$

the coefficient of π^{-1} of

$$\xi \frac{d\eta}{d\pi} = \left(\sum_{-\infty \ll n < \infty} a_n \pi^n\right)\left(\sum_{-\infty \ll n < \infty} n b_n \pi^{n-1}\right)$$
$$= \sum_{-\infty \ll n < \infty} c_n \pi^n$$

is given by

$$c_{-1} = \sum_{m+n=0} n a_m b_n .$$

THEOREM A.4.

$$\operatorname{Res}_P(\xi\, d\eta) = c_{-1} = \sum_{m+n=0} n a_m b_n .$$

PROOF. For a sufficiently large N, let

$$\xi' = \sum_{-\infty \ll n < N} a_n \pi^n \quad \text{and} \quad \eta' = \sum_{-\infty \ll n < N} b_n \pi^n .$$

From Lemma A.11, we obtain

$$\operatorname{Res}_P(\xi\, d\eta) = \operatorname{Res}_P(\xi'\, d\eta') .$$

Hence, we may assume that $a_n = b_n = 0$ for sufficiently large n's. Therefore, it is sufficient to consider the case where

$$\xi = \pi^m , \qquad \eta = \pi^n , \qquad \xi\, d\eta = n\pi^{m+n-1}\, d\pi .$$

Then Lemma A.9 and Lemma A.10 imply

$$\operatorname{Res}_P(\pi^{m+n-1}\, d\pi) = \begin{cases} 0, & m+n \neq 0, \\ n(P)1_k = 1_k, & m+n = 0, \end{cases}$$

completing the proof.

From this theorem, the coefficient of π^{-1} of the expansion of

$$\xi \frac{d\eta}{d\pi}, \qquad \xi, \eta \in K_P$$

is independent of the choice of a prime element π in K_P (see Theorem 2.16).

If the ground field k is algebraically closed, then any place P on an algebraic function field K over k satisfies $n(P) = 1$. Hence, we can then compute the residue $\operatorname{Res}_P(\xi\, d\eta)$ by the above formula. When k is a perfect field (e.g., a finite field), one can extend the above result to the case $n(P) > 1$. We will leave the proof of this case to the reader as an exercise.

Let L be another algebraic function field over k such that L is an extension field of K, i.e.,

$$k \subseteq K \subseteq L .$$

Then L/K is obviously a finite extension. Let P be the projection onto K of a place Q on L. Then we clearly have

$$K_P \subseteq L_Q .$$

THEOREM A.5. *For $\xi \in L_Q$ and $y \in K_P$, we have*

$$\operatorname{Res}_Q(\xi,\, dy) = \operatorname{Res}_P(T_{L_Q/K_P}(\xi)\, dy) .$$

PROOF. Using $K_P \otimes_{K_P} L_Q = L_Q$, Theorem A.3 implies the above.

LEMMA A.12. *Let K and L be as above, and for a place P on K, let $Q_1, \dots, Q_g$ be all the extensions on L of P. Then as algebras over K,*

$$K_P \otimes_K L \cong L_{Q_1} \oplus L_{Q_2} \oplus \cdots \oplus L_{Q_g}.$$

PROOF. From the inclusions K_P, $L \subseteq L_{Q_i}$, $i = 1, 2, \dots, g$, the homomorphism

$$f\colon K_P \otimes_K L \to L_{Q_1} \oplus \cdots \oplus L_{Q_g}$$

is induced. From Theorem 1.9 and Remark 2, §3, Chapter 2, as vector spaces over K_P, both sides in the above have the same (finite) dimension. On the other hand, the approximation theorem (Theorem 1.1) implies that $\operatorname{Im} f$ is dense in $L_{Q_1} \oplus \cdots \oplus L_{Q_g}$. For the linear map f over K_P, $\operatorname{Im} f$ is a subspace of $L_{Q_1} \oplus \cdots \oplus L_{Q_g}$. Since K_P is complete, the ontoness of f. Hence, f is an isomorphism.

THEOREM A.6. *Let K, L, P, $Q_1, \dots, Q_g$ be as in the above. Then for $x \in L$ and $y \in K$, we have*

$$\sum_{i=1}^{g} \operatorname{Res}_{Q_i}(x\,dy) = \operatorname{Res}_P(T_{L/K}(x)\,dy).$$

PROOF. From the above lemma,

$$\sum_{i=1}^{g} T_{L_{Q_i}/K_P}(x) = T_{L/K}(x).$$

Hence, Theorem A.5 implies Theorem A.6. We can also prove this theorem by applying Theorem A.3 to $K_P \otimes_K L \cong L_{Q_1} \oplus \cdots \oplus L_{Q_g}$. Note that the value of $\operatorname{Res}_{Q_i}(x\,dy)$ is the same whether $x\,dy$ is regarded as an element of $\Omega_{L_{Q_i}/k}$ or an element of $\Omega_{L/K}$. Similarly, $\operatorname{Res}_P(T_{L/K}(x)\,dy)$ retains the same value.

3.2. Let K be an algebraic function field over k as before. We will consider the adele ring $\widetilde{K}$ of K. See Chapter 2, §4. Since $k \subseteq K \subseteq \widetilde{K}$, $\widetilde{K}$ is a vector space over k. Furthermore, since the open sets $\widetilde{L}(A)$ defining the topology of $\widetilde{K}$ are k-subspaces of $\widetilde{K}$, $\widetilde{K}$ is a linear topological space over k. By the definition, $\widetilde{L}(A)$ is a direct product of $\mathfrak{p}_P^{m_P}$, where $m_P = -\nu_P(A)$, and where P runs through all the places P on K. Each $\mathfrak{p}_P^{m_P}$ is l.c. Hence, $\widetilde{L}(A)$ is also l.c., and as a consequence, the adele ring $\widetilde{K}$ is an l.l.c. linear topological space.

LEMMA A.13. *The k-vector space K is a discrete subspace of $\widetilde{K}$, and the quotient space $\widetilde{K}/K$ is l.c.*

PROOF. For a place P on K, we have

$$K \cap \widetilde{L}(P^{-1}) = \{0\}$$

by Theorem 2.2. Hence, K is discrete. For a divisor A on K, let $\mathfrak{L}'_A$ be the totality of linear maps

$$f: \widetilde{K}/(\widetilde{L}(A) + K) \to k.$$

Then $\mathfrak{L}'_A$ coincides essentially with $\mathfrak{L}_A$ in Chapter 2, §4. Hence, we have

$$\dim \mathfrak{L}'_A < +\infty.$$

Therefore, the dual space $\widetilde{K}/(\widetilde{L}(A) + K)$ is also finite dimensional over k. Hence $\widetilde{K}/(\widetilde{L}(A) + K)$ is l.c. On the other hand, $\widetilde{L}(A)$ is l.c. Since the isomorphism $(\widetilde{L}(A) + K)/K \cong \widetilde{L}(A)/\widetilde{L}(A) \cap K$ implies that $(\widetilde{L}(A) + K)/K$ is also l.c., $\widetilde{K}/K$ is l.c.

Whereas $\widetilde{K}$ is an l.l.c. topological ring and $K \subseteq \widetilde{K}$, $\widetilde{K}$ is a continuous K-module. Therefore, one can define the residue $\mathrm{Res}_{\widetilde{K}}(\omega)$ of a differential ω in $\Omega_{K/k}$.

LEMMA A.14. *We have*

$$\mathrm{Res}_{\widetilde{K}}(\Omega_{K/k}) = 0.$$

PROOF. Since K is a K-submodule of $\widetilde{K}$, from Theorem A.1, A.2, and Lemma A.13 we obtain

$$\mathrm{Res}_{\widetilde{K}}(\Omega_{K/k}) = \mathrm{Res}_{\widetilde{K}/K}(\Omega_{K/k}) + \mathrm{Res}_K(\Omega_{K/k}) = 0.$$

THEOREM A.7 (residue theorem). *For an arbitrary differential ω belonging to $\Omega_{K/k}$, we have*

$$\mathrm{Res}_P(\omega) = 0$$

for all but a finite number of places P. Furthermore, we have

$$\sum_P \mathrm{Res}_P(\omega) = 0.$$

PROOF. It suffices to prove the theorem for the case where $\omega = x\,dy$, x, $y \in K$. We clearly have

$$\nu_P(x) \geq 0 \quad \text{and} \quad \nu_P(y) \geq 0$$

for almost all P. Let $P_1, \ldots, P_h$ be the exceptional places for the above inequalities. Then Lemma A.11 implies

$$\mathrm{Res}_P(\omega) = 0 \quad \text{for } P \neq P_1, \ldots, P_h,$$

proving the first half of this theorem. Next, let

$$M = \{\bar{a} = (a_P) \in \widetilde{K};\ a_{P_i} = 0 \text{ for } i = 1, \ldots, h\}.$$

Then M is an ideal of $\widetilde{K}$, and

$$\widetilde{K} = K_{P_1} \oplus \cdots \oplus K_{P_h} \oplus M.$$

Let E be the identity element of the divisor group of K, and let

$$U = M \cap \widetilde{L}(E).$$

Then U is an l.c. open subspace of M and satisfies

$$xU \subseteq U \quad \text{and} \quad yU \subseteq U.$$

Hence, by Lemma A.8, we obtain

$$\mathrm{Res}_M(\omega) = 0.$$

Therefore, from the corollary of Theorem A.1,

$$\begin{aligned} \mathrm{Res}_{\widetilde{K}}(\omega) &= \mathrm{Res}_{P_1}(\omega) + \cdots + \mathrm{Res}_{P_h}(\omega) + \mathrm{Res}_M(\omega) \\ &= \mathrm{Res}_{P_1}(\omega) + \cdots + \mathrm{Res}_{P_h}(\omega) \\ &= \sum_P \mathrm{Res}_P(\omega). \end{aligned}$$

By Lemma A.14, $\mathrm{Res}_{\widetilde{K}}(\omega) = 0$ holds, completing the latter half of the proof as well.

3.3. Let k be algebraically closed. For an algebraic function field K over k, by Theorem 2.1, K contains a separating element x, i.e., $x \notin k$ such that $K/k(x)$ is a finite separable extension.

LEMMA A.15. *As an element of* $\Omega_{K/k}$, *we have*

$$dx \neq 0.$$

PROOF. There is a place P on the rational function field $k(x)$ such that $\nu_P(x) = 1$ and $n(P) = 1$ (see Example 2, §1, Chapter 1). Since $K/k(x)$ is a finite separable extension, there exists an element y in K such that

$$T_{K/k(x)}(y) = x^{-1}.$$

For all the extensions $Q_1, \ldots, Q_g$ on K of P, we obtain $\sum_{i=1}^g \mathrm{Res}_{Q_i}(y\,dx) = \mathrm{Res}_P(x^{-1}\,dx) = 1_k$ from Theorems A.4 and A.6. Hence, $dx \neq 0$ must hold.

THEOREM A.8. *The dimension of the* K*-module* $\Omega_{K/k}$ *over* K *is* 1:

$$\dim_K \Omega_{K/k} = 1.$$

PROOF. Let x be a separating element for K/k and let y be an arbitrary element of K. Let the irreducible equation for x and y be given by

$$g(x, y) = 0, \qquad g(X, Y) \in k[X, Y].$$

Then we get

$$g_x(x, y)\,dy + g_y(x, y)\,dy = 0.$$

Since y is a separable algebraic element for $k(x)$, we have $g_y(x, y) \neq 0$. Therefore, we may write

$$dy = w\,dx, \qquad \text{where } w = -g_x g_y^{-1} \in K.$$

Since y is arbitrary, we obtain

$$\Omega_{K/k} = K\,dx.$$

From Lemma A.15, $dx \neq 0$ holds so that we can conclude $\dim_K \Omega_{K/k} = 1$.

In the above proof, $K/k(x, y)$ is a separable extension. If $y \notin k$ is not a separating element for K/k, then $k(x, y)/k(y)$ has to be an inseparable finite extension. Hence, $g_x(x, y) = 0$ and $g_y(x, y) \neq 0$ must hold. Then we have

$$dy = 0.$$

In other words, an element x in K $(x \notin k)$ is a separating element if and only if $dx \neq 0$.

REMARK. One need not assume that k is algebraically closed in the above as long as there exists a separating element for K/k. When K does not possess a separating element over k, it is known that

$$\dim_K \Omega_{K/k} > 1.$$

Let $H_{K/k}$ be the totality of Hasse differentials as defined in Chapter 2, §5. For an element x in K, denote the Hasse differential of x by $d'x$ to distinguish it from the element dx of $\Omega_{K/k}$. Then $H_{K/k}$ is a K-module, and

$$\begin{array}{ccc} D\colon K & \longrightarrow & H_{K/k} \\ \cup\!\!\!\shortmid & & \cup\!\!\!\shortmid \\ x & \longmapsto & d'x \end{array}$$

is a k-derivation of K. By the remark in 2.1, we have an epimorphism

$$\begin{array}{ccc} f\colon \Omega_{K/k} & \longrightarrow & H_{K/k} \\ \cup\!\!\!\shortmid & & \cup\!\!\!\shortmid \\ dx & \longmapsto & d'x. \end{array}$$

Since $H_{K/k}$ and $\Omega_{K/k}$ are both 1-dimensional over K, by Theorem 2.18 and Theorem A.8, f is an isomorphism. Thus, a differential $y\,dx$ in $\Omega_{K/k}$ and a Hasse differential $y\,d'x$ are essentially the same. Furthermore, from Theorem A.4, the residue of $y\,dx$ at a place P coincides with the residue of $y\,d'x$ at P. Hence, Theorem A.7 gives another proof of Theorem 2.19. Note that Theorem A.5 is a generalization of the important formula (5.12) for the proof of Theorem 2.19. Notice that Theorem 2.16, which was fundamental for the definition of the residue of a Hasse differential, is immediately obtained from Theorem A.4.

As the reader observed in the above, as for applications to an algebraic function field, the theory of differentials and residues in the Appendix does not essentially differ from the theory in Chapter 2, §5. One should realize, however, that the various results obtained in Chapter 2 by clever and technical computations are natural and easy consequences of the general theorem in this Appendix. Furthermore, one can prove the Riemann-Roch theorem (Theorem 2.13) from the duality theorem for linear topological spaces by using the fact that the adele ring $\widetilde{K}$ is an l.l.c. space. In fact, based on the theory of linear topological spaces, one can build, with much more clarity, the whole algebraic theory of algebraic function fields as given in Chapter 2.

Subject Index

Abel's theorem, 219
Abelian
  function, 221
  function field, 221
Abstract Riemann surface, 94
Addition theorem, 250
Additive function, 196
Algebraic
  additive formula, 230
  function, 161
  function field, 33
Analytic
  function, 94
  function field, 95
  function field, of a closed Riemann surface, 156
  manifold, 220
  map, 91
  path, 100
  path, in the strict sense, 100
Analytical
  coordinates, 91
  domain, 91
  variable, 91
Approximation theorem, 6
Automorphism group, of a Riemann surface, 128

Canonical
  basis, 47
  class, 60
  divisor, 17
  form (of a Riemann surface), 130
Closed Riemann surface, 147
Coefficient field, 33
Complementary class, 61
Complete field, 14
Completion, 14
Complex multiplication, 263
Continuous curve (on a Riemann surface), 99
Covering
  manifold, 123
  map, 123
  Riemann surface, 123

Degree
  of a divisor, 42
  of a place, 36
Denominator of a divisor, 43
Different, Differente, 73, 232
Differential, 97, 269
  class, 60
  divisor, 60
  of the first kind, 61
  of the second kind, 76
  of the third kind, 76
Dimension, 48
Discontinuous topological transformation group, 124
Division equation, 262
Divisor, 42
  class, 42, 44
  class group, 44

Elementary Abelian function, 225
Elliptic
  function, 242
  function field, 79, 242
  integral, 251
Equivalent
  R-function system, 89
  valuations, 2
Exchange law of variable and parameter, 191
Expansion theorem, 14
Exponential valuation, 10
Extension, 44
  of a prime divisor, 20

Finitepotent map, 265
Fundamental
  domain, 188
  group, 100, 124
  system of periods, 177

Galois group, 236
Gauss, Green's theorem, 107

Genus, 52, 169
Gradient, 106
Group
of automorphisms, 94
of motions, 140

H-differential (Hasse differential), 65
Harmonic vector field, 117
Hensel, xvi
Hilbert space, 112
Homotopic, 100
Homotopy
group, 52
space, 126
Hurwitz's formula, 232
Hyperelliptic function field, 86, 242

Idèle, 53
ring, 53
Integral
along a path, 101
character, 205
divisor, 43
Integrating function, 104
Intersection number, 170
Invariant of an elliptic function field, 84, 247
Isomorphism, 93
of Riemann surfaces, 93

Jacobi's ϑ-function, 257
Jacobian variety, 220

Legendre relation, 253
Linear topology, 267
Local H-differential, 63
Locally uniformizing variable, 91

Measurable function on a Riemann surface, 97
Metric belonging to a valuation, 11
Modular
group, 247
transformation, 247
Multiple, 48
Multiplicative function, 198
Multiplier, 198

Non-Euclidean metric, 140
Norm, 109
Normal divisor, 52
Normal polygon, 143
Normalized valuation, 1
Numerator of a divisor, 43

Order, 35, 94, 98
of a differential, 63
Ostrowski, 4

$\mathfrak{p}$-adic valuation, 8
Path on a Riemann surface, 99
Period, 175, 196
group, 214, 217
matrix, 215
Periodic vector group, 214, 217
Place, 35
Poincaré group, 124
Pole, 35, 63, 98
Prime
divisor, 2
ideal of a prime divisor, 4
Principal
divisor, 43
divisor group, 43
Projection, 20
Puiseux series, 40

Ramification index, 20
Rational function field, 78
Regular, 94
Relative degree, 20
Representation function, 206
Residue, 65, 99, 271, 277
field, 4
theorem, 67
R-function system, 89
Riemann
matrix, 215
relation, 178
subsurface, 91
surface, 89, 147
Riemann's
formula, 74
inequality, 178
sphere, 91
Riemann-Roch theorem, 60

Schmidt, F. K., xvii
Separating element, 34
Simply connected, 124
Riemann surface, 130
Singularity set, 223
Standard subdivision, 173
Subdivision path, 174

ϑ-function, 227
2-dimensional closed manifold, 167
Theorem of Jacobi, 224
Theta-Fuchsian series, 211

Umkehrproblem, 231
Underlying space, 89
Unramified
Abelian extension, 240
extension, 232

Valuation, 1
ring, 4

Value group, 1
Vector field, 105

Weierstrass, xii
  canonical form, 82
  ℘-function, 246
  point, 76

Zero, 35, 63, 94, 98
Zeta-Fuchsian series, 211

Recent Titles in This Series

(Continued from the front of this publication)

83 **N. I. Portenko,** Generalized diffusion processes, 1990

82 **Yasutaka Sibuya,** Linear differential equations in the complex domain: Problems of analytic continuation, 1990

81 **I. M. Gelfand and S. G. Gindikin, Editors,** Mathematical problems of tomography, 1990

80 **Junjiro Noguchi and Takushiro Ochiai,** Geometric function theory in several complex variables, 1990

79 **N. I. Akhiezer,** Elements of the theory of elliptic functions, 1990

78 **A. V. Skorokhod,** Asymptotic methods of the theory of stochastic differential equations, 1989

77 **V. M. Filippov,** Variational principles for nonpotential operators, 1989

76 **Phillip A. Griffiths,** Introduction to algebraic curves, 1989

75 **B. S. Kashin and A. A. Saakyan,** Orthogonal series, 1989

74 **V. I. Yudovich,** The linearization method in hydrodynamical stability theory, 1989

73 **Yu. G. Reshetnyak,** Space mappings with bounded distortion, 1989

72 **A. V. Pogorelev,** Bendings of surfaces and stability of shells, 1988

71 **A. S. Markus,** Introduction to the spectral theory of polynomial operator pencils, 1988

70 **N. I. Akhiezer,** Lectures on integral transforms, 1988

69 **V. N. Saliĭ,** Lattices with unique complements, 1988

68 **A. G. Postnikov,** Introduction to analytic number theory, 1988

67 **A. G. Dragalin,** Mathematical intuitionism: Introduction to proof theory, 1988

66 **Ye Yan-Qian,** Theory of limit cycles, 1986

65 **V. M. Zolotarev,** One-dimensional stable distributions, 1986

64 **M. M. Lavrent′ev, V. G. Romanov, and S. P. Shishat·skiĭ,** Ill-posed problems of mathematical physics and analysis, 1986

63 **Yu. M. Berezanskiĭ,** Selfadjoint operators in spaces of functions of infinitely many variables, 1986

62 **S. L. Krushkal′, B. N. Apanasov, and N. A. Gusevskiĭ,** Kleinian groups and uniformization in examples and problems, 1986

61 **B. V. Shabat,** Distribution of values of holomorphic mappings, 1985

60 **B. A. Kushner,** Lectures on constructive mathematical analysis, 1984

59 **G. P. Egorychev,** Integral representation and the computation of combinatorial sums, 1984

58 **L. A. Aĭzenberg and A. P. Yuzhakov,** Integral representations and residues in multidimensional complex analysis, 1983

57 **V. N. Monakhov,** Boundary-value problems with free boundaries for elliptic systems of equations, 1983

56 **L. A. Aĭzenberg and Sh. A. Dautov,** Differential forms orthogonal to holomorphic functions or forms, and their properties, 1983

55 **B. L. Roždestvenskiĭ and N. N. Janenko,** Systems of quasilinear equations and their applications to gas dynamics, 1983

54 **S. G. Kreĭn, Ju. I. Petunin, and E. M. Semenov,** Interpolation of linear operators, 1982

53 **N. N. Čencov,** Statistical decision rules and optimal inference, 1981

52 **G. I. Èskin,** Boundary value problems for elliptic pseudodifferential equations, 1981

51 **M. M. Smirnov,** Equations of mixed type, 1978

50 **M. G. Kreĭn and A. A. Nudel′man,** The Markov moment problem and extremal problems, 1977

(See the AMS catalog for earlier titles)